D1295417

Fundamental Principles of Molecular Modeling

Edited by

Werner Gans

Free University of Berlin
Berlin, Germany

Anton Amann

ETH Zurich
Zurich, Switzerland

and

Jan C. A. Boeyens

University of the Witwatersrand
Johannesburg, South Africa

Plenum Press • New York and London

Library of Congress Cataloging-in-Publication Data

Fundamental principles of molecular modeling / edited by Werner Gans,
Anton Amann, and Jan C.A. Boeyens.
      p.   cm.
    "Proceedings of an international workshop on fundamental
principles of molecular modeling, held August 20-25, 1995, in
Skukuza, Kruger Park, South Africa"--T.p. verso.
    Includes bibliographical references and index.
    ISBN 0-306-45305-3
    1. Molecules--Models--Congresses.   I. Gans, W., 1949-   .
II. Amann, A., 1956-   . III. Boeyens, J. C. A. (Jan C. A.)
QD480.F86   1996
541.2'2'011--dc20   96-15801
                                                              CIP

Proceedings of an international workshop on Fundamental Principles of Molecular Modeling,
held August 20 – 25, 1995, in Skukuza, Kruger Park, South Africa

ISBN 0-306-45305-3

PREFACE

The concept of molecular shape emerged quite some time before quantum mechanics was used for the description of processes at the atomic and molecular level. The experimental methods for investigating molecular structure are numerous and have been refined continuously in the last decades, also due to progress in electronic data processing. Many molecular structures are now accessible in databases, and this availability facilitates the study of whole classes of compounds. Large and fast computers are, on the other hand, very helpful in investigating molecular structure theoretically by a procedure that is usually called molecular mechanics. It is based on classical mechanics, but quantum mechanical refinements are possible, and it can be applied to quite large structures, for which a purely quantum mechanical treatment by *ab initio* methods is impossible.

The fact that molecules can be modeled by methods using classical ingredients suggests that molecular shape is a concept independent of quantum mechanics, as was alluded to at the beginning. On the other hand, quantum mechanics shows that there are small molecules such as ammonia that do not have a molecular shape at all in the ground state. Quantum mechanics contains a wealth of pure states, some of which are strange from the point of view of molecular shape and which are, with very few exceptions, e.g., ammonia, not realized in nature.

Resonance structures of molecule with hydrogen bridges (intra- or intermolecular) are another interesting example of a fuzzy molecular shape that should be treated on a rigorous quantum mechanical level.

A further modification of the classical concept of molecular shape arises for molecules in a crowded environment, e.g., in a crystal, the medium in which crystallographers determine molecular structures. Structures determined in crystals have little to do with the structures of molecules in the gas phase, since the X-ray structures strongly depend on the way the molecules are packed into the crystal. Besides, a proper quantum theory of X-ray crystallography is still lacking.

This volume contains the proceedings of the "*Small Molecules Indaba.** Fundamental Principles of Molecular Modelling," an international workshop organized by the South African Crystallographic Society in collaboration with the Small Molecules Commission of the International Union of Crystallography, held at Skukuza, Kruger Park, South Africa, 20–25 August 1995.

Werner Gans
for the Editors

---

* Indaba is an African term to describe a meeting to analyze a difficult problem from all angles.

CONTENTS

Fundamentals of Molecular Modelling 1
  J.C.A. Boeyens

Molecular Shape 11
  B.T. Sutcliffe

New Tests of Models in Chemical Binding – Extra-Mechanical Effects and
    Molecular Properties 41
  J.F. Ogilvie

Can Quantum Mechanics Account for Chemical Structures? 55
  A. Amann

Environmental Factors in Molecular Modelling 99
  J.C.A. Boeyens

Knowledge Acquisition from Crystallographic Databases: Applications in
    Molecular Modelling, Crystal Engineering and Structural Chemistry 105
  F.H. Allen

Hydrogen Bonding Models: Their Relevance to Molecular Modeling 119
  P. Gilli, V. Ferretti, and G. Gilli

Molecular Electrostatic Properties from X-Ray Diffraction Data 143
  T. Koritsánszky

Modeling of Structural and Spectroscopic Properties of Transition Metal
    Compounds 167
  P. Comba

Conformational Analysis of Long Chain Seco-Acids Used in Woodward's Total
    Synthesis of Erythromycin A – Conformational Space Search as the Basis
    of Molecular Modeling 189
  E. Ōsawa, E. Deretey, and H. Gotō

Packing Molecules and Ions into Crystals 199
  L. Glasser

On the Isostructurality of Supramolecules: Packing Similarities Governed by
    Molecular Complementarity 209
  A. Kálmán

The Phenomenon of Conglomerate Crystallization. Part 44. Counterion Control of
    Crystallization Pathway Selection 223
    I. Bernal, X. Xia, and F. Somoza

Contributors 245

Index 247

Fundamental Principles of
Molecular Modeling

FUNDAMENTALS OF MOLECULAR MODELLING

Jan C.A. Boeyens

Centre for Molecular Design
Department of Chemistry
University of the Witwatersrand
Johannesburg

INTRODUCTION

Molecular modelling is an important growth industry in chemistry, practised by many young scientists, and more often because of the appeal of computers than a desire to understand the fundamentals of the subject. There is a danger that the cowboys of the keyboard could turn a serious pursuit into disrepute and a circus of pretty pictures, unless the balance is restored by noting the scientific importance of the basic principles. From another point of view molecular mechanics has established itself as the only reliable computational technique that allows of unbiased modelling of molecular properties, albeit without recourse to fundamental theory.

These two caricatures are equally unfortunate and both are responsible for the common conclusion that molecular modelling represents little more than a rather frivolous dabbling into a serious subject, to be replaced in good time by a more difficult, but a theoretically pure and respectable analysis[1]. This is not a fair assessment, but also not one that is readily re-futed,without addressing some hard questions about the meaning of concepts such as molecular shape, chemical bonding, and their formulation in terms of basic theory.

Under the assumption that consistent quantum-mechanical analysis of chemical systems automatically reveals the same three-dimensional structure of molecules that can be determined uniquely by X-ray diffraction or molecular spectroscopy, the previous assessment is indeed valid and molecular mechanics should be avoided because of its approximate nature. However, it is worrisome to note that the real situation is considerably more complex and that the basic theory predicts almost no details pertaining to electronic structure, spectroscopy, shape or bonding of an isolated molecule.

It is not even clear that the two most sturdy legs that theoretical chemistry stands upon can be reconciled. The quantum theory of X-ray crystallography remains an open question, and in fact all known methods of structure determination could stand some closer scrutiny.

Most structural studies are done by the methods of diffraction, spectroscopy and ab initio calculation. These techniques have certainly contributed tremendously to a rapid growth in the understanding of chemistry, but have also spawned some poor assumptions that should be analyzed more thoroughly to understand the real basis of the molecular structure hypothesis.

Fundamental Principles of Molecular Modeling
Edited by Werner Gans *et al.*, Plenum Press, New York, 1996

DIFFRACTION METHODS

Diffraction is almost universally accepted as the nearly infallible method for molecular and/or crystal structure determination. In the minds of many chemists "crystal structure" implies complete characterization of a material at the level of atomic resolution. The distinction between diffraction in the gas phase, amorphous materials, powders and single crystals is often not appreciated. The magic phrase "X-ray/electron/neutron diffraction", by itself is enough. A brief outline of diffraction techniques as they feature in structure analysis, could therefore be helpful to clarify many of the remarks that follow.

The fundamental phenomenon underlying diffraction is the scatter of radiation by matter. Neutron and electron beams are considered as radiation in this context. The radiation quantum (e.g. X-ray) excites a unit of matter (e.g. electron) which on relaxation emits radiation at the same wavelength, but scattered in all directions. Scattered waves from various electrons of the same atom interfere, so that the total atomic scattering (called the scattering factor) is a function of the radial atomic electron density distribution and the angle of scatter, with respect to the direction of incidence. The core electrons which are more tightly bound, scatter more effectively at high angles, whereas the more loosely bound valence electrons only scatter at small angles.

Scattered waves from neighbouring atoms interfere in exactly the same way and unless the atoms are ordered as in a crystal, the total diffraction pattern is a function of the radial distribution of scattering density (atoms) only. This is the mechanism whereby diffraction patterns arise during gas-phase electron diffraction, scattering by amorphous materials, and diffraction by dissolved species. The only information contained in this type of diffraction pattern is a function describing the radial distribution of scattering centres. For a pure substance in the gas phase this is strictly an intramolecular distribution. All possible interatomic distances are represented, but it contains no conformational information.

It is therefore impossible to determine molecular structures from gas-phase electron-diffraction data, or from the diffraction patterns of amorphous materials or solutions. The only procedure that works is to compare the radial distribution function with the corresponding function, generated by some assumed spatial arrangement of atoms. In many cases sufficient chemical knowledge is available to argue convincingly that the postulated arrangement actually occurs, but this can only be negated and never confirmed by diffraction. A completely structureless group of atoms which maintains a constant average separation between given kinds of atom, also has the correct radial distribution to fit the diffraction data.

Another diffraction technique which is rapidly gaining in popularity as a method of structure determination is X-ray powder diffraction. In this case one deals with a randomly oriented collection of single crystallites, which is midway between the total disorder of the gas phase and the perfect order of an ideal single crystal. Each crystallite produces a three-dimensional diffraction pattern as described for single crystals below, but because of their random orientation the directional properties are destroyed by the superposition of diffraction patterns in all possible orientations. It follows that any powder pattern can be generated by three-dimensional tumbling of the single-crystal diffraction pattern. In other words, it can be generated from a known single-crystal molecular structure. The real problem however, is to achieve the inverse transformation, for which no methods are known. Only limited structure refinement is feasible, provided it gets a kick-start in the form of a well-behaved trial structure.

X-RAY CRYSTALLOGRAPHY

This is the most widely applied and respected diffraction technique for crystal/molecular structure determination. Its success is atttested to by the dozen or more Nobel prizes earned by its practitioners. It is the final arbiter in many arguments about molecular conformation, bonding parameters, phase relationships and physical properties of solids. In a sense it has fallen victim to its own success.

Single-crystal analysis differs from other diffraction methods through the measurement of a diffraction pattern resolved in three dimensions, and generated from an oriented single crystal. It is the only diffraction technique where a two-way mathematical relationship exists between the observed diffraction pattern and the structure of the scatterer, which is the electron density, as distributed throughout the crystal. The key assumption is that a single crystal has three-dimensional translational symmetry, which reduces the problem to a study of the scattering density of a unit cell, rather than the entire crystal. The relationship that exists between the scattering density and the diffraction pattern is the Fourier transform

$$\rho(xyz) = (1/V) \sum_{hkl} F(hkl) \exp[-2\pi i(hx + ky + lz)] \qquad (1)$$

which means that the density (ρ) at any point (x, y, z) in the unit cell can be computed immediately by summing over all structure factors (F), measured at the diffraction points identified by the integers h, k, l. The only problem is that a diffraction intensity, $I(hkl) = |F(hkl)|^2$, rather than $F = |F| \exp(2\pi i\phi)$ is measured experimentally. This means that the phase angle (ϕ) between the primary and scattered beams cannot be measured. Ingenuity in overcoming this phase problem is what makes crystallographers famous.

It is noted that phases can be calculated if all atomic positions in the unit cell are known, by using an alternative expression for the structure factor

$$F(hkl)_j = \sum_{atoms} f_{(j)} \exp[2\pi i(hx_{(j)} + ky_{(j)} + lz_{(j)})]$$

where the $f(j)$ are the atomic scattering factors mentioned before. To succeed, this would require an inspired guess of both stucture and position. In the early days of crystallography this so-called trial-and-error method was the only procedure available. However, things have changed dramatically and today the phase problem can be overcome almost routinely by statistical methods, based on mathematical relationships between phases. The details are not important for this discussion, but the consequence is that any chemist with a crystal and a computer program can solve a structure. The drawback is that an assessment of the results becomes rather hard on somebody not familiar with the input and criteria used in the analysis. One compensation is that the structure refinement procedure and statistical analysis of the results have also been automated.

The experiment consists of intensity measurement at all accessible diffraction maxima according to index (hkl), on a spherical surface centred at the crystal. The aim is to find that unique electron distribution in the unit cell that produces a calculated diffraction pattern to match the observed in all respects. It is important to note that the Fourier transform (1) relates the entire diffraction pattern to the total electron density function, and vice versa. This is the crucial difference between structure determination by diffraction and spectroscopic techniques. The assignment of characteristic spectral features to molecular fragments has no counterpart in diffraction, which yields either all or nothing.

The analysis is often simplified by the presence of symmetry in the unit cell. This reduces the problem to finding the density in an asymmetric unit only. The total unit cell density is then generated by the space-group symmetry operations, readily deduced from the diffraction symmetry. The measure of fit between the two sets of observed and calculated intensities is a reliable guide for assessing the suitability of a trial structure, which can be a fragment only. Starting from the atomic positions of the fragment, phases can be calculated for all hkl and combined with the observed intensities in a Fourier synthesis of electron density to show up the positions of missing atoms. This is how a structure is determined crystallographically without prior assumptions. It is synthesized from the data. Once appropriate positions for all atoms are available, the fit between calculated and observed data is optimized as a function of atomic positions by least-squares procedures. The goodness of fit is commonly assessed in terms of the ratio,

$$R_{hkl} = \sum |F_{(obs)} - F_{(calc)}|_{hkl} / \sum F_{(obs)}$$

which in the hands of non-experts and the minds of many chemists has become the only criterion for evaluating crystallographic results. This is dangerous.

An obvious factor that should be taken into account is the degree of over determination, or the ratio between the number of observations and the number of variable parameters in the least-squares problem. The number of observations depends on many factors, such as the X-ray wavelength, crystal quality and size, X-ray flux, temperature and experimental details like counting time, crystal alignment and detector characteristics. The number of parameters is likewise not fixed by the size of the asymmetric unit only and can be manipulated in many ways, like adding parameters to describe complicated modes of atomic displacements from equilibrium.

A second over-worked quantity, featured in all crystallographic papers, is the estimated standard deviation on derived bond parameters. These values are obtained from the estimated standard deviations of the optimized atomic coordinates, which derive from the least-squares covariance matrix. It is a measure of the internal consistency of the least-squares problem only, and does not take any physical factors into account. It is therefore utterly misleading to view these quantities as referring to the absolute values of bond lengths or angles, but this is how they are interpreted by chemists worldwide.

The single most important factor in crystallographic analysis is crystal quality. The diffraction patterns of perfect crystals, like whiskers however, often consist of spikes that are too sharp to measure by conventional methods. The broad peaks usually measured are caused by crystal imperfections, collectively called crystal mosaicity. This means that strict translational symmetry does not extend beyond small domains in a single crystal. These domains are slightly out of alignment with neighbouring domains, stacking into a three-dimensional mosaic of randomly misaligned blocks. In order to interpret the diffraction intensities it is necessary to assume that the mosaic blocks scatter independently and without interference to add up as if all blocks had the same orientation. This is a convenient theory, but not necessarily an infallible one. Most crystals have an arrangement somewhere between the two extremes and some measure of interference between mosaic blocks is inevitable and difficult to compensate for. Any averaging over misaligned mosaic blocks must then have an influence on measured structure parameters. A standard deviation of $0.0001 \, \text{Å}$ in bond length then probably is an optimistic estimate. However, chemists have been conditioned so well that crystallographically measured variances of this magnitude, and even from different crystals, are used routinely as the basis for subtle theoretical analyses or to define modelling parameters. The absolute lower limit probably exceeds this value by two orders of magnitude, excluding the systematic bias.

Another complication is that the molecules in different blocks, or even neighbouring unit cells can have different orientations. If this happens in an orderly fashion the size of the unit cell is increased by an integral factor and a superstructure appears. Not if the alternative orientations appear at random. The size of the unit cell then remains the same and the different molecular images are crystallographically superimposed in the same space, with coincident atomic positions only occurring by accident. The situation is no worse than trying to find a solution from the composite diffraction pattern of a twinned crystal. The real problem comes when the superimposed images are largely in register, with many pairs of atoms in exact or near overlap. The resulting composite image can then easily be interpreted as that of a single molecule with severe thermal motion. Should this yield to least-squares refinement the deception becomes complete. There are probably dozens of wrong structures in the literature, because of this effect. Large thermal displacement parameters and inconsistent bond properties are the signals to look out for.

One moiety that almost invariably appears to be disordered in crystals is the perchlorate ion. There are numerous examples in the literature, particularly because of the well-known tendency of perchlorate to promote the crystallization of many complex anionic species. The perchlorate groups often occur at different sites in the same crystal with different degrees of disorder, which seems to depend on the volume of the available space between anions. In a small cavity it could even be ordered, with well-defined tetrahedral geometry. The larger the void, the more severe is the disorder. It is almost as if it has the ability to fill any free volume, within limits, to capacity, by adapting a smeared-out structure to match the shape and size of the available space. This raises the interesting possibility that a free perchlorate group is essentially structureless and assumes a tetrahedral shape only when confined into a tight spot. This could be an illustration of the Woolley mechanism[2] whereby molecular shape is an induced property that becomes apparent only in condensed phases.

If structure and shape are not intrinsic properties of free molecules and only emerge in response to environmental pressure, it reopens the whole question of just how reliable the structure is, that one obtains by crystallographic analysis. It could be argued that the electron-density transform (1), formulated in terms of space coordinates only, would always project a static ensemble average, even from a situation that fluctuates periodically at the unit cell level. Such fluctuation would add a fourth element of translational symmetry in the time coordinate, but which remains undetected unless a coherent radiation source is used to record the diffraction pattern as a function of time. This situation is no different from the calculation of a two-dimensional transform to produce a flat projection of the structure that contains no information about the third dimension. The possibility that the crystallographer looks at a time-average projection with the appearance of a rigid arrangement only, can therefore not be discounted. It would mean that the crystallographic structure is an artefact, imposed on the data by the assumption that limits translational symmetry to three space dimensions.

SPECTROSCOPIC METHODS

Spectroscopic measurements at virtually all accessible wavelenghts have been used at some stage to obtain molecular-structure information, but most of these are now of historical interest only. The most important methods still in use for structural studies are microwave spectroscopy, nuclear magnetic resonance and spectroscopy with polarized light, which provides direct information on chirality. It is important to note that all of these techniques suffer from the same problem as diffraction techniques, other than single-crystal crystallography: There is no reciprocal relationship between atomic coordinates and the spectrum. Once a conformation has been selected however, microwave spectra translate into internal coordinates, more accurate by orders of magnitude, than crystallographic results. Another advantage of microwave spectroscopy is that it provides information about (isolated) molecules in the gas phase.

The microwave experiment studies rotational structure at a given vibrational level. The spectra are analyzed in terms of rotational models of various symmetries. The vibration of a diatomic molecule is, for instance, approximated by a Morse function and the rotational frequencies are related to a molecular moment of inertia. For a classical rigid diatomic molecule this reduces to $I\mu r^2$, and since the reduced mass is known, the equilibrium bond length is calculated directly, assuming zero centrifugal distortion.

For a non-linear triatomic molecule three parameters are needed to specify the (assumed fixed) geometry, but only two moments of inertia can be measured. The situation becomes rapidly worse for larger molecules. Innovations like isotopic substitution have been used to good effect, but the situation where sufficient data exist for a unique solution has never been realized. It is also of interest to note that the quantum-mechanical models for molecular vibration and rotation are derived by assuming the potential-energy functions of classical structures. No provision is made for non-classical behaviour, like the inversion of ammonia. One faces the same dilemma as before: The spectrum may be consistent with an assumed molecular conformation, but without excluding many other non-classical possibilities.

Nuclear magnetic resonance is the most versatile of all spectroscopic techniques because it's used for the widest range of materials including liquids, solids and solutions, but like all other spectroscopic techniques it cannot be used independently to determine molecular structures[3]. More than any other spectroscopic technique it allows determination of molecular symmetry and the observation of unexpected chemistry in solution, but the literature abounds with 'confident' assignments based on chemical-shift arguments that turned out to be wrong[4].

The technique is based on the very simple principle that nuclei with spin have an associated magnetic moment which interacts with an applied field to produce level splittings. Radiation absorbed at the resonance frequency and causing transition between the spin levels is characteristic of the atom, its environment and the strength of the applied magnetic field. The resonance frequency is therefore sensitive to coupling with neighbouring spins through bonds, through space, and through chemical reorganization.

All of these effects have been studied in sufficient detail to allow reconstruction of the environment of each atom by analyzing all spectral features.

The higher the field strength the more information does the spectrum carry, but the more difficult does it become to disentangle the many overlapping multiplets. It is now standard procedure to spread the spectrum into an extra dimension and carbon atoms in an organic specimen can be sorted by number of attached protons, by their mobility, or by the chemical shift of spins to which they are coupled. However, in the final analysis a postulated structure against which to test the fragments of information is still required. Even then it's only the dominant conformation in the case of solution studies that is measured against the postulate. Many minor conformations and fluctuations are routinely overlooked. Despite its enormous power it is still true that any NMR structure determination remains tentative until confirmed by crystallographic results.

AB-INITIO CALCULATION

In the very early days of wave mechanics there existed the reasonable expectation that a comprehensive theory of chemistry could be developed by the solution of molecular wave equations. With the advent of fast electronic computers this expectation turned into believe that numerical solution of molecular wave equations was imminent and only a question of suitable computer hardware. In the meantime however, ingeneous approximate methods were devised and featured as "ab initio" theoretical chemistry. The uninitiated may believe that a molecular wave equation is actually solved by this procedure, which unfortunately is not the case.

To arrive at a theoretical model of electron density in real, chemically meaningful systems, requires many assumptions. The first essential approximation is to accept the solutions for hydrogen as the model for all other atoms. The angular dependence is in fact accepted to be identical for all atoms, assuming a central field and spherical symmetry for all. This is where chemical theory has been most uncritical, decorating each atom with a set of orbitals, each with its own pair of electrons. This qualitative use of molecular orbital theory to explain chemical reactivity and reaction mechanisms is a mine-field[5], but not of immediate concern in the present discussion that focusses more on the prediction of molecular shape by quantitative

6

calculation of the ab-initio type. The procedure is based on a technique, known as the Linear Combination of Atomic Orbitals, and used to produce molecular wavefunctions[6].

The biggest problem remains the computational effort required to reach convergence and to define the criterion for truncation of the basis set. The number of basis terms seldom exceeds the number of electron pairs. Basis functions are therefore chosen to resemble hydrogenic wavefunctions. It provides some rational basis for selecting terms and in practice they can be built up from mathematically simple functions like Slater-Type Orbitals or Gaussian Functions. The molecular Hamiltonian, operating on these improvised wavefunctions produces molecular energies which can be optimized in a variational procedure. At convergence, the coefficients of the atomic orbitals in the linear expansion constitute the ab-initio "solution".There is no attempt to predict molecular structure, which is assumed in order to initiate the calculation. The final result, after geometry optimization, returns the same assumed conformation, with only minor adjustment to individual internal parameters. In the most advanced procedure the approximate wavefunctions, obtained after the first optimization are used to calculate derivatives of the total energy with respect to nuclear displacement[7]. A force field is constructed from these and used to further optimize the geometry by iterative minimization methods. It is of interest to note how this procedure is the same as molecular mechanics, starting from an assumed structure and an empirical force field[8].

Although quantum-mechanical calculations cannot be used to predict molecular shape they have another important function, calculating the electron-density distribution in a molecule of known structure. The purpose of this is to relate chemistry to changes in molecular electron density. The effects are too subtle for detection by X-ray diffraction which maps the complete density function, and unless the core densities can be separated out the fluctuations due to chemical effects cannot be identified. It is therefore attempted to simulate the scattering from atomic cores by ab-initio calculation and to subtract this contribution from the total experimental density. The difference is called deformation density. This is a very treacherous procedure, subject to many imponderables.

The most serious problem is the quality of the best ab-initio analyses, also with respect to the calculation of atomic scattering factors. One deals with molecules in a condensed crystal phase and the simulation is attempted in terms of free-atom wavefunctions. It is therefore necessary to correct for thermal smearing, lattice vibrations, crystal-field effects and distortions due to confinement. There are no reliable guidelines and in practice special parameters to represent these factors are introduced and refined by least squares. The separation between core and valence density is therefore not as clean as one is often led to believe. The choice of basis set is also not objective and it's not inconceivable that the analysis can be tainted considerably by expectation.

ADVICE TO THE STRUCTURAL CHEMIST

X-ray crystallographers, molecular spectroscopists and quantum chemists have all engaged individually in public relations exercises, for many years and with considerable success. So successfully indeed that the average chemist needs no convincing that each molecule has an intrinsic structure which can be measured, both for isolated molecules and in condensed phases like solutions or single crystals. Even when for some reason, an experimental analysis is not feasible, the structure can be correctly predicted by computations based on the first principles of quantum theory. Since the three approaches are different in principle and design one often has three independent witnesses in agreement, leaving little room for doubt.This has not been a conscious effort to mislead chemists, but rather a bit of oversell too eagerly accepted by the customer. The situation can be put into perspective by noting that:

1. There is no proof that an isolated molecule has a rigid characteristic shape.

2. It is not possible to determine the structure of a molecule in the gas phase or in solution, experimentally.

3. The three-dimensional Fourier transform of crystallography does not necessarily contain the full translational symmetry of a crystal.

4. The R-factor of a crystallographic analysis is not the only measure of the reliability of a reported structure.

5. Crystallographic standard deviations are a measure of internal consistency only and not an absolute measure.

6. Undetected disorder can be mistaken for unusual molecular conformations in the absence of independent molecular modelling.

7. Molecular structures and shapes cannot be predicted or calculated by quantum mechanics.

8. Deformation densities are not an objective representation of the electronic effects of chemical bonding.

Indeed, there is a whole new body of structural information flowing from correlation studies using the Cambridge Structural Database and showing that the conformation of any molecule in a crystal depends critically on its environment[9]. These authors comment that crystal-packing forces stabilize low-lying points on the potential-energy surface, besides just the global minimum. With a slight change in emphasis it can also be argued that the global minimum in each case depends on the environment and that a sharp minimum does not exist for an isolated molecule. For this reason molecules in solution appear more flexible, with several conformations contributing to the observed spectra. It is even more pronounced in the gas phase and probably goes beyond flexibility towards an absence of fixed shape[10].

MOLECULAR MECHANICS

This enquiry has revealed no quantum-mechanical concept or operation that corresponds to the classical idea of molecular structure. The only direct experimental evidence that molecules have the three-dimensional shapes usually ascribed to them, comes from X-ray crystallography, but not without a question mark. The mathematical analysis of the diffraction data is designed so as to transform the information into a three-dimensional distribution which could be the static projection of a fluctuating, or even non-classical distribution, periodic·in time. The question of whether a molecule has a structure is therefore the same as asking whether a molecule is a classical or a non-classical entity. The answer may however, depend on the relative size and environment of the molecule. The same molecule that behaves non-classically under certain conditions, could behave classically in a different environment. This raises two interesting questions: How to identify the quantum/classical limit and how to describe a molecule in each of the two situations. The second question is the easier to address and one feels that non-classical molecules should be described quantum mechanically and classical molecules classically.

The quantum-mechanical problem, intuitively presumed to apply to small molecules only, will remain a problem for some time to come, but the modelling of classical molecular structures has largely been solved in terms of molecular mechanics. Its value in application to practical problems has been demonstrated convincingly in many instances, but the empirical nature of the method is theoretically less satisfying. The way forward is not in anticipation of eventually replacing molecular mechanics by quantum methods, but in discovering the fundamental principles underlying its practical successes.

The force field represents the equilibrium electron density and should be derived from that. The density, in turn, is a function of the nuclear distribution, suggesting an approach like the Hartree self-consistent-field procedure. The necessary steps are to find atomic wavefunctions that generate the electron-density function without recourse to least-squares refinement of ad hoc parameters, followed by the derivation of a transferable force field from the density gradients. That would be as close as one may hope to get to the calculation of molecular structure from first principles.

REFERENCES

1. N.L. Allinger *in:* "Theoretical and Computational Models for Organic Chemistry," S.J. Formosinho, I.G. Csizmadia and L.G. Arnaut, eds., Nato ASI Series C339, Kluwer, Dordrecht (1991)
2. R.G. Woolley, *J. Am. Chem. Soc.* 100:1073(1978)
3. J.K.M. Sanders, E.C. Constable, and B.K. Pearce, "Modern NMR Spectroscopy," Oxford University Press, 2nd ed., Oxford (1993)
4. Compare: D.A.H. Taylor, *Tetrahedron* 43:2779 (1978)
5. J.F. Ogilvie *in:* "Conceptual Trends in Quantum Chemistry," E.S. Kryachko and J.L. Calais, eds., Kluwer, Dordrecht (1994), p. 171
6. W.G. Richards and D.L. Cooper, "Ab Initio Molecular Orbital Calculations for Chemists," Clarendon Press, Oxford, 2nd ed. (1983)
7. J.E. Boggs *in:* "Theoretical Models of Chemical Bonding," Z.B. Maksić, ed., Vol. 1:185 (1990)
8. A.Y. Meyer *in:* "Theoretical Models of Chemical Bonding," Z.B. Maksić, ed., Vol. 1:214 (1990)
9. R. Taylor and F.H. Allen *in:* "Structure Correlation," H. Bürgi and J.D. Dunitz, eds., Volume 1, VCH Verlagsgesellschaft, Weinheim (1994)
10. A. Amann, *S. Afr. J. Chem.* 45:29 (1992)

MOLECULAR SHAPE

Brian T. Sutcliffe

Department of Chemistry
University of York
York YO1 5DD, England

INTRODUCTION

That a molecule has a shape has been a central idea in theoretical chemistry since about 1874. It has become so deeply embedded in chemical thinking that students take it for granted from their earliest years and regard it as a completely uncontroversial idea. Before examining the idea further, we explain how the word "shape" is going to be interpreted in the context of this paper.

The view taken is that the notion of molecular shape makes sense in contexts in which one can state in a meaningful way that the components of the molecule form a geometrical figure. Thus it is generally considered that it is meaningful to describe the methane molecule as shaped like a regular tetrahedron with hydrogen atoms at the apices and the carbon atom at the centroid, or the DNA molecule as a double stranded helix whose strands are made up of nucleotides. There exist many other possible examples but the important idea is that the molecule should be helpfully describable in terms of a three-dimensional figure, composed of sub-units with specifiable form. That is, they should have a structure: that structure should define, in terms of substructures, a three-dimensional shape. In the case of the methane molecule, the sub-structures (the atoms) are conventionally treated as points, labelled with appropriate letters of the alphabet, somewhat like a classical figure in Euclidean geometry. In the case of DNA, the substructures have various conventional representations, each used depending on the context. If the substructure units are atoms then it is conventional to join certain of them by lines, regarded as chemical bonds; the idea of certain bonded units in a molecule is clearly one intimately related to aspects of molecular shape but it is not always central in describing complicated inorganic systems such as clays, or complicated biological molecules. We end the description of the notion of shape by offering a criterion of meaningfulness in this context. All that is required of an idea in order to be meaningful is that it should serve in fruitful exchanges between groups of informed workers in the field to describe and to account for their common experience. As this criterion is essentially that for meaningfulness in a Wittgensteinian language game, it has a respectable pedigree.

In the first part of this paper we develop the idea of shape in an historical context; in the second part we put it in the context of modern quantum mechanical thinking. The first part is concerned largely with the development of the idea of a chemical bond: without this idea,

Fundamental Principles of Molecular Modeling
Edited by Werner Gans *et al.*, Plenum Press, New York, 1996

the original ideas of a three-dimensional chemical structure could not have developed. In this account the idea of a bond is not considered as central to the idea of shape. In the second part of this paper we consider to what extent a reductionist perspective is possible in attempting to account for molecular shape in the context of the quantum mechanical description of an isolated molecule. Less formally, we ask the question: "If one knew nothing about the history of the idea of molecular shape and only knew about Schrödinger's equation for the isolated molecule, would one ever get round to attributing a shape to a molecule?"

A HISTORY OF THE IDEA OF MOLECULAR STRUCTURE

In his Tilden lecture[1], Coulson dated the advent of the idea of chemical structure from the work of Frankland, Odling, Couper and Crum Brown undertaken between 1850 and 1860. We begin a bit earlier than that with the publication in 1661 of Robert Boyle's book "The Sceptical Chymist". In this book he enunciated the axiom that only what can be demonstrated to be undecomposable constituents of bodies are to be regarded as elements. Offering a distinction between chemical compounds and mixtures, he characterised a compound as the result of combination of two constituents so as to form a third that has properties different from either separate constituent. Boyle's theoretical position begs questions, but associated as it was, with a firm grasp of the need for careful experiment, it provided the basis of an operational programme that was pursued fruitfully throughout Europe in the eighteenth century. During the implementation of this programme (particularly in the work of Lavoisier) there arose the discoveries of the law of constant combining proportions and, later, the law of multiple proportions in the context of which Dalton in 1808 proposed his atomic theory.

Dalton's theory was centred on the ideas that every element comprises homogeneous indestructible atoms of which the mass is constant and that chemical compounds are formed by the union of the atoms of different elements, in the simplest numerical proportions. It is difficult now for us to appreciate the great imaginative bound made by Dalton in asserting that compounds were to be understood in terms of combinations of the atoms of elements, for there were at that time no known physical forces that seemed capable of accounting for this combination. In making this assertion Dalton therefore began in some ways the separation of chemistry from physics. The idea that atoms have a latent power (called first atomicity and later valency) of combining with other atoms in a manner that seemed peculiar to chemical change had no counterpart in the physicist's view of atoms.

The historical development of views of chemical combination until the development of a theory of molecular structure after 1860 is extremely complicated, many events in it remain in dispute among scholars. Laurent appears to be the first person in 1846, to use the word molecule in a way that is recognisable today. Laurent[2] supposed that an atom was the smallest quantity of an element that can be present in a compound. He further supposed that a molecule was the smallest quantity that can be employed in order to produce a compound. He understood the molecular weight (molar mass) of an element or compound to pertain to the quantity that, under like conditions, occupies the same volume as two atoms of hydrogen; the latter quantity he considered as a molecule of hydrogen. He thus arrived at what are considered correct empirical formulae such as Cl_2, O_2, N_2 HCl and so on, and correspondingly correct molar masses.

The extent to which Laurent's ideas influenced chemical thinking is uncertain. These ideas are regarded as becoming really influential only after Cannizzaro in 1857 re-interpreted in a convincing way Avogadro's work of 1811. (An English translation Cannizzaro's paper is found in "The Question of the Atom" which is Vol. IV of *The History of Modern Physics, 1800-1950*[3].)

However that may be, by 1852 the English chemist Edward Frankland was clearly think-ing about the concept of molecular structure for in a paper he proposed the idea of saturation capacity of an element, an idea central to development of the notion of valency and hence of molecular structure.

Frankland's early work was on inorganic compounds, he seems to have understood that what is called variable valency was a possibility, but he observed that the variability was strictly limited. He wrote that "no matter what the character of the uniting atoms may be, the combining power of the attracting element, if I may be allowed the term, is always satisfied by the same number of atoms".

Most historians of chemistry now agree that Frankland was responsible for inventing the chemical bond. According to Russell[4] the first formal statement of the idea of a bond appears in Frankland's article in *J. Chem. Soc.* **19**, 377-8 (1866). In that article Frankland wrote,

> "By the term *bond*, I intend merely to give a more concrete expression to what has received various names from different chemists, such as an atomicity, an atomic power, and an equivalence. A monad is represented as an element having one bond, a dyad as an element having two bonds, *etc.* It is scarcely necessary to remark by this term I do not intend to convey the idea of any material connection between the elements of a compound, the bonds actually holding the atoms of a chemical compound being, as regards their nature much more like those which connect the members of our solar system."

The idea of representing a bond by means of a straight line joining the atomic symbols, we probably owe to Alexander Crum Brown, who lived from 1832 until 1922. With due acknowl-edgement, Frankland adopted Crum Brown's representation (which put circles round the atom symbols) but by 1867 the circles had been dropped and more or less modern chemical notation became widespread. This adoption occurred, almost certainly because of the ease with which isomers of hydrocarbons could be easily represented and enumerated within this approach, as shown especially in the work of Cayley[5].

What is notable is Frankland's extreme caution "to avoid any speculation as to the na-ture of the tie which enables an element thus to attach itself to one or more atoms of other elements"[4]. The historical reasons for this caution are not hard to fathom for there had been much bitter argument about the nature of chemical combination since the early years of the nineteenth century. An early view was that chemical combination arose from electrical forces. We fairly attribute to Berzelius the early development of this important idea, but the rise of organic chemistry, in which the combination was not obviously of an electrical kind, led to the eclipse of his approach. The theories of types and of radicals both bid to replace it. Af-ter 1850 the idea of atoms having autonomous valences developed; according to this idea (to the development of which he contributed much) Frankland[4] invented the chemical bond. His caution about specifying precisely what was a bond, probably reflected his desire not to give hostage to fortune in a way that could support those adherents of either theory of chemical combination. This caution might also reflect philosophical controversies about the existence of unobservable entities, such as atoms and molecules, that was then a lively part of the discus-sions current in scientific and learned circles. We recall that the distinguished Anglo-American mathematician, Sylvester, wrote an article in 1878[6] with the aim to show how graphical for-mulae for molecules could be realised in purely algebraic terms. (Sylvester called his algebra "the algebra of binary quantics" but we call it the theory of binary or bilinear forms.) The reasons for his choice can, at least in part, be inferred from the following quotation

> "Chemical graphs, at all events, for the present are to be regarded as translations into geometrical forms of trains of priorities and sequences having their proper *habitat* in the sphere of order and existing quite outside the world of space. Were

it otherwise, we might indulge in some speculation as to the directions of the lines of emission or influence or radiation or whatever else the bonds might then be supposed to represent as dependent on the manner of the atoms entering into combination to form chemical substances. Such not being the case ..."

The perhaps unworthy suspicion does enter one's mind that Sylvester founded and was first editor-in-chief of the *American Journal of Mathematics* in which his article appeared, to relieve himself of such a diatribe. This suspicion is supported somewhat by the following, in which sententious thoughts are expressed.

"In regard of atomicity theory [*read* 'valency theory'], all these modes of colligation are identical, and the supposition that there is any real difference between them, or that figures in space are distinguishable from figures in a plane (as I heard suggested might be the case by a high authority at a meeting of the British Association for the Advancement of Science, where I happened to be present), is a departure from the cautious philosophical views embodied in the theory as it came from the hands of its illustrious authors and continued to be maintained by their sober-minded successors and coadjutors, and affords an instructive instance of the tendency of the human mind to worship, as if of self-subsistent realities, of the symbols of its own creation."

The idea of directed valence existed four or so years by the time that Sylvester wrote this and he was not alone in opposing van't Hoff and le Bel. In particular Kolbe (with whom Frankland actually published a paper on valence in 1857) remained vociferous in his opposition to the idea until his death in 1884. (An account of Kolbe's extreme reaction to the paper of van't Hoff on the tetrahedrally directed bonds in carbon, is found in the entertaining book by Klotz[7] and Kolbe's position is examined, in context, in Russell's book[4].)

In his paper Sylvester at first attributed the idea of a bond to Kekulé and corrected himself only in a footnote towards the end of the first appendix. Frankland however remained alert for his priority, and wrote a letter to Sylvester, which Sylvester published, to rectify matters. At the end of this letter[8], Frankland says

"I trust that you will go on with the consideration of chemical phenomena from a mathematical point of view, for I am convinced that the future progress of chemistry, as an exact science, depends very much indeed upon the alliance of mathematics ..."

I cannot decide whether Frankland had his tongue in his cheek when he wrote that, and no other chemist seems to have thought much of Sylvester's ideas either.

Frankland's ideas (as developed by Kekule in particular) led to the theory of structural isomerism. Eventually at the hands of van't Hoff and le Bel the structural formulae assumed three-dimensional form to explain optical activity and what came to be called geometrical or stereo- isomerism. Thus by 1879 the modern molecule as an entity with a definite shape had emerged as a chemical concept, *sui generis*. The graphical notation and ideas of Frankland were also further developed by Werner after 1890 in his coordination theory of inorganic complexes. (An entertaining account of the role and some of the history of structural representations in Chemistry can be found in[9] and see also[10].)

There was no agreement about whether this shape was that of a object in the real world or simply in the world of chemical ideas and the argument was often conducted with great vituperation and bitterness (see, for example[11,12]).

However this may be, developments in both physics and chemistry from about 1880 up to 1916 seemed to most people to settle the question firmly in favour of the reality of molecules as objects existing "out there". Among experiments that seemed to compel the realist view were those of Perrin on Brownian motion and of the Braggs on x-ray diffraction patterns of alkali halides. Results of both these sets of work seemed to indicate the reality of atoms and hence, by implication, that of molecules. Less well remembered now perhaps, was the work of Bjerrum who in 1912 attained a somewhat imperfect but still very plausible interpretation of the infrared spectrum of CO_2 by treating the molecular model as a semi-rigid rotor[13,14].

Although this work seemed to dictate molecular structure to be a reality with a molecule regarded as a geometrical figure in three dimensions of which the defining points could be considered as mass points, there was still no coherent theory of even atomic structure, let alone molecular structure. Such a theory was on the point of development following Thomson's "discovery" or "invention" (the term used depended on ones philosophical position) of the electron. By 1904 Thomson had developed a rather primitive theory of molecular structure involving electrons and Lewis had begun work (published only in 1916) on his theory. Lewis's theory, which was to become known as the electronic theory of valency, is still widely taught to chemistry students. The history of its development is found in Lewis's monograph[15] and an account of it at its high point is found in Sidgwick's book of 1927[16].

This theory once more placed electrical effects at the centre of an explanation of chemical combination and was much admired by chemists with a theoretical turn of mind. It seemed to account for molecular structure in a way that provided a potential link to physics. As pairs of dots and crosses taken to denote electrons in the standard Lewis symbolism might be interpreted as chemical bonds, it was hoped that, given suitable equations of motion for the electrons, chemical structure could be understood in physical terms. The basic idea was that the nuclei were the passive mass points defining the molecular shape and the electrons were somehow the active participants in causing chemical combination.

As Woolley[17] has noted, this view was adopted into the work done within the old quantum theory to derive equations of motion for electrons in molecules, but with relatively little success. However, this work influenced the way in which the "new" quantum mechanics was used after 1927 in attempts to describe molecules. In particular, in all work according to the old quantum theory it had been assumed that nuclear motion could be treated classically and that the chemist's molecular model could be used to describe the "equilibrium geometry" of a molecule. These notions were indeed at the heart of early attempts to account for molecular spectra in the old quantum theory.

Heitler and London applied the new quantum mechanics to the molecule H_2 in 1927 against this background. Instead of being concerned about nuclear motion they simply concentrated on applying quantum mechanics to the electrons at a fixed nuclear separation to evaluate whether the dissociation energy was calculated reasonably well. In fact the work of Heitler and London, though pioneering, is imperfect at a technical level: they estimated one integral and used perturbation theory. Wang[18] provided the first full account of H_2 in quantum-mechanical terms; his paper published early in 1928 reports a variational approach, with all integrals calculated *ab initio*. The equilibrium internuclear distance was evaluated by plotting the potential energy as a sum of quantum-mechanical electronic energy and classical nuclear repulsion energy; extensive comparisons were made between calculated and experimental results. It seemed that a theory of molecular structure and hence of molecular shape had been found.

In summary of the preceding discussion, to describe a molecule as having a definite shape became established following the work of van't Hoff and le Bel published in 1874. The initial basis thereof was in the explanation of optical activity and so it was not necessary to decide whether the molecule was a real object in real three-dimensional space or not. As the century progressed realism prevailed, for it seemed the most useful way to account for the vast majority of experimental results involving molecules, particularly those arising from the developing "physical" techniques such as spectroscopy, calorimetry and x-ray diffraction. No theory of chemical structure was compatible with the physicists' developing views on the structure of matter until the first tentative steps were made by Lewis to incorporate the electron into a chemical theory. Developed very fruitfully during the first quarter of the twentieth century, Lewis's ideas established the chemical view of the molecule as being composed of nuclei responding to electronic interactions to form a molecule of a particular shape. On the advent of quantum mechanics this view was incorporated into the earliest calculations. In these, the electrons were treated according to quantum mechanics whereas the nuclei made a classical response to the electronic distribution. Although the work of Born and Oppenheimer on nuclear motion, published in 1927[19], is commonly supposed to be central to thinking about the way in which nuclei were treated, a survey of the literature until 1935 shows that their paper was seldom if ever mentioned; when it was mentioned, its arguments were used as justification *a posteriori* for what was being done anyway. The way that molecules became treated in quantum mechanics arose from the ideas that chemists had about molecules and was not strictly compelled by quantum mechanics.

A QUANTUM MECHANICAL PERSPECTIVE ON MOLECULAR STRUCTURE

If all that were known about a molecule was that its structure was implicitly contained in the solutions to Schrödinger's equation for a specified number of nuclei and electrons, then it would be extremely puzzling to decide how to proceed to find the required solutions. The underlying symmetries of the problem, considered to be that of a stationary state, are those of invariance under uniform translations of all particle variables, rigid rotation-reflections of all particle variables in any origin and permutation of variables of identical particles. From the latter invariance, because of particle statistics, one can determine that only states of limited number are allowed; otherwise the invariances seem unhelpful in identifying molecules among the solutions. A spherical symmetry implied by the rotation-reflection invariance would seem to militate against any structure with a geometrical shape, for which, at most, invariance of point-group only would be expected. There is also the perennial problem that if a given Schrödinger equation describes the molecule of choice, it also describes equally all reactants from which that molecule can be formed and all products into which that molecule can decompose. Hence to adopt a radically reductionist perspective would be absurd, for that would reduce to nil the chances of finding molecular structure in solutions of the Schrödinger equation. To be too inclusive in assumptions made about solution structure would be imprudent, for fear that these might simply incorporate what one seeks to demonstrate. It is reasonable to ask, therefore, what might be the minimal assumption that could be made, that might give a chance to molecular structure to emerge from solutions to the problem, but without biasing the outcome too strongly.

Here the minimal assumption is that molecular structure has the best chance of emerging from an approach to solutions of Schrödinger's problem when the electronic and nuclear motions are decoupled as far as is possible. We emphasise that this is an assumption - other views are perfectly possible. Some workers take the view that molecular structure cannot possibly emerge from solutions for the isolated molecule problem and can emerge only from the molecule considered as interacting either with other molecules or with the vacuum field. A recent article[20] views molecular structure from such a perspective. The defence of the present assumption consists only of its minimalist equipment and its congruence with the way in which a molecule is generally considered to be a discrete entity. This assumption becomes justified or not, according to the extent to which it provides a satisfactory picture of the molecule as having structure and hence shape whilst satisfying all requirements of a solution of Schrödinger's equation. To express this in more concrete terms, we regard the picture as satisfactory if we can identify in it, without breaking any basic quantum-mechanical rules, a potential-energy hypersurface of some kind on which we can characterise a minimum to be interpreted as endowing the assemblage of nuclei with a geometrical shape. The picture, if it emerges at all, arises from an approximate wavefunction. But this condition is no source of anxiety, provided that the approximate wavefunction can be made exact in principle, again without breaking a rule.

The plan of the paper henceforth is to deploy all the technical apparatus that can be mustered to provide a rigorous formal account of decoupling and then at the end to try to recover the molecule. If the molecule seems to disappear in what follows, the reason is that as explained above, our policy is not to involve it too soon. However, we shall advert to the idea of a potential-energy hypersurface for nuclear motion because that invention is specifiable in terms of quantum mechanics as it develops.

A standard approach to nuclear motion

The quantum-mechanical approach to decoupling electronic from nuclear motions in quantum mechanics is generally attributed to Born and Oppenheimer[19], although it seems to have been something that was in the air at the time, for the earliest papers in which the perception is evident, predate the publication of their paper. The current approach to separation stems from work of Born in the early nineteen-fifties that is most easily accessible as appendix VIII in the book by Born and Huang[21].

The procedure of this approach (to be called *standard*) is that variables in the laboratory frame form of the molecular hamiltonian are divided into two sets, one consisting of L variables, x_i^e, describing the electrons and the other of H variables, x_i^n, describing the nuclei. There are $N = L + H$ particles altogether; when to distinguish types of particle is unnecessary each of N particles in the collection is labeled in the laboratory frame as x_i; $i = 1, 2, \ldots, N$ with mass m_i and charge $Z_i e$. The charge number Z_i is a positive integer for a nucleus and minus one for an electron. In a neutral system the charge-numbers sum to zero; for convenience we think of x_i as a column matrix of three cartesian components $x_{\alpha i}$, $\alpha = x, y, z$ and think of x_i collectively as a matrix x with dimensions 3 by N.

The wave function for the full problem is then assumed writable as

$$\Psi(\mathbf{x}) = \sum_p \Phi_p(\mathbf{x}^n)\psi_p(\mathbf{x}^n, \mathbf{x}^e) \tag{1}$$

in which $\psi_p(\mathbf{x}^n, \mathbf{x}^e)$ is supposed to be a solution of an electronic problem and is commonly assumed to be a solution of the clamped nuclei electronic hamiltonian

$$\hat{H}^{cn}(\mathbf{a}, \mathbf{x}^e) = -\frac{\hbar^2}{2m}\sum_{i=1}^{L}\nabla^2(\mathbf{x}_i^e) - \frac{e^2}{4\pi\varepsilon_0}\sum_{i=1}^{H}\sum_{j=1}^{L}\frac{Z_i}{|\mathbf{x}_j^e - \mathbf{a}_i|} + \frac{e^2}{8\pi\varepsilon_0}\sum_{i,j=1}^{N}{}'\frac{1}{|\mathbf{x}_i^e - \mathbf{x}_j^e|} \tag{2}$$

This hamiltonian one obtains from that fixed in the laboratory simply by assigning a values $\mathbf{a}_i$ to the nuclear variable $\mathbf{x}_i^n$; hence arises the designation *clamped nuclei* for this form. Within the electronic problem each nuclear position $\mathbf{a}_i$ is treated as a parameter; for solution of the entire problem, the electronic wave function must be available for all values of these parameters. The energy obtained from the solution of this problem depends also on the nuclear parameters and is generally called the electronic energy. We consider the potential energy hypersurface to be formed by adding the electronic energy to the classical nuclear repulsion.

Examining (2), one finds that the wavefunctions in its solutions are not quite adequate to fill the part of $\psi_p(\mathbf{x}^n, \mathbf{x}^e)$ in (1). In the first place the clamped nuclear positions in (2) define the reference frame in which the electronic problem is specified. Viewed from the laboratory frame the hamiltonian with clamped nuclei is invariant under all uniform translations and all rigid rotations and rotation-reflections of the reference frame so defined. This means that the electronic energy can depend only on the $3H - 6$ coordinates that define (locally, at least) the geometry of the nuclei and not on all the $3H$ nuclear coordinates that are assumed present in $\psi_p(\mathbf{x}^n, \mathbf{x}^e)$.

But even were the solutions of (2) adequate, the expansion form of (1) can have only a formal validity, since the molecular hamiltonian is invariant under uniform translations of the laboratory frame. This means that the centre of molecular mass moves through space like a free particle and the states of a free particle are not quantised. Thus the full molecular problem as formulated has a ubiquitously continuous spectrum and it is not possible to disentangle discrete states from this continuum. But discrete states must be characterised if there is to be any hope of recognising a molecule. Further, the molecular hamiltonian is also invariant under rigid rotations of the laboratory-fixed frame and though property presents no spectral problems, a problem arises how this invariance should be treated to maintain compatibility with the problem involving the clamped nuclei. The problem of the continuous spectrum can be solved, much as it is in classical mechanics, by separating out the centre of mass motion from the full problem. We show that this separation is problematic when we seek to undertake it so as to keep a clear notational distinction between electrons and nuclei in order to connect with the hamiltonian for clamped nuclei and hence with the usual idea of a potential energy hypersurface.

Removal of translational motion

Schrödinger's hamiltonian to describe the molecule as a system of N particles in a laboratory frame is

$$\hat{H}(\mathbf{x}) = -\frac{\hbar^2}{2} \sum_{i=1}^{N} m_i^{-1} \nabla^2(\mathbf{x}_i) + \frac{e^2}{8\pi\varepsilon_o} \sum_{i,j=1}^{N} {}' \frac{Z_i Z_j}{x_{ij}} \tag{3}$$

in which the separation between particles is defined according to

$$x_{ij}^2 = \sum_{\alpha} (x_{\alpha j} - x_{\alpha i})^2 \tag{4}$$

and α is summed over x y and z; the notation is otherwise standard. This will be taken as the full molecule hamiltonian.

To remove the centre of mass motion from this hamiltonian we transform coordinates according to

$$(\mathbf{t}\, \mathbf{X}_T) = \mathbf{x}\, \mathbf{V} \tag{5}$$

Here $\mathbf{t}$ is a matrix of size 3 by $N - 1$ and $\mathbf{X}_T$ is a matrix of size 3 by 1. $\mathbf{V}$ is a matrix of size N by N that, according to the structure of the left side of (5), has a special last column whose elements are

$$V_{iN} = M_T^{-1} m_i, \qquad\qquad M_T = \sum_{i=1}^{N} m_i \tag{6}$$

Hence $\mathbf{X}_T$ is the standard coordinate of the centre of mass.

$$\mathbf{X}_T = M_T^{-1} \sum_{i=1}^{N} m_i \mathbf{x}_i \tag{7}$$

As the coordinates $\mathbf{t}_j, j = 1, 2, \ldots, N-1$ are to be translationally invariant, we require on each remaining column of $\mathbf{V}$ that

$$\sum_{i=1}^{N} V_{ij} = 0, \qquad\qquad j = 1, 2, \ldots, N-1 \qquad (8)$$

and it is easy to see that (8) forces $\mathbf{t}_j \to \mathbf{t}_j$ as $\mathbf{x}_i \to \mathbf{x}_i + \mathbf{a}$ for all i.

The coordinates $\mathbf{t}_i$ are independent if the inverse transformation

$$\mathbf{x} = (\mathbf{t} \mathbf{X}_T) \mathbf{V}^{-1} \qquad (9)$$

exists. When we write the column matrix of the cartesian components of the partial derivative operator as $\partial / \partial \mathbf{x}_i$ then the coordinates transformed according to (5) give

$$\frac{\partial}{\partial \mathbf{x}_i} = m_i M_T^{-1} \frac{\partial}{\partial \mathbf{X}_T} + \sum_{j=1}^{N-1} V_{ij} \frac{\partial}{\partial \mathbf{t}_j} \qquad (10)$$

Hence the hamiltonian (3) in the new coordinates becomes

$$\hat{H}(\mathbf{t}, \mathbf{X}_T) = -\frac{\hbar^2}{2M_T} \nabla^2(\mathbf{X}_T) - \frac{\hbar^2}{2} \sum_{i,j=1}^{N-1} \mu_{ij}^{-1} \vec{\nabla}(\mathbf{t}_i) . \vec{\nabla}(\mathbf{t}_j) + \frac{e^2}{8\pi\varepsilon_o} \sum_{i,j=1}^{N} {}' \frac{Z_i Z_j}{f_{ij}(\mathbf{t})}. \qquad (11)$$

Here

$$\mu_{ij}^{-1} = \sum_{k=1}^{N} m_k^{-1} V_{ki} V_{kj} \qquad\qquad i, j = 1, 2, \ldots, N-1 \qquad (12)$$

and f_{ij} is just x_{ij} as given by (4) but expressed as a function of $\mathbf{t}_i$. Thus

$$f_{ij}(\mathbf{t}) = (\sum_{\alpha} (\sum_{k=1}^{N-1} ((\mathbf{V}^{-1})_{kj} - (\mathbf{V}^{-1})_{ki}) t_{\alpha k})^2)^{1/2} \qquad (13)$$

In (11) $\vec{\nabla}(\mathbf{t}_i)$ are conventional grad operators expressed in cartesian components of $\mathbf{t}_i$ and the first term represents the kinetic energy of the centre of mass. As the centre of mass variable is absent from the potential term, the effect of the centre of mass becomes separated completely; then the full solution has the form

$$T(\mathbf{X}_T)\Psi(\mathbf{t}) \qquad (14)$$

in which $\Psi(\mathbf{t})$ is a solution to the problem specified in the latter two terms of (11) which will be denoted collectively by $\hat{H}(\mathbf{t})$ and called the translationally invariant hamiltonian.

It is convenient to express the angular momentum operator in terms of X_T and the $\mathbf{t}_i$. The total angular momentum operator is written as

$$\hat{L}(\mathbf{x}) = \frac{\hbar}{i} \sum_{i=1}^{N} \hat{\mathbf{x}}_i \frac{\partial}{\partial \mathbf{x}_i} \qquad (15)$$

19

in which $\hat{\mathbf{L}}(\mathbf{x})$ and $\frac{\partial}{\partial \mathbf{x}_i}$ are column matrices of cartesian components and the skew-symmetric matrix $\hat{\mathbf{x}}_i$ is

$$\hat{\mathbf{x}}_i = \begin{pmatrix} 0 & -x_{zi} & x_{yi} \\ x_{zi} & 0 & -x_{xi} \\ -x_{yi} & x_{xi} & 0 \end{pmatrix} \tag{16}$$

The matrix $\hat{\mathbf{x}}_i$ is written also in terms of the generators of infinitesimal rotations

$$\mathbf{M}^x = \begin{pmatrix} 0 & 0 & 0 \\ 0 & 0 & 1 \\ 0 & -1 & 0 \end{pmatrix} \qquad \mathbf{M}^y = \begin{pmatrix} 0 & 0 & -1 \\ 0 & 0 & 0 \\ 1 & 0 & 0 \end{pmatrix} \qquad \mathbf{M}^z = \begin{pmatrix} 0 & 1 & 0 \\ -1 & 0 & 0 \\ 0 & 0 & 0 \end{pmatrix} \tag{17}$$

so that

$$\hat{\mathbf{x}}_i = \sum_\alpha x_{\alpha i} \mathbf{M}^{\alpha T} \tag{18}$$

A symbol for a variable capped with a caret is henceforth used to denote a skew-symmetric matrix as defined by (18).

Transforming to coordinates $(\mathbf{t}\mathbf{X}_T)$ gives

$$\hat{\mathbf{L}}(\mathbf{x}) \rightarrow \frac{\hbar}{i} \hat{\mathbf{X}}_T \frac{\partial}{\partial \mathbf{X}_T} + \frac{\hbar}{i} \sum_{i=1}^{N-1} \hat{\mathbf{t}}_i \frac{\partial}{\partial \mathbf{t}_i} \tag{19}$$

The second term is henceforth denoted $\hat{\mathbf{L}}(\mathbf{t})$ and called the translationally invariant angular momentum.

For an expansion such as (1) to be valid as a solution to the problem specified by $\hat{H}(\mathbf{t})$ and hence for the idea of a potential-energy hypersurface to be well founded, this hamiltonian must have bound-states. Whether or not the states are found is problematic. An eigenfunction of a bound state is square-integrable, has a negative energy and lies below the onset of the continuous spectrum of the system. The location of the onset of the continuous spectrum of a system can be determined, in principle, with the aid of a theorem developed by Hunziker, van Winter and Zhislin. Accounts of this, the HVZ theorem, can be found in in Reed and Simon[22] and in Thirring[23]; this theorem essentially asserts that the infimum of the continuous spectrum is the least energy at which the system can break into two non-interacting clusters. For atoms the infimum is obviously at the first ionisation energy and so is (in hartree) 0 for the hydrogen atom, $-\frac{1}{2}$ for the helium atom and so on. The intricate part of the problem is to determine if there is any state below this infimum and if so, the numbers.

For a system with a single nucleus — that is, for an atom rather than a molecule — if it is electrically positive or neutral, it was shown first by Zhislin[24] and later by Uchiyama[25], that there are a countably infinite number of bound states. Accounts of these proofs are accessible in[23] and [26]. If the system is negative then it has at most a finite number of bound states as, again, was first shown by Zhislin[27]. If the nuclei number more than one but are held clamped then the spectral properties of the system are similar to those of an atom[28], whereas if the nuclei are allowed to move, then general results are few. If a molecule becomes either too positively or too negatively charged, then it lacks bound states[29,30]. For a neutral system Simon[31] argued very persuasively that only if the infimum of the continuous spectrum is determined by disso-ciation into a pair of oppositely charged ionic clusters, will a neutral molecule have an infinite number of bound states. If the clusters are neutral, then there will be, at most, only a finite number of bound states. Vugal'ter and Zhislin[32] were able to show rigorously that Simon's belief about the neutral clusters was well founded and Evans et al.[33], were able to show that his belief about the charged clusters was too. An examination of tables of experimental values of electron affinities and ionisation energies leads to the conclusion that it is very unlikely that any diatomic molecule has an infinite number of bound states. This observation is not inconsistent with spectroscopic experience. The awkward problem is to know whether a neutral system has a bound state. Ordinary chemical experience makes it seem likely that there are some atomic combinations that lack a bound state but, so far, there are no rigorous results that enable it to be said that a particular kind of neutral system has no bound states. As for showing that a system which might have molecular solutions, actually has bound states, the most that has been rigorously proved, is that the hydrogen molecule has at least one bound state[34]. It is necessary, therefore in order to make progress, to summon up courage and go ahead as if for all problems of interest, there were a number of bound states.

This emphasis on bound states should not obscure the fact that even with the hamiltonian free of translations, there will be among its solutions continuum states resulting from the relative motion of two or more fragments. Any such solutions correspond to reactants that produce or products that result from, the putative molecule. The energies of continuum states exceed those of the bound states (if any) and thus produce no formal problems like that of the translation continuum. Such states can be treated, formally at least, in an expansion, by stipulating that the sum includes integration over the continuum. In further expansions this interpretation is assumed where appropriate, but in practice such integration cannot be achieved.

We can understand why the separation of translation is problematic for the identification of electrons and nuclei. In the hamiltonian from which translation has been removed, the inverse mass terms μ_{ij}^{-1} and the form of the potential functions f_{ij} depend intimately on the choice of V and this is essentially arbitrary. Because the hamiltonian depends on only $N-1$ variables these cannot, except in the most conventional of senses, be considered particle coordinates; the non-diagonal nature of μ_{ij}^{-1} and the peculiar form of f_{ij} also preclude any simple particle interpretation of this hamiltonian. Thus to identify electrons and nuclei after this separation is made is troublesome.

Distinguishing electronic and nuclear motions

Because the spectrum of the hamiltonian $\hat{H}(\mathbf{t})$ is independent of the choice of V, the way in which it is chosen would be immaterial if it were possible to construct exact solutions. As it is necessary in practice to use an approximation scheme, it is rational to choose V adapted to the scheme. Here the aim is to design the approximate wave functions so as to decouple maximally electronic and nuclear motions. The electronic part of the approximate wave function should ideally consist of solutions of an electronic hamiltonian which is as much like the hamiltonian (2) as possible and whose eigenvalues can be identified with electronic energies as functions of the nuclear coordinates. The nuclear part of the approximation should, again ideally, consist of solutions to a problem composed of the nuclear motion kinetic energy operator expressed entirely in terms of coordinates that arise from the original nuclear variables alone together with a potential that consists of a sum of electronic and potential energy of nuclear repulsion.

It seems reasonable to require that a subset $\mathbf{t}^n$ of the $\mathbf{t}$ for the nuclei be expressible entirely in terms of original nuclear coordinates. Thus analogously to (5)

$$(\mathbf{t}^n\mathbf{X}) = \mathbf{x}^n\mathbf{V}^n \tag{20}$$

Here $\mathbf{t}^n$ is a matrix of order 3 by $H-1$ and $\mathbf{X}$ is one of order 3 by 1. $\mathbf{V}^n$ is a square matrix of order H of which the last column is special, with elements

$$V_{iH}^n = M^{-1}m_i, \qquad\qquad M = \sum_{i=1}^{H} m_i \tag{21}$$

so that $\mathbf{X}$ is the coordinate of the centre of nuclear mass. The elements in each of the first $H-1$ columns of $\mathbf{V}^n$ each sum to zero, precisely as in (8), to ensure translational invariance.

The comparable electronic coordinates will have to involve the original nuclear coordinates so that (20) becomes generalised to

$$(\mathbf{t}^e \mathbf{t}^n \mathbf{X}) = (\mathbf{x}^e \mathbf{x}^n) \begin{pmatrix} \mathbf{V}^e & \mathbf{0} \\ \mathbf{V}^{ne} & \mathbf{V}^n \end{pmatrix} \tag{22}$$

in which $\mathbf{t}^e$ is a 3 by L matrix. It is not possible to choose $\mathbf{V}^{ne}$ to be a null matrix and to satisfy simultaneously the requirements of translational invariance as specified by (8) while leaving the whole matrix non-singular. Given that it exists, the inverse of (22) may be written as

$$(\mathbf{x}^e \mathbf{x}^n) = (\mathbf{t}^e \mathbf{t}^n \mathbf{X}) \begin{pmatrix} (\mathbf{V}^e)^{-1} & \mathbf{0} \\ \mathbf{B} & (\mathbf{V}^n)^{-1} \end{pmatrix} \tag{23}$$

in which

$$\mathbf{B} = -(\mathbf{V}^n)^{-1} \mathbf{V}^{ne} (\mathbf{V}^e)^{-1} \tag{24}$$

Because for present purposes these transformations are too general, the forms of $\mathbf{V}^{ne}$ and of $\mathbf{V}^e$ are restricted so that that the electronic coordinates are invariant under any permutation of identical nuclei and also such that they change in the usual manner under permutation of the original electronic variables. The implementation of this discussed in detail in[35] and[36] where it is shown that it is necessary to require that all columns of $\mathbf{V}^{ne}$ be identical, with a typical column denoted $\mathbf{v}$. To make a definite choice for the elements of $\mathbf{v}$ is also convenient namely,

$$v_k = -M^{-1} m_k, \tag{25}$$

$\mathbf{V}^e$ is chosen as a unit matrix so that $\mathbf{t}_i^e$ are the electronic coordinates referred to the centre of nuclear mass. This choice eliminates cross-terms between the electronic and nuclear coordinates in the kinetic energy operator and causes the elements of (24) to vanish. With these restrictions a general permutation of identical particles can be written as

$$\mathcal{P}(\mathbf{t}^e \mathbf{t}^n \mathbf{X}_T) = (\mathbf{t}^e \mathbf{t}^n \mathbf{X}_T) \begin{pmatrix} \mathbf{P}^e & \mathbf{0} & \mathbf{0} \\ \mathbf{0} & \mathbf{H} & \mathbf{0} \\ \mathbf{0} & \mathbf{0} & 1 \end{pmatrix} \tag{26}$$

in which

$$(\mathbf{H})_{ij} = ((\mathbf{V}^n)^{-1} \mathbf{P}^n \mathbf{V}^n)_{ij} \qquad\qquad i,j = 1, 2, \ldots, H-1 \tag{27}$$

The matrix $\mathbf{H}$ is not in general in standard permutation form neither is it orthogonal even though it a has determinant ± 1 according to the sign of $|\mathbf{P}^n|$.

The form of the derivative operators remains unchanged from that given in the previous section but the partition made here does enable a more specific structure to be given to them with parts attributable to the types of particle.

Thus the derivative operator (10) becomes distinguished to consist of two parts

$$\frac{\partial}{\partial x_i^e} = m M_T^{-1} \frac{\partial}{\partial \mathbf{X}_T} + \frac{\partial}{\partial t_i^e} \tag{28}$$

$$\frac{\partial}{\partial x_i^n} = m_i M_T^{-1} \frac{\partial}{\partial \mathbf{X}_T} - M^{-1} m_i \sum_{j=1}^{L} \frac{\partial}{\partial t_j^e} + \sum_{j=1}^{H-1} V_{ij}^n \frac{\partial}{\partial t_j^n} \tag{29}$$

The hamiltonian arising from the latter two terms in (11) expands into three parts

$$\hat{H}(\mathbf{t}) \rightarrow \hat{H}^e(\mathbf{t}^e) + \hat{H}^n(\mathbf{t}^n) + \hat{H}^{en}(\mathbf{t}^n, \mathbf{t}^e) \tag{30}$$

Here

$$\hat{H}^e(\mathbf{t}^e) = -\frac{\hbar^2}{2\mu} \sum_{i=1}^{L} \nabla^2(\mathbf{t}_i^e) - \frac{\hbar^2}{2M} \sum_{i,j=1}^{L}{}' \vec{\nabla}(\mathbf{t}_i^e) \cdot \vec{\nabla}(\mathbf{t}_j^e) + \frac{e^2}{8\pi\varepsilon_0} \sum_{i,j=1}^{L}{}' \frac{1}{|\mathbf{t}_j^e - \mathbf{t}_i^e|} \tag{31}$$

with

$$\mu^{-1} = m^{-1} + M^{-1} \tag{32}$$

and

$$\hat{H}^n(\mathbf{t}^n) = -\frac{\hbar^2}{2} \sum_{i,j=1}^{H-1} \mu_{ij}^{-1} \vec{\nabla}(\mathbf{t}_i^n) \cdot \vec{\nabla}(\mathbf{t}_j^n) + \frac{e^2}{8\pi\varepsilon_0} \sum_{i,j=1}^{H}{}' \frac{Z_i Z_j}{f_{ij}(\mathbf{t}^n)} \tag{33}$$

in which μ_{ij}^{-1} is defined just as in (12) but in terms of only the nuclear masses and using $\mathbf{V}^n$. Similarly $f_{ij}(\mathbf{t}^n)$ is defined just as in (13) but using only $\mathbf{t}_i^n$ and $(\mathbf{V}^n)^{-1}$.

Finally

$$\hat{H}^{en}(\mathbf{t}^n, \mathbf{t}^e) = -\frac{e^2}{4\pi\varepsilon_0} \sum_{i=1}^{H} \sum_{j=1}^{L} \frac{Z_i}{f'_{ij}(\mathbf{t}^n, \mathbf{t}^e)} \tag{34}$$

in which f'_{ij} is the distance between electron and nucleus and so is the modulus

$$|\mathbf{x}_i^n - \mathbf{x}_j^e| = |\sum_{k=1}^{H-1} \mathbf{t}_k^n (\mathbf{V}^n)_{ki}^{-1} - \mathbf{t}_j^e| \tag{35}$$

The angular momentum operator comprising the second term in (19) becomes

$$\hat{\mathbf{L}}(\mathbf{t}^n, \mathbf{t}^e) = \frac{\hbar}{i} \sum_{i=1}^{H-1} \hat{\mathbf{t}}_i^n \frac{\partial}{\partial \mathbf{t}_i^n} + \frac{\hbar}{i} \sum_{i=1}^{L} \hat{\mathbf{t}}_i^e \frac{\partial}{\partial \mathbf{t}_i^e} \tag{36}$$

and has this form, regardless of precise choices made for $\mathbf{t}^e$ and $\mathbf{t}^n$.

It is always possible to choose the $\mathbf{t}^e$ and $\mathbf{t}^n$ to make the respective kinetic energy terms "diagonal" that is, contain only the purely quadratic terms. This can be done for the electronic variable, for example, by an equivalent of the Radau choice[37] using the centre of nuclear mass as the unique heavy centre. Although any such choice undoubtedly simplifies the form of the kinetic energy operator, it does so only at the cost of making the interparticle separation terms like (35), and hence the form of the potential, more complicated. Such complications would cause difficulties in identifying of the hamiltonian for the clamped nuclei and so no such choices will be made here.

Although at this stage one could plausibly attempt a solution $\Psi(\mathbf{t})$ to the problem specified by $\hat{H}(\mathbf{t})$ in the form

$$\Psi(\mathbf{t}) = \sum_p \Phi_p(\mathbf{t}^n) \psi_p(\mathbf{t}^n, \mathbf{t}^e) \tag{37}$$

because one can reasonably believe that the problem some solutions expressible in terms of square integrable functions, the sum of (31) and (34) is still not equivalent to the required electronic hamiltonian. The reason is that the operator formed from this sum depends on $3H - 3$

23

nuclear coordinates rather than on the $3H - 6$ that the hamiltonian (2) depends on. To circumvent this difficulty, we must, as indicated earlier, account for the invariance of the full problem under rigid rotations and rotation-reflections in such a way that the remaining nuclear variables "carry" all such motions. This is done by defining a coordinate frame that rotates in a defined way with the system. This is usually called defining a frame fixed in the body and in the present case, the definition must involve only nuclear variables.

The hamiltonian fixed in the body

For a system with more than two nuclei one can transform the coordinates $\mathbf{t}^n$ such that rotational motion can be expressed in terms of three orientation variables, with the remaining motions expressed in terms of variables (commonly called internal coordinates) which are invariant under all orthogonal transformations of the $\mathbf{t}^n$. For $H = 2$ only two orientation variables are required and this case (the diatomic molecule) is rather special and is excluded from all subsequent discussion. To construct the frame fixed in the body it is supposed that the three orientation variables are specified by means of an orthogonal matrix $\mathbf{C}$, the elements of which are expressed as functions of three eulerian angles $\phi_m, m = 1, 2, 3$ which are orientation variables. We require that the the matrix $\mathbf{C}$ is specified entirely in terms of the H nuclear variables and so there will be just $3H - 6$ internal variables for the nuclei.

Thus the nuclear cartesian coordinates $\mathbf{t}^n$ are considered related to a set $\mathbf{z}^n$ by

$$\mathbf{t}^n = \mathbf{C}\mathbf{z}^n \tag{38}$$

Since $\mathbf{z}^n$ are fixed in the body, not all their $3H - 3$ components are independent, for there must be three relations between them. Hence components of $\mathbf{z}_i^n$ must be writable in terms of $3H - 6$ independent internal coordinates $q_i, i = 1, 2, \ldots, 3H - 6$. Some of the q_i maybe components of $\mathbf{z}_i^n$ but generally q_i are expressible in terms of scalar products of the $\mathbf{t}_i^n$ (and equally of the $\mathbf{z}_i^n$) since scalar products are the most general constructions that are invariant under orthogonal transformations of their constituent vectors.[*]

The electronic variables fixed in the body are then defined in terms of the above transformation by

$$\mathbf{z}_i = \mathbf{C}^T \mathbf{t}_i^e \qquad\qquad i = 1, 2, \ldots, L \tag{39}$$

in which there superscript on the electronic variables fixed in the body has been dropped. These equations *define* the electronic variable fixed in the body and thus *any* orthogonal transformation of $\mathbf{t}^n$ (including inversion) leaves them, by definition, invariant.

It can be shown[35] that the required expressions for the partial derivatives of the new variables with respect to those free of translation are

$$\frac{\partial \phi_m}{\partial t_{\alpha i}^n} = (\mathbf{C}\Omega^i \mathbf{D})_{\alpha m} \qquad\qquad \frac{\partial \phi_m}{\partial t_{\alpha i}^e} = 0 \tag{40}$$

$$\frac{\partial q_k}{\partial t_{\alpha i}^n} = (\mathbf{C}\mathbf{Q}^i)_{\alpha k} \qquad\qquad \frac{\partial q_k}{\partial t_{\alpha i}^e} = 0 \tag{41}$$

The equivalent derivatives of the internal electronic coordinates are

$$\frac{\partial z_{\gamma j}}{\partial t_{\alpha i}^n} = (\mathbf{C}\Omega^i \hat{\mathbf{z}}_j)_{\alpha \gamma} \qquad\qquad \frac{\partial z_{\gamma j}}{\partial t_{\alpha i}^e} = \delta_{ij} C_{\alpha \gamma} \tag{42}$$

---

[*] If only proper orthogonal transformations are considered the scalar triple products are also invariants but they change sign under improper operations. This and related matters are discussed in[38].

Here the elements of $\mathbf{Q}^i$ and the elements of the matrix Ω^i can be shown to depend only on internal variables. The elements of the matrices $\mathbf{C}$ and $\mathbf{D}$ can likewise be shown to be functions of only the eulerian angles. Equations (40), (41), and (42) are clearly expressions of jacobian matrix elements for transformation from the $(\phi, \mathbf{q}, \mathbf{z})$ to the $(\mathbf{t}^n, \mathbf{t}^e)$.

It is therefore possible to write derivatives of $\mathbf{t}_i^n$ coordinates in terms of the orientation and the internal coordinates as

$$
\frac{\partial}{\partial \mathbf{t}_i^n} = \mathbf{C}(\Omega^i \mathbf{D} \frac{\partial}{\partial \phi} + \mathbf{Q}^i \frac{\partial}{\partial \mathbf{q}} + \Omega^i \sum_{j=1}^{L} \hat{\mathbf{z}}_j \frac{\partial}{\partial \mathbf{z}_j}) \tag{43}
$$

and derivatives of $\mathbf{t}_i^e$ as

$$
\frac{\partial}{\partial \mathbf{t}_i^e} = \mathbf{C} \frac{\partial}{\partial \mathbf{z}_i} \tag{44}
$$

in which $\partial/\partial \phi$ and $\partial/\partial \mathbf{q}$ are column matrices of three and $3N - 6$ partial derivatives respectively and $\partial/\partial \mathbf{t}_i$ and $\partial/\partial \mathbf{z}_i$ are column matrices of three partial derivatives.

According to the formal change of variable in (38) or (39) it follows that

$$
\hat{\mathbf{t}}_i = |\mathbf{C}| \mathbf{C} \hat{\mathbf{z}}_i \mathbf{C}^T \tag{45}
$$

in which $\mathbf{t}$ and $\mathbf{z}$ can be nuclear or electronic variables and in which $|\mathbf{C}|$ is either plus or minus unity according to whether $\mathbf{C}$ corresponds to a proper or improper rotation. Using this relation together with the orthogonality relations for the jacobian and its inverse[36], we obtain for the angular momentum operator as given by the second term on the right in (19)

$$
\hat{\mathbf{L}}(\mathbf{t}) = -\frac{\hbar}{i} |\mathbf{C}| \mathbf{C} \mathbf{D} \frac{\partial}{\partial \phi} \tag{46}
$$

There is at this stage an element of choice for the sign of the angular momentum expressed in terms of the eulerian angles: we choose

$$
\hat{\mathbf{L}}(\phi) = \frac{\hbar}{i} \mathbf{D} \frac{\partial}{\partial \phi} \tag{47}
$$

and it is possible to show that components of the operators with this choice obey the *standard commutation conditions*. With this choice the angular-momentum eigenfunctions $|JMk\rangle$ have standard properties defined in Brink and Satchler[39] or in Biedenharn and Louck[40]. Explicitly, if the elements of $\mathbf{C}$ are expressed according to the conventional eulerian angle choice[39,40], then it is easy to show that

$$
|JMk\rangle = \left(\frac{2J+1}{8\pi^2}\right)^{\frac{1}{2}} (-1)^k \mathcal{D}^{J*}_{M-k}(\phi) \tag{48}
$$

in which $\mathcal{D}^J$ is the Wigner matrix[39,40]. The functions $|JMk\rangle$ are commonly called *symmetric-top* eigenfunctions. A more extended discussion of these matters can be found in section (3.8) of reference[40] and also in[41] and[42].

It can be shown[40,43] that, whatever the parametrisation of $\mathbf{C}$, the appropriate Wigner $\mathcal{D}^1$ matrix can be written as

$$
\mathcal{D}^1 = \mathbf{X}^\dagger \mathbf{C} \mathbf{X} \tag{49}
$$

25

with

$$\mathbf{X} = \begin{pmatrix} -1/\sqrt{2} & 0 & 1/\sqrt{2} \\ -i/\sqrt{2} & 0 & -i/\sqrt{2} \\ 0 & 1 & 0 \end{pmatrix} \tag{50}$$

provided that $C_{\alpha\beta}$ is ordered $\alpha, \beta = x, y, z$ and that indices on $\mathcal{D}^1$ run $+1, 0, -1$ across each row and down each column.

The elements of the general matrix $\mathcal{D}^J$ are then obtained (see[43] appendix 2) by repeated vector-coupling of the elements of $\mathcal{D}^1$.

It is thus possible to get expressions for the eigenfunctions of angular momentum directly in terms of elements of $\mathbf{C}$. (For details, see[40] section 6.19.) This possibility is extensively exploited in Louck's derivation[44] of Eckart's hamiltonian[45].

The transformation of the nuclear part of the kinetic energy operator from (33) into the coordinates ϕ and $\mathbf{q}$ is long and tedious, but the final result can be stated directly; as the derivation is mechanical, simply involving letting (43) operate on itself and summing over i and j, there is no need to go into details. The resulting operator is

$$\hat{K}(\phi, \mathbf{q}, \mathbf{z}) + \hat{K}(\mathbf{q}, \mathbf{z}) \tag{51}$$

in which

$$\hat{K}(\phi, \mathbf{q}, \mathbf{z}) = \frac{1}{2} \left(\sum_{\alpha\beta} M_{\alpha\beta} \hat{L}_\alpha \hat{L}_\beta + \hbar \sum_{\alpha} (\lambda_\alpha + 2(\mathbf{M}\hat{\mathbf{I}})_\alpha) \hat{L}_\alpha \right) \tag{52}$$

with $\hat{\mathbf{I}}$ a column matrix of cartesian components of dimension 3 by 1

$$\hat{\mathbf{I}} = \frac{1}{i} \sum_{i=1}^{L} \hat{\mathbf{z}}_i \frac{\partial}{\partial \mathbf{z}_i} \tag{53}$$

and

$$\hat{K}(\mathbf{q}, \mathbf{z}) = -\frac{\hbar^2}{2} \left(\sum_{k,l=1}^{3H-6} G_{kl} \frac{\partial^2}{\partial q_k \partial q_l} + \sum_{k=1}^{3H-6} \tau_k \frac{\partial}{\partial q_k} \right) + \frac{\hbar^2}{2} \left(\sum_{\alpha\beta} M_{\alpha\beta} \hat{l}_\alpha \hat{l}_\beta + \sum_{\alpha} \lambda_\alpha \hat{l}_\alpha \right) \tag{54}$$

The complete kinetic energy operator may be written as

$$\hat{K}(\mathbf{z}) + \hat{K}(\mathbf{q}, \mathbf{z}) + \hat{K}(\phi, \mathbf{q}, \mathbf{z}) \tag{55}$$

The first term in (55) arises trivially from the kinetic energy part of (31) simply by replacing the $\mathbf{t}^e$ by the $\mathbf{z}$ and so is

$$\hat{K}(\mathbf{z}) = -\frac{\hbar^2}{2\mu} \sum_{i=1}^{L} \nabla^2(\mathbf{z}_i) - \frac{\hbar^2}{2M} \sum_{i,j=1}^{L} {}' \vec{\nabla}(\mathbf{z}_i) . \vec{\nabla}(\mathbf{z}_j) \tag{56}$$

The potential energy operator is

$$V(\mathbf{q}, \mathbf{z}) = \frac{e^2}{8\pi\varepsilon_0} \sum_{i,j=1}^{L} {}' \frac{1}{|\mathbf{z}_j - \mathbf{z}_i|} + \frac{e^2}{8\pi\varepsilon_0} \sum_{i,j=1}^{H} {}' \frac{Z_i Z_j}{f_{ij}(\mathbf{z}^n)} - \frac{e^2}{4\pi\varepsilon_0} \sum_{i=1}^{H} \sum_{j=1}^{L} \frac{Z_i}{f'_{ij}(\mathbf{z}^n, \mathbf{z}^e)} $$

or

$$V(\mathbf{q}, \mathbf{z}) = V^e(\mathbf{z}) + V^n(\mathbf{q}) - V^{ne}(\mathbf{q}, \mathbf{z}) \tag{57}$$

Here f'_{ij} is the distance between electron and nucleus and so (see (35)) is the modulus

$$|\mathbf{x}^n_i - \mathbf{x}^e_j| = |\sum_{k=1}^{H-1} \mathbf{z}^n_k(\mathbf{q})(\mathbf{V}^n)^{-1}_{ki} - \mathbf{z}_j| \tag{58}$$

and f_{ij} is defined as explained below (33) but with $z^n_{\alpha k}(\mathbf{q})$ replacing $t^n_{\alpha k}$.

The matrix $\mathbf{M}$ is an inverse generalised inertia tensor defined as

$$\mathbf{M} = \sum_{i,j=1}^{H-1} \mu^{-1}_{ij} \boldsymbol{\Omega}^{i^T} \boldsymbol{\Omega}^j \tag{59}$$

and $\mathbf{G}$ is given by

$$\mathbf{G} = \sum_{i,j=1}^{H-1} \mu^{-1}_{ij} \mathbf{Q}^{i^T} \mathbf{Q}^j \tag{60}$$

In the term linear in angular momentum

$$\lambda_\alpha = \frac{1}{i}(\nu_\alpha + 2\sum_{k=1}^{3H-6} W_{k\alpha}\frac{\partial}{\partial q_k}) \tag{61}$$

with

$$\mathbf{W} = \sum_{i,j=1}^{H-1} \mu^{-1}_{ij} \mathbf{Q}^{i^T} \boldsymbol{\Omega}^j \tag{62}$$

and

$$\nu_\alpha = \sum_{i,j=1}^{H-1} \mu^{-1}_{ij} (\sum_\beta ((\boldsymbol{\Omega}^{i^T} \mathbf{M}^\beta \boldsymbol{\Omega}^j)_{\beta\alpha} + \sum_{l=1}^{3H-6} Q^i_{\beta l}\frac{\partial}{\partial q_l}\Omega^j_{\beta\alpha})) \tag{63}$$

As the term (61) is associated with Coriolis coupling, no coordinate system can be found in which it vanishes.

In the term linear in derivatives of the q_k

$$\tau_k = \sum_{i,j=1}^{H-1} \mu^{-1}_{ij} (\sum_\beta ((\boldsymbol{\Omega}^{i^T} \mathbf{M}^\beta \mathbf{Q}^j)_{\beta k} + \sum_{l=1}^{3H-6} Q^i_{\beta l}\frac{\partial}{\partial q_l}Q^j_{\beta k})) \tag{64}$$

It is possible to choose a coordinate system in which this term vanishes.

It is clear from both the preceding and from general group-theoretical arguments, that the eigenfunctions $\Psi(\mathbf{t})$ from (14) can be written in the form

$$\Psi(\mathbf{t}) \to \Psi^{J,M}(\phi,\mathbf{q},\mathbf{z}) = \sum_{k=-J}^{+J} \Phi^J_k(\mathbf{q},\mathbf{z})|JMk\rangle \tag{65}$$

in which the function of internal coordinates on the right side cannot depend on M because, in the absence of a field, the energy of the system does not depend on M.

Hence one can eliminate angular motion from the problem and write an effective hamiltonian within any (J,M,k) rotational manifold that depends on only the internal coordinates. From an examination of this effective hamiltonian, the origin of the clamped nuclei hamiltonian may most fruitfully be sought. Before going further, we discuss the jacobian of the transformation and its domain of validity.

As the transformation (22) is linear its jacobian is simply a constant that can be ignored: the transformation from the $\mathbf{t}_i^e$ to the $\mathbf{z}_i$ is essentially a constant orthogonal one with a unit jacobian. The transformation from the $\mathbf{t}_i^n$ to the eulerian angles and the internal coordinates is non-linear and has a jacobian $|\mathbf{J}|^{-1}$ where $\mathbf{J}$ is the matrix constructed from the nuclear terms in (40) and (41). The non-linearity is a topological consequence of any transformation that allows rotational motion to be separated[46] and there is always some conformation of the particles that causes the jacobian to vanish. Clearly where the jacobian vanishes, the transformation is undefined. This failure manifests itself in the hamiltonian by the presence of terms which diverge unless, acting on the wavefunction, they vanish. This can occur either by cancellation or by the wavefunction itself being vanishingly small in the divergent region. These and related matters are discussed in more detail elsewhere[47].

The origin of these divergences is not physical: they arise simply as a consequence of the choice of coordinates. A particular choice can obviously preclude the description of a possible physical state of a system. Thus suppose that a triatomic is described according to Eckart's approach[45,48] with the equilibrium geometry specified as bent. In this case the jacobian vanishes when the internal coordinates correspond to a linear geometry. The problem then becomes ill conditioned for states with large amplitude angular deformations. Such states are physically reasonable, merely they cannot be described according to this formulation.

The important point is that the non-linear transformation cannot be globally valid. As it has only local validity, one can at most derive a local hamiltonian that is valid within a particular domain. According to general topological considerations[46] one can construct a sequence of transformations which have common ranges of validity sufficient for passage from one to another to cover the whole space.

To remove the rotational motion we write (55) as

$$\hat{K}_I(\mathbf{q}, \mathbf{z}) + \hat{K}_R(\phi, \mathbf{q}, \mathbf{z}) \tag{66}$$

in which the first term $\hat{K}_I$ consists of the first two terms in (55). The matrix elements with respect to angular functions of operators that depend on only q_k and z_i are trivial. Thus

$$\langle J'M'k' \mid \hat{K}_I + V \mid JMk \rangle = \delta_{J'J} \delta_{M'M} \delta_{k'k} (\hat{K}_I + V) \tag{67}$$

In what follows explicit allowance for the diagonal requirement on J and M is assumed; the indices are suppressed to save writing. Similarly the fact that the integration implied is over ϕ only is left implicit.

To treat the second term in (66) is much more complicated; as no new features or principles arise from its inclusion, it is not considered further, details are available[36]. Here it can be imagined that only states with $J = 0$ are being discussed.

Within any rotational manifold eigensolutions of the effective hamiltonian given by (67) are invariant to orthogonal transformations and these functions are used to consider the separation of electronic and nuclear motion.

Separating electronic and nuclear motions

Returning to (65), we see that an expansion like (37) that applies to the internal motion in the problem and that is hence an approximate solution to the effective hamiltonian in (67) is expressed in terms of a sum of products of the form

$$\Phi_{kp}^J(\mathbf{q}) \psi_p(\mathbf{q}, \mathbf{z}) \tag{68}$$

in which p labels the electronic state and the sum is over p. The function $\psi_p(\mathbf{q}, \mathbf{z})$ is assumed known, as $|JMk\rangle$ is assumed known; the effective hamiltonian for nuclear motion is obtained in terms of matrix elements of the effective hamiltonian for internal motion between $\psi_p(\mathbf{q}, \mathbf{z})$ with respect to variables $\mathbf{z}$, just as the effective internal motion hamiltonian itself is expressed in terms of matrix elements of the hamiltonian between the $|JMk\rangle$ with respect to ϕ. The effective nuclear motion hamiltonian then contains the electronic state labels p as parameters, in much the same way that the effective hamiltonian for internal motion contains labels k of angular momentum. The analogy between the two derivations is simply formal. There is no underlying symmetry in the effective nuclear problem, nor is the sum over p of definite extent as is the sum over k.

Hunter[49] showed (at least for the case $J = 0$) that the *exact* wavefunction can be written as a single product of this form. However in Hunter's form ψ is not determined as the solution of an electronic problem but is obtained as a conditional probability amplitude by a process of integration and is to be associated with a marginal probability amplitude Φ to constitute a complete probability amplitude. The work of Czub and Wolniewicz[50] indicates that it would be very difficult to use this scheme to define, *ab initio*, a potential in terms of which nuclear motion functions could be calculated. Unless the full function is known it seems impossible to determine its parts factored in this way. For all practical purposes we must use the standard approach. In the original formulation it was stipulated that the set of known functions, $\psi_p(\mathbf{q}, \mathbf{z})$, were to be looked on as exact solutions of a problem such as

$$(\hat{K}(\mathbf{z}) + V^e(\mathbf{z}) - V^{ne}(\mathbf{q}, \mathbf{z}))\psi_p(\mathbf{q}, \mathbf{z}) = E_p(\mathbf{q})\psi_p(\mathbf{q}, \mathbf{z})$$

that is

$$\hat{H}^{\text{elec}}(\mathbf{q}, \mathbf{z})\psi_p(\mathbf{q}, \mathbf{z}) = E_p(\mathbf{q})\psi_p(\mathbf{q}, \mathbf{z}) \tag{69}$$

Because in this equation there are no terms that involve derivatives with respect to q_k, there is no development with respect to $\mathbf{q}$ in $E_p(\mathbf{q})$ or $\psi_p(\mathbf{q}, \mathbf{z})$. Thus the $\mathbf{q}$ act here simply as parameters that can be chosen at will.

It is not essential for what follows to require ψ_p to be eigenfunctions of $\hat{H}^{\text{elec}}$. A reasonably concise and useful form is obtained simply by requiring that

$$\int \psi_{p'}^*(\mathbf{q}, \mathbf{z})\psi_p(\mathbf{q}, \mathbf{z})d\mathbf{z} \equiv \langle \psi_{p'}|\psi_p \rangle_{\mathbf{z}} = \delta_{p'p} \tag{70}$$

with the above abbreviation to denote integration over all $\mathbf{z}$ only,

$$\langle \psi_{p'}|\hat{H}^{\text{elec}}|\psi_p \rangle_{\mathbf{z}} = \delta_{p'p}E_p(\mathbf{q}) \tag{71}$$

The requirements (70) and (71) are met in a simple and practical way by requiring ψ_p to be solutions of a linear variation problem with matrix elements determined by integration over $\mathbf{z}$ alone, for each and every value assigned to $\mathbf{q}$. One can argue that an account according to linear variation is more convincing than the direct one, because the assumed basis can be extended to include functions capable of providing an L^2 approximation to the continuum. In such an account, the continuum states can be approximately included.

The effective hamiltonian for nuclear motion, only depending on the $\mathbf{q}$, is expressed in terms of matrix elements of the hamiltonian just as before, between pairs of functions such as (65) but with internal coordinate parts (68) integrated over both $\mathbf{z}$ and angular factors. This procedure yields an equation rather like (67) but with coupling between different electronic states, labeled with p. In deriving it, one recalls that the product rule must be used when considering the effect of derivative operators with respect to the q_k because *both* terms in the product (68) depend on variables $\mathbf{q}$.

The term analogous to (67) becomes

$$\langle JMk'p' \mid \hat{K}_I + V \mid JMkp \rangle_{\mathbf{z}} = \delta_{p'p}\delta_{k'k}(\hat{K}_H + E_p(\mathbf{q}) + V^n(\mathbf{q})) + \delta_{k'k}\gamma_{p'p}(\mathbf{q}) \tag{72}$$

in which the designation of the variables of angular integration is left implicit as before, as has the diagonal requirement on J and M. The term $\hat{K}_H$ consists of the first group of terms from (54), namely the nuclear kinetic energy terms. The last term in (72) is given by

$$
\gamma_{p'p}(\mathbf{q}) = \frac{\hbar^2}{2} \left(\sum_{\alpha\beta} \langle \psi_{p'} | \hat{l}_\alpha \hat{l}_\beta | \psi_p \rangle_{\mathbf{z}} M_{\alpha\beta} + \sum_\alpha \langle \psi_{p'} | \hat{l}_\alpha | \psi_p \rangle_{\mathbf{z}} \lambda_\alpha \right.
$$

$$
- \sum_{k,l=1}^{3H-6} G_{kl} \left(\langle \psi_{p'} | \frac{\partial^2}{\partial q_k \partial q_l} | \psi_p \rangle_{\mathbf{z}} + \langle \psi_{p'} | \frac{\partial}{\partial q_k} | \psi_p \rangle_{\mathbf{z}} \frac{\partial}{\partial q_l} + \langle \psi_{p'} | \frac{\partial}{\partial q_l} | \psi_p \rangle_{\mathbf{z}} \frac{\partial}{\partial q_k} \right)
$$

$$
\left. + \sum_{k=1}^{3H-6} \left(\frac{2}{i} \langle \psi_{p'} | (\mathbf{W}\hat{\mathbf{l}})_k \frac{\partial}{\partial q_k} | \psi_p \rangle_{\mathbf{z}} - \tau_k \langle \psi_{p'} | \frac{\partial}{\partial q_k} | \psi_p \rangle_{\mathbf{z}} \right) \right) \tag{73}
$$

If the last term in (72) is ignored, what is left seems to be a hamiltonian for nuclear motion of the right sort for present purposes. The sum $E_p(\mathbf{q}) + V^n(\mathbf{q})$ is clearly a potential in the nuclear variables and the kinetic energy operator depends on only the nuclear variables. To place the potential in its conventional context, we try to relate the hamiltonian (2) to the electronic hamiltonian (69).

An electronic hamiltonian for clamped nuclei

The explicit form of an electronic hamiltonian $\hat{H}^{\text{elec}}$, that arises from (69) with (56), (57) and (58) is

$$
\hat{H}^{\text{elec}}(\mathbf{q}, \mathbf{z}) = -\frac{\hbar^2}{2\mu} \sum_{i=1}^L \nabla^2(\mathbf{z}_i) - \frac{e^2}{4\pi\varepsilon_0} \sum_{i=1}^H \sum_{j=1}^L \frac{Z_i}{f'_{ij}(\mathbf{q},\mathbf{z})} + \frac{e^2}{8\pi\varepsilon_0} \sum_{i,j=1}^L{}' \frac{1}{|\mathbf{z}_j - \mathbf{z}_i|}
$$

$$
- \frac{\hbar^2}{2M} \sum_{i,j=1}^L{}' \vec{\nabla}(\mathbf{z}_i) . \vec{\nabla}(\mathbf{z}_j) \tag{74}
$$

Here, from (58)

$$
f'_{ij}(\mathbf{q},\mathbf{z}) = |\mathbf{x}_i^n - \mathbf{x}_j^e| = \left| \sum_{k=1}^{H-1} \mathbf{z}_k^n(\mathbf{q}) (\mathbf{V}^n)_{ki}^{-1} - \mathbf{z}_j \right| \tag{75}
$$

If the correspondence $\mathbf{x}_i^e \to \mathbf{z}_i$ is made in (2) then it matches (74) fairly closely. The last term in (74) has no matching term in (2); neither does μ match m though this latter is unimportant for it is simply a matter of scale.

How this discrepancy should be managed is a matter of judgement. We suggest that it is reasonable to *redefine* $\hat{H}^{\text{elec}}$ as consisting of all but the last term in (74) and to extend the definition of $\gamma_{p'p}(\mathbf{q})$ in (73) so that the operator there includes the term neglected here. That is

$$
\gamma_{p'p}(\mathbf{q}) \to \gamma_{p'p}(\mathbf{q}) - \frac{\hbar^2}{2M} \sum_{i,j=1}^L{}' \langle \psi_{p'} | \vec{\nabla}(\mathbf{z}_i) . \vec{\nabla}(\mathbf{z}_j)) | \psi_p \rangle_{\mathbf{z}} \tag{76}
$$

This course of action seems reasonable for the integral in (76) resembles the first term in (73), and like that term is here multiplied by a factor involving reciprocals of nuclear masses.

One might expect the added term to be the smallest there, because it involves the reciprocal of the total nuclear mass. Its inclusion in the diagonal terms produces at most a constant energy shift dependent on nuclear mass for any electronic state. If (73) is so modified and (74) equivalently truncated all that remains is to establish the electron-nucleus attraction term in (2) matches that in (75). We suppose that the choice $\mathbf{x}^n = \mathbf{a}$ has been made and a set of constant translationally invariant coordinates $\mathbf{t}^n(\mathbf{a})$ has been defined according to (20). Then

$$\mathbf{b}_i = \sum_{k=1}^{H-1} \mathbf{t}_k^n(\mathbf{a})(\mathbf{V}^n)_{ki}^{-1} + \mathbf{X}(\mathbf{a}) \tag{77}$$

in which $\mathbf{X}(\mathbf{a})$ is the centre of nuclear mass for the conformation chosen. This set of constant translationally invariant coordinates, $\mathbf{b}_i$ generates a constant matrix $\mathbf{C}(\mathbf{b})$ that in turn generates according to (38) a set of constant cartesian coordinates $\mathbf{z}^n(\mathbf{q}(\mathbf{b}))$. The constant internal coordinates $\mathbf{q}(\mathbf{b})$ are generated in terms of scalar products of the constant translationally invariant coordinates. Then

$$\sum_{k=1}^{H-1} \mathbf{z}_k^n(\mathbf{q}(\mathbf{b}))(V^n)_{ki}^{-1} = \mathbf{C}^T(\mathbf{b})(\mathbf{b}_i - \mathbf{X}(\mathbf{a})) = \mathbf{c}_i \tag{78}$$

and (75) becomes

$$|\mathbf{a}_i - \mathbf{x}_i^e| = |\mathbf{c}_i - \mathbf{z}_i| \tag{79}$$

Although the left side fails to match the right on making the correspondence $\mathbf{x}_i^e \to \mathbf{z}_i$, at a deeper level a match can be made. This reason is that the set of all $\mathbf{c}_i$ define a geometrical object that differs from that defined by the set of all $\mathbf{a}_i$ by at most a uniform constant translation and a constant rigid rotation, so that the terms for attraction yield identical expectation values with respect to integration over the appropriate free variable $\mathbf{x}_i^e$ or $\mathbf{z}_i$. Thus, without loss of generality $\mathbf{c}_i$ can be replaced by $\mathbf{a}_i$ on the right side of (79) if it is done for all i.

From an alternative viewpoint, it is clearly possible to choose $\mathbf{a}_i$ in a set such that $\mathbf{X}(\mathbf{a}) = 0$ and such that $\mathbf{C}(\mathbf{b}) = \mathbf{E}_3$; if this is done then the matching is evident at once, as $\mathbf{c}_i = \mathbf{a}_i$.

With (76) in place of (73) in defining $\gamma_{p'p}(\mathbf{q})$, the form of (74) that may be used in (69) is

$$\hat{H}^{\text{elec}}(\mathbf{q}(\mathbf{a}), \mathbf{z}) = -\frac{\hbar^2}{2\mu} \sum_{i=1}^{L} \nabla^2(\mathbf{z}_i) - \frac{e^2}{4\pi\varepsilon_0} \sum_{i=1}^{H} \sum_{j=1}^{L} \frac{Z_i}{|\mathbf{a}_i - \mathbf{z}_j|} + \frac{e^2}{8\pi\varepsilon_0} \sum_{i,j=1}^{L}{}' \frac{1}{|\mathbf{z}_j - \mathbf{z}_i|} \tag{80}$$

The hamiltonian (2) becomes mapped exactly onto this form according to the correspondences $m \to \mu$ and $\mathbf{x}_i^e \to \mathbf{z}_i$.

There remains a disjunction between the hamiltonian (80) and the hamiltonian (2). The hamiltonian (2) is well defined in its own right and its solution can be undertaken without any thought for a particular coordinate system. It would also be possible to obtain distinct solutions at nuclear conformations that differed from one another by only a uniform translation or a rigid rotation. The hamiltonian (80) is defined assuming a definite embedding and a particular choice of translationally and rotationally invariant variables. This specification is implicit, and (80) would have the same form whatever choices were made. But in order to match (2) with (80) correctly, it is important to require that the possible solutions of (2) that differ from one another by only nuclear translations or rotations, be excluded from consideration. Although such solutions would be proper, in practice they are obtained only by accident or oversight. Such conformations must be excluded from any discussion of the potential-energy hypersurface not simply because of the matching requirement, but since their inclusion would mean that the mapping from $(\phi, \mathbf{q}, \mathbf{z}, \mathbf{X})$ to $(\mathbf{x}^n, \mathbf{x}^e)$ becomes many to one.

This discussion places the conventional clamped nuclei hamiltonian into the context of a computational strategy for the full problem. Given that it is straightforward, according to the arguments used above, to show that $V^n(\mathbf{q}(\mathbf{b}))$ in (72) is precisely the conventional nuclear repulsion, it also places the conventional form of the potential-energy function in context.

RECOGNISING THE MOLECULE

We return to seek the molecule and to try to establish its structure and shape, if this is at all possible.

Any scheme of body fixing must have regions in which the definitions fail because the jacobian vanishes. However as internal coordinates for the problem can always be written in terms of scalar products of a set of translationally invariant coordinates, the internal coordinates remain well defined even when a particular scheme of body fixing fails. In principle one may move from one embedding scheme to another keeping a single set of internal coordinates. It is therefore sufficient to assume this condition and to concentrate on identifying the molecule in terms of internal coordinates. For this purpose in this approach, we must assume that there are real systems in which the coupling term $\gamma_{p'p}(\mathbf{q})$ as given in (76) is properly defined and small, so that (72) can be sensibly treated as an operator diagonal in the electronic state label; if it is supposed that the $\psi_p(\mathbf{q}, \mathbf{z})$ are eigenfunctions of the electronic hamiltonian (69) then it is easy to show that the first derivative terms in (73) can be written as

$$\langle \psi_{p'} | \frac{\partial}{\partial q_k} | \psi_p \rangle_{\mathbf{z}} = (E_{p'}(\mathbf{q}) - E_p(\mathbf{q}))^{-1} \langle \psi_{p'} | \frac{\partial V^{ne}}{\partial q_k} | \psi_p \rangle_{\mathbf{z}} \tag{81}$$

In so far as this is a valid approach it is seen that the coupling term must be divergent whenever the two electronic energy hypersurfaces touch or intersect unless the integral on the right hand side of (81) vanishes strongly.

There is no reason to believe that the electronic wave function is generally such as to cause the right side of (81) to vanish nor is there reason to believe that the electronic wave function itself will vanish either here or where the jacobian vanishes. The electronic functions alone will not generally vanish because they contain internal coordinates (and hence the nuclear conformations) simply as parameters and these coordinates are definable independently of the embedding choice.

The electronic function enters the full problem only as part of a product function such as (68): it is sufficient then that the matrix elements of the full hamiltonian between such products not diverge. To ensure this condition the functions $\Phi_{kp}^f(\mathbf{q})$ for the particular coordinate system fixed in the body must be chosen to force any potentially divergent matrix element to vanish. Whether this is possible is clearly contingent - it must be checked in any concrete case. For present purposes it will be assumed that it is possible at least in some cases. In such cases the diagonal term from (76) may be incorporated into the potential terms in (72) and this equation solved as an uncoupled equation. If it has bound state solutions then they are prime candidates for recognition as molecules. For the time being we shall assume that suitable nuclear motion functions can be chosen to satisfy the constraints described above if the effective problem specified by (72) is otherwise well-posed.

For it to be well-posed the potential must be an analytic function of internal coordinates. Schmelzer and Murrell investigated this question in an attempt to determine internal coordinates invariant under permutations of identical particles[51] and further work was done by Collins and Parsons[52]. It is instructive to consider the question somewhat indirectly. The set of internuclear distances forms a proper set of internal coordinates and in the general case of

a system with H nuclei, there are $H(H-1)/2$ of these of which only $3H-6$ can be independent. When there are only either three or four nuclei then there are the same number of internuclear distances as there are independent coordinates; the internuclear distances can serve as
internal coordinates. However when there are five or more nuclei, the number of internuclear
distances exceeds the number of independent internal coordinates: they form a redundant set.
If they are to be used, an independent subset of them must be chosen. But one can construct
two (or more) distinct figures for the nuclear geometry in which all the chosen independent
internuclear distances are the same. An example in the case of five nuclei is given in the paper by Collins and Parsons cited above[52]. In such cases the potential energy function cannot
be an analytic function of the internal coordinates. This observation does not, of itself, show
that for systems containing five or more nuclei it is impossible to find internal coordinates in
which the potential energy is an analytic function, but clearly, if they exist one has to seek them
rather carefully. We know that the conventionally defined potential energy is described independently of any of body fixing, and so a global description in terms of coordinates which are
not independent is possible. It might be sufficient to construct a set of internal coordinates in
which the potential energy is an analytic function within a limited range. It might be possible
to connect this range with the domain in which the jacobian is non-vanishing or in which the
coupling terms do not diverge: again this would be a matter for investigation in a particular
case.

Putting aside this problem for the moment we consider the behaviour of both internal
coordinates and the eulerian angles under the permutation of of identical nuclei. We must be
able to specify the behaviour of our trial functions under permutations of identical nuclei so
that we may properly account for the particle statistics. Because of choices made in deriving
equation (26), the permutation of electrons is standard and need not be explicitly considered
here.

Let the (redundant) set of $(H-1)^2$ scalar products of $\mathbf{t}_i^n$ be denoted by the square matrix
$\mathbf{S}$, of dimension $H-1$. According to (27), a permutation

$$\mathcal{P}\mathbf{t}^n = \mathbf{t}^n\mathbf{H} = \mathbf{t}'^n \tag{82}$$

so that

$$\mathbf{S}' = \mathbf{H}^T\mathbf{S}\mathbf{H} \tag{83}$$

Making explicit the functional dependencies, (38) becomes

$$\mathbf{t}^n = \mathbf{C}(\phi)\mathbf{z}^n(\mathbf{q}) \tag{84}$$

and using (82) and (83) two expressions for permuted translationally invariant coordinates
may be obtained. The first follows at once from (84) and (82)

$$\mathbf{t}'^n = \mathbf{t}^n\mathbf{H} = \mathbf{C}(\phi)\mathbf{z}^n(\mathbf{q})\mathbf{H} \tag{85}$$

this relation yields $\mathbf{t}_i'^n$ as functions of ϕ and $\mathbf{q}$.

Alternatively, the eulerian angles and the internal coordinates can be expressed directly
as functions of $\mathbf{t}^n$ and hence of $\mathbf{t}'^n$ according to

$$\phi_m(\mathbf{t}^n) = \phi_m(\mathbf{t}'^n\mathbf{H}^{-1}) = \phi_m'(\mathbf{t}'^n) = \overline{\phi}_m(\phi, \mathbf{q}) \tag{86}$$

and

$$q_k(\mathbf{S}) = q_k(\mathbf{H}^{-T}\mathbf{S}'\mathbf{H}^{-1}) = q_k'(\mathbf{S}') = \overline{q}_k(\mathbf{q}) \tag{87}$$

33

Whereas the effect of a permutation on q_k can produce at most a function of the q_k, the effect of a permutation on ϕ_m can produce a function of both ϕ_m and q_k. If the permuted internal coordinates and eulerian angles are now used in (84) the resulting expression becomes for the permuted translationally invariant variables

$$\mathbf{t}'^n = \mathbf{C}(\overline{\phi}(\phi,\mathbf{q}))\mathbf{z}^n(\overline{\mathbf{q}}(\mathbf{q})) \tag{88}$$

so that

$$\mathbf{t}'^n = \overline{\mathbf{C}}(\phi,\mathbf{q})\overline{\mathbf{z}}^n(\mathbf{q}) \tag{89}$$

When one equates (85) and (89) one obtains

$$\overline{\mathbf{z}}^n = \overline{\mathbf{C}}^T \mathbf{C}\mathbf{z}^n\mathbf{H} \tag{90}$$

As this expression can be a function of internal coordinates at most the orthogonal matrix $\overline{\mathbf{C}}^T\mathbf{C}$ must have elements that are at most functions of internal coordinates. With this matrix denoted $\mathbf{U}$ (the coordinates are original ones, the variables are henceforth not explicitly given) it follows that

$$\overline{\mathbf{z}}^n = \mathbf{U}\mathbf{z}^n\mathbf{H} \tag{91}$$

and

$$\overline{\mathbf{C}} = \mathbf{C}\mathbf{U}^T \tag{92}$$

Thus the permuted variables are related to those unpermuted. There exists such a relationship for every distinct permutation and thus the matrices should carry a designation to indicate which permutations is being considered, but that practice would be to overload the notation in a way that is unnecessary here and is not done.

As these relationships are established, we proceed to treat the effects of a permutation on the various parts of the wavefunction. To avoid overloading the notation the convention adopted is to write the change (91) as

$$\mathbf{z}^n \to \mathbf{U}^T\mathbf{z}^n\mathbf{H}^{-1} \tag{93}$$

while (92) becomes

$$\mathbf{C} \to \mathbf{C}\mathbf{U} \tag{94}$$

when considering how a function changes upon transformation of variables.

We consider first the form of $|1Mk\rangle$ given by (48) and (49)

$$|1Mk\rangle = \left(\frac{3}{8\pi^2}\right)^{\frac{1}{2}} (\mathbf{X}^T\mathbf{C}\mathbf{X})_{Mk} \tag{95}$$

As elements of $\mathcal{D}^J$ are obtained obtained by repeated vector coupling of the elements of $\mathcal{D}^1$, a similar process is possible for $|JMk\rangle$ from $|1Mk\rangle$; thus it is sufficient to know how $|1Mk\rangle$ transforms in order to know the general result. With (94) for the change in $\mathbf{C}$ it follows from (95) that

$$|1Mk\rangle \to \left(\frac{3}{8\pi^2}\right)^{\frac{1}{2}} (\mathbf{X}^T\mathbf{C}\mathbf{U}\mathbf{X})_{Mk}$$

$$= \left(\frac{3}{8\pi^2}\right)^{\frac{1}{2}} (\mathbf{X}^T \mathbf{C} \mathbf{X} \mathbf{X}^\dagger \mathbf{U} \mathbf{X})_{Mk}$$

$$= \sum_{n=-1}^{+1} |1Mn\rangle \mathcal{D}_{nk}^1(\mathbf{U}) \tag{96}$$

Thus the change induced in the general function under $\mathcal{P}$ is

$$|JMk\rangle \rightarrow \sum_{n=-J}^{+J} |JMn\rangle \mathcal{D}_{nk}^J(\mathbf{U}) \tag{97}$$

In this equation $\mathcal{D}^J(\mathbf{U})$ is the matrix containing elements of $\mathbf{U}$ in exactly the same way that $\mathcal{D}^J$ contains elements of $\mathbf{C}$. A precise account of how this is to be done is given in[40], section 6.9. If $\mathbf{U}$ is a constant matrix, $\mathcal{D}^J(\mathbf{U})$ is a constant matrix and (97) represents simply a linear combination. If $\mathbf{U}$ is a unit matrix then $|JMk\rangle$ is invariant. This coupling of rotations by permutations can mean that certain rotational states are not allowed by Pauli's principle; this result is important in assigning statistical weights to rotational states.

What can we say precisely about the change induced in the q_k under the permutation? As the internal coordinates are expressible entirely in terms of scalar products[†] and as scalar products of $\mathbf{t}_i^n$ are identical to the scalar products of $\mathbf{z}_i^n$, the change is that given in (87), namely

$$\mathbf{q}(\mathbf{S}) \rightarrow \mathbf{q}(\mathbf{H}^{-T}\mathbf{S}\mathbf{H}^{-1}) \tag{98}$$

in which the notation of (93) is used and in which $\mathbf{S}$ is regarded as a function of q_k. This result has no general form; the best that can be stated is that a permutation of nuclei induces a general change

$$\Phi_k^J(\mathbf{q}, \mathbf{z}) \rightarrow \Phi'^J_k(\mathbf{q}, \mathbf{z}) \tag{99}$$

in which the precise nature of the transformation depends on the permutation, the chosen form of the internal coordinates and the chosen functional form. Thus the general change induced in (65) by $\mathcal{P}$ is

$$\Psi^{J,M}(\phi, \mathbf{q}, \mathbf{z}) \rightarrow \sum_{k=-J}^{+J} \sum_{n=-J}^{+J} \mathcal{D}_{nk}^J(\mathbf{U}) \Phi'^J_k(\mathbf{q}, \mathbf{z}) |JMn\rangle$$

$$= \sum_{n=-J}^{+J} \overline{\Phi}_n^J(\mathbf{q}, \mathbf{z}) |JMn\rangle \tag{100}$$

This expression is very difficult to handle because not only is $\mathbf{U}$ difficult to determine but also one must be found for each distinct permutation of identical nuclei; in a problem of any size there will are many such permutations. To choose a body-fixing matrix $\mathbf{C}$ that is invariant under all permutations of identical particles can be done by choosing $\mathbf{C}$ to be the matrix that diagonalises the translationally invariant instantaneous inertia tensor. This approach was adopted in the first attempts to fix a frame in a molecule[53,54] and has been used subsequently[55]. If this choice is made the resulting hamiltonian is inappropriate because the jacobian for the transformation vanishes in regions of physical interest. For ammonia this occurs at what is usually considered its equilibrium geometry. For this and other reasons Eckart developed his

---

[†]For this reason internal coordinates are invariant under inversion, which simply causes the $\mathbf{t}_i^n$ to change sign. Thus only the nuclear permutation group, not the permutation-inversion group, is relevant here.

body-fixing prescription[45] that is generally taken as the basis for the interpretation of molecular spectra. Here, as mentioned above, a molecule is assumed defined by a minimum in the potential hypersurface and the geometry of that minimum defines the molecular structure. Historically it was not possible to calculate such a minimum, instead a geometrical structure was assumed and the spectrum interpreted in terms of that structure. The potential can be expanded in terms of displacement coordinates and the equations of motion solved to a particular order of the expansion, commonly just to second order, leading to the familiar harmonic model. Undoubtedly this procedure is practicable and spectra are thereby interpreted successfully. The reason for this success is that molecular structure is imposed on the solution rather than arising naturally from it.

The embedding defined in Eckart's approach is generally invariant only under the subset of permutations of identical nuclei that correspond to point-group operations on the equilibrium geometry figure, discussed in detail elsewhere[43,56]. We should expect that in Eckart's approach generally, permutations would mix rotational and internal coordinates. This obviously poses problems for the standard view of the separation of rotational and vibrational motion. These matters, discussed further elsewhere[36,37], are neglected here because we assume that $J = 0$ and problems of this kind cannot arise in such states.

We proceed to consider the internal-motion part of the problem with trial wavefunctions of product kind as in (68). Any function or operator that is expressible in terms of inter-particle distances and that involves all of these distances is clearly invariant under any permutation of identical particles, and the electronic hamiltonian is obviously invariant under permutations of identical nuclei. We expect therefore, that the electronic wave function is similarly invariant. Although this condition applies to the exact solution, it does not to the usual sort of approximate electronic wavefunction. Imagine a MO calculation performed on, for example, ethene in the LCAO approximation. It would be natural to choose the AO basis η_i to be the same on each identical nucleus. Under a permutation of identical nuclei, nuclear variables in the laboratory frame transform as a standard permutation representation. As these variables enter the electronic wavefunction as the positioning parameters in the AO, the AO basis transforms in blocks of identical orbitals according to the inverse of that standard permutation representation. Thus a standard block, written as a row matrix, behaves as

$$\mathcal{P}\eta \to \eta \mathbf{P}^T \tag{101}$$

Under this permutation the matrix of one-electron integrals $h_{ij} = \langle \eta_i | \hat{h} | \eta_j \rangle$ is transformed as

$$\mathbf{h} \to \mathbf{P} \mathbf{h} \mathbf{P}^T \tag{102}$$

and similarly for the two electron integrals. Only if the permutation in (101) has the same effect as a point group operation is the matrix in (102) invariant; at a general geometry in a calculation potential energy, there is no point-group symmetry. The electronic energy calculated with the original MO coefficients and the transformed integrals are not then the same as the original energy, because the transformed wavefunction differs from the original one. The problem arises because nuclei are identified in this calculation, thus the permutational symmetry of the problem is generally broken. This point seems first to have been noted in print by Berry[57]; some difficulties arising from it are discussed elsewhere[38]. For present purposes we note that in order to calculate rigorously an electronic energy one must use a function that belongs to the totally symmetric representation of the overall permutation group of identical nuclei in the problem. Only in this way can the the electronic energy belong to the totally symmetric representation and hence match the behaviour of the potential for nuclear repulsion and with it form a properly invariant potential energy hypersurface. Not to ignore nuclear spin and statistics, one simply requires them taken properly into account in the solution to the nuclear problem.

One can achieve such permutational invariance in the electronic energy using a wave function on a single centre, the centre of nuclear mass, but the results of such calculations are generally so poor as to make them worthless for quantitative discussion. In principle one might project, from a chosen trial wave function of the typical kind, a function with the required invariance but in practice no such calculations are undertaken, instead the invariance of the hamiltonian is recognised and one assumes that the electronic energy associated with a typical wavefunction has that value for all wavefunctions that refer to the permuted identical nuclear variables. The problem from such a viewpoint is discussed in[58]. The construction of potential-energy functions that are invariant under permutations then amounts to using these typical values in functional forms that are suitably invariant. In this context, the natural variables in which to work are internuclear distances, even though they do form a redundant set.

Why typical solutions having broken symmetry turn out to be so effective has been a vexing puzzle since the beginning of molecular quantum mechanics. Much work has followed the publication in 1963 of a paper[59] by Longuet-Higgins in which permutations were classified as *feasible* or *unfeasible*; hence it is necessary to consider only the (typically small) set of feasible permutations in a given problem. A summary of much relevant work pertinent to molecular spectroscopy is reviewed in Ezra[43] and Bunker[60] and in a more general context by Kaplan[61] and in the monograph edited by Maruani and Serre[62].

The idea of unfeasibility rests on the notion that the permutation is a real motion of particles in the potential energy determined from an electronic structure calculation. If between the typical geometry and the permuted geometry there is a high barrier, the permutation is unfeasible. That a permutation is a real motion of particles is an incongruous idea from a mathematical point of view, as is the idea of unfeasibility, outside the approximation of the wave function by single product and hence a single uncoupled potential function. We speculate within the present context that if the usual orbital electronic wavefunction for a typical geometry is projected so that it has the proper permutational invariance, the elements in the projected function that are small and make only a small contribution to the electronic energy, can be neglected without great loss in energy. The permutations that produce these negligible terms can perhaps properly be called unfeasible. Although calculations might not settle this matter it would be interesting to attempt some and to discover what emerged.

CONCLUSIONS

We have sought to relate the idea of molecular structure, that is of a molecule having a definite shape, to solutions to the full problem. The relation is more problematic than is commonly thought. Therefore the idea of molecular structure is problematic from the standpoint of quantum mechanics. We do not urge that the idea should be abandoned, but recommend that when it is deployed in a quantum mechanical context, it should be deployed with care and sensitivity. Perhaps the chemical idea of a structured molecule is just not derivable from quantum mechanics as there is no logical reason why one theory should be reducible through another theory.

ACKNOWLEDGEMENTS

I should like to thank John Ogilvie very much for reading the manuscript critically and for helpful comments on its presentation.

1. C.A. Coulson, *J. Chem. Soc.* 2069 (1955).
2. E. von Meyer, "A History of Chemistry from Earliest Times to the Present Day," (transl. G. McGowan) Macmillan, London (1898).
3. Mary Jo Nye, ed., "The Question of the Atom," Tomash Publishers, Los Angeles (1984).
4. C.A. Russell, "The History of Valency," Leicester University Press, Leicester (1971).
5. A. Cayley, *Philos. Mag.* 67:444 (1874).
6. J.J. Sylvester, *American Journal of Mathematics* 1:64 (1878).
7. I.M. Klotz, "Diamond Dealers and Feather Merchants. Tales from the Sciences," Birkhauser, Boston (1986).
8. E.W. Frankland, *American Journal of Mathematics* 1:345 (1878).
9. R. Hoffmann and P. Laszlo, *Angew. Chem. Int. Ed. Eng.* 30:1 (1991).
10. B.T. Sutcliffe, *Physics Bull.* 360 (1977).
11. B.T. Sutcliffe, *J. Molec. Struct. (Theochem)* 259:29 (1992).
12. B.T. Sutcliffe, *Int. J. Quant. Chem.*, to appear (1995).
13. N. Bjerrum *in:* "Nernst-Festschrift," 90, Verlag von Knapp, Berlin (1912).
14. N. Bjerrum, *Verhandl. deut. physik. Ges.* 16:737 (1914).
15. G.N. Lewis, "Valence and the Structure of Atoms and Molecules," Chemical Catalog Co., New York (1923).
16, N.V. Sidgwick, "The Electronic Theory of Valency," Oxford University Press, Oxford (1927).
17. R.G. Woolley, *Adv. Phys.* 25:27 (1976).
18. S.C. Wang, *Phys. Rev.* 31:579 (1928).
19. M. Born and J.R. Oppenheimer, *Ann. der Phys.* 84:457 (1927).
20. E.B. Davies, *J. Phys. A. Math. and Gen.* 28:4025 (1995).
21. M. Born and K. Huang, "Dynamical Theory of Crystal Lattices," Oxford University Press, Oxford (1955).
22. M. Reed and B. Simon, "Methods of Modern Mathematical Physics, IV, Analysis of Operators," Academic Press, New York (1978).
23. W. Thirring, "A Course in Mathematical Physics, 3, Quantum Mechanics of Atoms and Molecules," transl. by E.M. Harrell, Springer-Verlag, New York (1981).
24. G.M. Zhislin, *Trudy Mosk. Mat. Obšč.* 9:82 (1960).
25. J. Uchiyama, *Pub. Res. Inst. Math. Sci. Kyoto* A 2:117 (1966).
26. B. Simon, "Quantum Mechanics for Hamiltonians Defined as Quadratic Forms," Princeton University Press, Princeton (1971).
27. G.M. Zhislin, *Theor. Math. Phys.* 7:571 (1971).
28. M.B. Ruskai and J.P. Solovej *in:* "Schrödinger Operators," Lecture Notes in Physics 403, 153, E. Balslev, ed., Springer-Verlag, Berlin (1992).
29. M.B. Ruskai, *Ann. Inst. Henri Poincaré* 52:397 (1990).
30. M.B. Ruskai, *Commun. Math. Phys.* 137:553 (1991).
31. B. Simon, *Helv. Phys. Acta* 43:607 (1970).
32. S.A. Vugal'ter and G.M. Zhislin, *Theor. Math. Phys.* 32:602 (1977).
33. W.D. Evans, R.T. Lewis and Y. Saito, *Phil. Trans. Roy. Soc. Lond.* A 338:113 (1992).
34. J.-M. Richard, J. Fröhlich, G.-M. Graf and M. Seifert, *Phys. Rev. Lett.* 71:1332 (1993).
35. B.T. Sutcliffe, *J. Chem. Soc., Faraday Transactions* 89:2321 (1993).
36. B.T. Sutcliffe *in:* "Conceptual Trends in Quantum Chemistry," E.S. Kryachko and J.L. Calais, eds., 53, Kluwer Academic, Dordrecht (1994).
37. F.T. Smith *Phys. Rev. Letts.* 45:1157 (1980).
38. B.T. Sutcliffe *in:* "Methods of Computational Chemistry 4," S. Wilson, ed., 33, Plenum Press, New York and London (1991).
39. D.M. Brink and G.R. Satchler, "Angular Momentum," 2nd ed., Clarendon Press, Oxford (1968).
40. L.C. Biedenharn and J.C. Louck, "Angular Momentum in Quantum Physics," Addison–Wesley, Reading, Mass. (1982).
41. J.M. Brown and B.J. Howard, *Mol. Phys.* 31:1517 (1976).
42. R.N. Zare, "Angular Momentum," Chap. 3.4, Wiley, New York (1988).
43. G. Ezra, "Symmetry Properties of Molecules," Lecture Notes in Chemistry 28, Springer-Verlag, Berlin (1982).
44. J.C. Louck, *J. Mol. Spec.* 61:107 (1976).
45. C. Eckart, *Phys. Rev.* 47:552 (1935).
46. B. Schutz, "Geometrical Methods of Mathematical Physics," Cambridge University Press, Cambridge (1980).
47. B.T. Sutcliffe *in:* "Theoretical Models of Chemical Bonding," Part 1, Z. Maksić, ed., 1, Springer-Verlag Berlin (1990).
48. J.K.G. Watson, *Mol. Phys.* 15:479 (1968).

49. G. Hunter, *Int. J. Quant. Chem.* 9:237 (1975).
50. J. Czub and L. Wolniewicz, *Mol. Phys.* 36:1301 (1978).
51. A. Schmelzer and J.N. Murrell, *Int. J. Quant. Chem.* 28:288 (1985).
52. M.A. Collins and D.F. Parsons, *J. Chem. Phys.* 99:6756 (1993).
53. C. Eckart, *Phys. Rev.* 46:487 (1934).
54. J.O. Hirschfelder and E. Wigner, *Proc. Nat. Acad. Sci.* 21:113 (1935).
55. B. Buck, L.C. Biedenharn and R.Y. Cusson, *Nucl. Phys.* A 317:215 (1979).
56. J.D. Louck and H.W. Galbraith, *Rev. Mod. Phys.* 48:69 (1976).
57. R.S. Berry, *Rev. Mod. Phys.* 32:447 (1960).
58. M.S. Reeves and E.R. Davidson, *J. Chem. Phys.* 95:6651 (1991).
59. H.C. Longuet-Higgins, *Molec. Phys.* 6:445 (1963).
60. P.R. Bunker, "Molecular Symmetry and Spectroscopy," Academic Press, London (1979).
61. I.G. Kaplan, "Symmetry of Many-Electron Systems," Academic Press, London (1975).
62. J. Maruani and J. Serre, eds., "Symmetries and Properties of Non-Rigid Molecules," Elsevier, Amsterdam (1983).

NEW TESTS OF MODELS IN CHEMICAL BINDING – EXTRA-MECHANICAL EFFECTS AND MOLECULAR PROPERTIES

J.F. Ogilvie

Department of Chemistry
University of the Witwatersrand
Johannesburg, South Africa

Abstract

We examine the validity of two models according to which we derive information about structural, dynamic, electric and magnetic properties of small molecules from analysis of frequency data of vibration-rotational spectra in absorption or emission. One model is the atomic approximation, according to which we associate with a particular atomic centre (atomic nucleus) electrons of number approximately equal to the protonic number of the nucleus; by this means we derive structural and dynamic information. Another model of the diatomic molecule is a rotating electric dipole, according to which parameters associated with extra-mechanical (adiabatic and nonadiabatic) effects yield information about electric and magnetic properties, namely the permanent electric dipolar moment and the rotational g factor. Spectral data of LiH and GeS are employed as tests of these models, on the basis of which evaluation of these properties proves practicable for other small molecules, for which illustrations are presented.

INTRODUCTION

Molecular spectra that consist of discrete lines measured at great resolution provide much information, initially in the forms of frequencies of line centres, integrated intensities of individual lines and shapes of lines, that with the aid of theories and models yield data about properties both of discrete molecules and of their collection in the macroscopic sample. If spectra are measured under conditions of temperature and total density such that intermolecular interactions are considered negligible, the line shape holds little interest. Frequencies may then be measured with great precision: the ratio of the uncertainty of measurement to the line frequency approaches commonly parts in 10^7 and in the best cases even parts in 10^{10}; according to Bohr's relation these frequencies that characterise the energies of the photons are proportional to differences of energies of various eigenstates of the molecule. Intensities are still subject to relatively poor precision: the ratio of the uncertainty of measurement to the integrated intensity typically approaches only parts in 10^2; these intensities are proportional to probabilities of transitions between eigenstates. If spectra of interest involve transitions within one particular electronic state, then all properties deduced from these spectra pertain primarily to that electronic state; accordingly this state becomes characterised. According to an explanation in terms of classical mechanics, such optical spectra of free molecules are associated with vibrational and rotational motions of nuclear centres and their associated electrons. In treating spectra in terms of such motions according to mechanics either classical or quantal, one introduces conventionally the idea of molecular structure as a fairly rigid arrangement of atomic centres in space of three dimensions: the position of a nucleus defines the location of that atomic centre, and the mass of a nucleus and of electrons of number about the protonic number of that nucleus so as to bestow almost electric neutrality define masses of the atomic centres. This model enables us to deduce parameters that describe structural and dynamic properties, or the mechanical effects, of the molecules within the classical model of its structure. To the extent that electrons fail to follow exactly one or other nucleus in their motions, this model is quantitatively inadequate. Spectral repercussions of this condition we attribute to extra-mechanical effects, classified as adiabatic or nonadiabatic for reasons to be explained. With the aid of a second model, that of a rotating electric dipole, we can analyse nonadiabatic rotational effects to derive parameters that pertain to electric and magnetic properties of the free molecule and to their collection within the macroscopic sample. In what follows we explain the application of these two models to a diatomic molecule; this term implies a collection of particles comprising two atomic nuclei and their associated electrons but not necessarily net electric neutrality of the molecular species. The reason that this applicability is limited to diatomic molecules is that sophistication of treatment of spectra is naturally in inverse proportion to the number of nuclei within a molecular carrier of those spectra. In analyses of vibration-rotational spectra of molecules containing three or more nuclei, formal separation of vibrational and rotational motions becomes increasingly difficult: vibrational angular momentum, vibrational and intricate vibration-rotational interactions greatly complicate these analyses. At present some progress is made for linear triatomic molecules, but at the expense of an analytic treatment that yields maximal physical insight.

After a summary of an applicable mathematical formalism, we test first the applicability of the notion that molecular structure is based on atomic centres, and then proceed to derive electric and magnetic properties based on the understanding of extra-mechanical effects.

THEORETICAL BASIS

For an assembly of two atomic nuclei A and B, designated by subscripts 'a' and 'b' and having masses M_a and M_b and charges $+Z_a$ and $+Z_b$ in units of the protonic charge respectively, and N electrons, each of mass m_e and charge -1 in the same units, the hamiltonian of minimal length contains terms for kinetic energies of electrons and of nuclei and for potential energies of coulombic attraction between nuclei and electrons and of repulsion between electrons and between the nuclei[1].

$$\mathcal{H} = \frac{-\hbar^2}{2m_e} \sum_{j=1}^{N} \nabla_j^2 - \frac{1}{2}\hbar^2 \sum_{l=a,b} \nabla_l^2/M_l - \frac{e^2}{4\pi\varepsilon_0} \left[\sum_{j=1}^{N} \sum_{l=a,b} \frac{Z_l}{r_{jl}} - \sum_{j=1}^{N} \sum_{k>j}^{N} \frac{1}{r_{jk}} - \frac{Z_a Z_b}{R} \right] \quad (1)$$

Because the only parameters in this hamiltonian are masses and charges of constituent particles of a molecule, eigenvalues of this expression enable no characterisation of a particular electronic state according to analysis of pertinent molecular spectra. For this reason we invoke a conventional classical notion of molecular structure according to which we distinguish a distance between two particular constituent particles, namely R as the instantaneous internuclear distance. We suppose that discrete spectra of interest imply a stable electronic state with a characteristic equilibrium internuclear distance R_e at which the internuclear potential energy is a minimum. Under these conditions we postulate the following effective hamiltonian to describe molecular motions relative to the origin of a coordinate system at the centre of molecular mass.

$$\mathcal{H}_{\text{eff}}(R) = \hat{P}[1 + \beta(R)]\hat{P}/(2\mu) + V(R) + [1 + \alpha(R)][J(J+1)]\hbar^2/(8\pi^2\mu R^2) \qquad (2)$$

The first term represents kinetic energy of atomic centres, the second denotes interatomic potential energy and the third takes into account centrifugal motion of atomic centres about the origin at the centre of molecular mass. The first and third terms contain explicitly the reduced atomic mass $\mu = M_a M_b/(M_a + M_b)$. The interatomic potential energy comprises three terms,

$$V(R) = V^{BO}(R) + V^{ad}(R) + V^{na}(R) \qquad (3)$$

of which the first is independent of mass.

We explain the various terms as follows. If the diatomic molecule were to consist of only two structureless 'atoms', being two point masses separated a distance R, the terms $\beta(R)$, $\alpha(R)$, $V^{ad}(R)$, $V^{na}(R)$ and any further contributions to the preceding two expressions would be absent; hence these terms take into account the fact that electrons fail to follow perfectly one or other nucleus during rotational and vibrational motions of nuclei. As vibrational potential energy depends in general not only on the distance between nuclei, in $V^{BO}(R)$, but also slightly on relative momenta of nuclei – hence on their individual masses, the term $V^{ad}(R)$ embodies this correction. All three terms $\beta(R)$, $\alpha(R)$ and $V^{ad}(R)$ contain implicitly a ratio of electronic (rest) mass m_e to a nuclear (or a reduced nuclear) mass to the power unity relative to adjacent terms[1], made explicit in subsequent equations, whereas the further terms $V^{na}(R)$ and others that might appear in an extended treatment[2] contain such a ratio to a power greater than unity. As this ratio of masses is much smaller than unity and as corresponding contributions to molecular energies from such further terms are negligible by comparison with present accuracy of almost all measurements of transition frequencies, we eliminate such terms from present consideration; in practice, retained terms thereby absorb effects of these neglected terms that might prove significant in current experiments. In a derivation of this effective hamiltonian[2], the radial function $V^{ad}(R)$ is shown to represent expectation values of various operators within an electronic state of interest, namely the electronic state within which the vibrational and rotational transitions occur, which is typically the ground electronic state; for electrically neutral molecular species to be discussed here for which the hamiltonians in equations 1 and 2 suffice this electronic state belongs to symmetry class $^1\Sigma^+$ or O^+. In contrast radial functions $\alpha(R)$ and $\beta(R)$, which represent inertia of electrons with respect to rotational and vibrational motions of nuclei respectively, represent matrix elements[2] of operators (analogous to those in the expectation values of $V^{ad}(R)$) between an electronic state of interest and excited electronic states, correspondingly belonging to symmetry classes $^1\Pi$ or 1 for $\alpha(R)$ and $^1\Sigma^+$ or O^+ for $\beta(R)$. For this reason effects within the same electronic state that $V^{ad}(R)$ represent are called adiabatic, whereas interactions between electronic states that $\alpha(R)$ and $\beta(R)$ represent formally generate nonadiabatic rotational and nonadiabatic vibrational effects, respectively. Although these adiabatic and nonadiabatic effects are mathematical artefacts due to special treatment of nuclear motion, not really physical in origin, some effects are related to significant phenomena, to be discussed.

For convenience, we introduce instead of R a reduced variable z[3,4] for displacement from equilibrium separation R_e, such that

$$z = 2(R - R_e)/(R + R_e) \qquad (4)$$

which possesses the valuable property of remaining finite throughout the range of molecular existence: for $0 < R < \infty$, $-2 < z < 2$. Henceforth with units of wavenumber for potential and other energies in conformity with spectral convention, the potential energy independent of nuclear mass becomes

$$V^{BO}(R) \to V^{BO}(z) = c_0 z^2 \left(1 + \sum_{j=1} c_j z^j\right) \tag{5}$$

and a correction for adiabatic effects is expressed

$$V^{ad}(R) \to V^{ad}(z) = (m_e/M_a) \sum_{j=1} u_j^a z^j + (m_e/M_b) \sum_{j=1} u_j^b z^j \tag{6}$$

Remaining pertinent radial functions are, for nonadiabatic rotational effects,

$$\alpha(R) \to \alpha(z) = (m_e/M_a) \sum_{j=0} t_j^a z^j + (m_e/M_b) \sum_{j=0} t_j^b z^j \tag{7}$$

and for nonadiabatic vibrational effects,

$$\beta(R) \to \beta(z) = (m_e/M_a) \sum_{j=0} s_j^a z^j + (m_e/M_b) \sum_{j=0} s_j^b z^j \tag{8}$$

Coefficients c_j for potential energy and u_j^a and u_j^b for adiabatic effects are in principle independent parameters, but coefficients t_j^a and t_j^b are more directly inter-related through other quantities, to be discussed; s_j^a and s_j^b are likewise inter-related.

With these terms in the hamiltonian, we express energies within a particular electronic state, or vibration-rotational terms of molecular eigenstates, according to an expression[5]

$$E_{vJ} = \sum_{k=0} \sum_{l=0} (Y_{kl} + Z_{kl}^{v,a} + Z_{kl}^{v,b} + Z_{kl}^{r,a} + Z_{kl}^{r,b})(v + \tfrac{1}{2})^k [J(J+1)]^l \tag{9}$$

in which explicit dependence of E_{vJ}, Y_{kl} and various Z_{kl} on nuclear masses in a particular isotopic species is suppressed. How these term coefficients Y_{kl} and Z_{kl} depend on radial coefficients $c_j, s_j^{a,b}, t_j^{a,b}$ and $u_j^{a,b}$ according to analytic relations is explained elsewhere[5]: in essential summary, coefficients Y_{kl} depend on reduced atomic mass μ, on R_e and on potential-energy coefficients c_j; coefficients Z_{kl}^v take into account additional vibration-rotational effects that depend on the individual mass of atomic centre A or B, M_a or M_b respectively, and depend too not only on c_j but also on s_j and u_j of A or B, whereas coefficients Z_{kl}^r that take into account further rotational effects of an individual atomic centre A or B depend on the mass of A or B, on c_j, and on s_j and t_j of A or B. According to this scheme, coefficients c_j with $j > 0$, s_j and t_j have typically magnitudes of order unity and are dimensionless, whereas c_0 and u_j have magnitudes of order B_e/γ^2 and dimensions conventionally of wavenumber; here γ is a dimensionless ratio of the limiting interval $2B_e$ between successive lines in a pure rotational band to the limiting interval ω_e between vibrational bands in a sequence, and takes values in a range $[0.0001, 0.026]$ for known molecules in their electronic ground states, with B_e the equilibrium rotational parameter. Relations for Y_{kl} in terms of c_j that are coefficients of z^j are equivalent to those reported by Dunham[6] in terms of a_j that are coefficients of x^j; $x = (R - R_e)/R_e$, which lacks a finite value as $R \to \infty$. The fact that vibration-rotational terms E_{vJ} containing only term coefficients Y_{kl} failed to represent accurately effects of mass scaling in molecular spectra of isotopic variants led van Vleck to delineate[7] adiabatic and nonadiabatic effects that are represented here in additional term coefficients Z_{kl}.

An alternative formula for vibration-rotational terms

$$E_{vJ} = \sum_{k=0} \sum_{l=0} U_{kl} \mu^{-(k/2+l)} [1 + m_e(\Delta_{kl}^a/M_a + \Delta_{kl}^b/M_b)](v + \tfrac{1}{2})^k [J(J+1)]^l \tag{10}$$

has an empirical basis[8], which is underpinned theoretically[1]. To understand the relationship between parameters in equations 9 and 10, we explain that term coefficients in each set Y_{kl} or various Z_{kl} are expressed as a sum of contributions; rapidly decreasing magnitudes of successive contributions generally ensure acceptable convergence. Thus

$$Y_{kl} = Y_{kl}^{(0)} + Y_{kl}^{(2)} + Y_{kl}^{(4)} + \dots \tag{11}$$

in which each successive contribution contains a further factor γ^2. The leading term $Y_{kl}^{(0)}$ contains the reduced atomic mass μ to a power $-(k/2+l)$; defined as $Y_{kl}^{(0)} \mu^{(k/2+l)}$, U_{kl} then becomes formally independent of mass. Parameters Δ_{kl} of atomic centre A or B, which are also formally independent of mass, are likewise evidently related to $Y_{kl}^{(2)}$ and to the sum of all Z_{kl} (or to at least their leading terms $Z_{kl}^{(0)}$). Further terms containing $(m_e/M_j)^2$, $j = a$ or b, etc. that might be included in equation 10 are not yet needed in relation to uncertainties of measurements of transition frequencies. As we accordingly truncate equation 10, we correspondingly truncate equation 11 after $Y_{kl}^{(2)}$ and the analogous relations for Z_{kl} after $Z_{kl}^{(0)}$, because the ratio of electronic and nuclear masses, m_e/M_j, $j = a$ or b, has a magnitude of the order of γ^2.

A physical but qualitative explanation of the nature of adiabatic and nonadiabatic effects follows. Although one formally considers vibrational and rotational motions of molecules to involve primarily nuclei – indeed some theoretical treatments are based rigorously on that premise, in practice only relatively imprecise values of nuclear masses are available for most nuclides; to match the precision of typical relevant spectral data one has perforce to use atomic masses. If electrons followed perfectly one or other nuclei (so as to maintain effective near neutrality of charge about each atomic centre), terms $V^{ad}(R)$, $\alpha(R)$ and $\beta(R)$ in equations 2 and 3 would be superfluous. One might imagine that electrons outside a region between nuclei can follow well the motion of the nearer nucleus because of strong electrostatic attraction; as electrons between nuclei would likewise be attracted to both nuclei, they might tend to follow neither nucleus. The effective mass of each atomic centre would then differ from that of the isolated neutral atom; terms $\alpha(R)$ and $\beta(R)$ thereby constitute corrections to these atomic masses to take account of this nominally physical effect, and the radial dependence reflects that effective atomic mass depends on internuclear distance. As 'valence' electrons are indistinguishable from electrons of any other purported kind, this explanation is qualitative but may assist one to appreciate the magnitude of adiabatic and nonadiabatic effects despite their artefactual nature.

According to the preceding argument, from spectra of a molecule containing atomic centres characterised with distinct atomic numbers one can derive information of two types – dependence of vibration-rotational terms on mass of each atomic centre separately (rather than on merely the reduced atomic mass), and further rotational effects. Terms within the hamiltonian that yield these components of eigenenergies number three – adiabatic, nonadiabatic rotational and nonadiabatic vibrational. For this reason one cannot in general evaluate separately these effects from only frequencies of transitions of diatomic molecules of multiple isotopic variants applicable to samples of which spectra are measured in the absence of an externally applied electric or magnetic field. There exists however a magnetic interaction of rotating molecules that lack net electronic spin or orbital angular momentum in their rotationless states (electronic state of symmetry class $^1\Sigma^+$ or O^+). Under conditions of appropriate alignment of the magnetic field, this Zeeman effect produces a splitting of lines, the extent of which is proportional to a molecular parameter called the rotational g factor, which is formally an expectation value of a particular vibration-rotational state[9]. In principle, on analysis of such data for many and varied vibration-rotational states one can generate precisely the radial function $\alpha(R)$, but in practice measurements of g_J are common for only the rotational state $J = 1$, and seldom for other than the vibrational ground state $v = 0$.

EVALUATION OF ADIABATIC AND NONADIABATIC EFFECTS FOR LiH

The diatomic species LiH, which exists as a stable molecular compound in the gaseous phase in equilibrium with the ionic crystalline substance at temperatures above about 800 K, presents in principle an excellent opportunity to examine adiabatic and nonadiabatic effects according to the preceding discussion: because isotopic variants of both Li (^{6}Li and ^{7}Li) and H (actually ^{1}H and ^{2}H, and prospectively even ^{3}H) are available and because these nuclei are relatively light, these effects might be expected to be more readily detected for this compound than for others. The rotational g factor is measured accurately for both ^{7}Li^{1}H and ^{7}Li^{2}H [10], but for only one rotational state in the vibrational ground state (i.e. $v = 0$, $J = 1$); a qualitative indication of the rotational variation of g_J is available from other measurements[11]. Available spectra include pure rotational transitions in the millimetre-wave and far infrared regions and vibration-rotational transitions in the mid infrared region corresponding to the progression $\Delta v = 1$, for four isotopic variants containing stable nuclei, as described in detail elsewhere[12]. Although these spectral data have smaller precisions than those for the best measurements on comparable compounds, they serve to illustrate the approximate orders of magnitudes of pertinent effects, provided that we combine them with calculated information about the radial dependence of the rotational g factor, as available experimental data[10,11] are too sparse.

To employ meaningfully these data to discern the sought magnitudes we consider the ratio of contributions to Z_{kl}^v and Z_{kl}^r of each nuclear type to Y_{kl}, and further we divide this ratio by the quotient of electronic and atomic masses. The reason for this approach is that according to a theoretical analysis, developed first by Born and Oppenheimer and later clarified and extended[13], we expect these adiabatic and nonadiabatic effects to exhibit such relative magnitudes involving the ratio of m_e and a nuclear (or atomic) mass M_a or M_b, as is clear from the comparison of equations 9 and 10. Accordingly, in table 1 we present the quantities $(Z_{kl}/Y_{kl})/(m_e/M_j)$, for various contributions to Z_{kl}, for various values of k and l for which data are available[14], and for two atomic centres, $j = $ Li and H, in ^{7}Li^{1}H.

The most important conclusion from these results is that ratios of components of extra-mechanical effects to mechanical effects have indeed magnitudes of the order of the ratio of the electronic to an atomic (or nuclear) mass, although there appears to be a significant trend of increased magnitude in the sequence with $k = 0$ and increasing l of LiH. Apart from exact cancellation of nonadiabatic vibrational effects in Z_{kl}^r with those in Z_{kl}^v for $k = 0$ previously noted[5], one discerns that adiabatic, nonadiabatic rotational and nonadiabatic vibrational contributions to extra-mechanical effects in LiH have comparable orders of magnitude, whereas those adiabatic effects that vary with internuclear distance in SiS have magnitudes too small to be discerned from available spectral data; adiabatic effects that do not vary with internuclear distance are undetectable from these spectral analyses. In the case of CO, adiabatic contributions are detectable but their magnitudes seem consistently smaller than those of nonadiabatic rotational and nonadiabatic vibrational contributions[15]. Thus for vibration-rotational states in the electronic ground state far from the limit of the least energy of dissociation to neutral atomic fragments in their appropriate electronic states, the approximation of atomic centres within a molecule is proved to have an extent of validity expected according to work of Born, Oppenheimer, Fernandez and others[13]. A classical model of a molecule as a fairly rigid arrangement of 'atoms' in space is hereby justified for vibration-rotational states of diatomic molecules well below the dissociation limit within the expected range of validity, although there remains no implication from these results that within molecules there exist atoms

Table 1. Ratio of adiabatic (ad), nonadiabatic rotational (nr) and nonadiabatic vibrational (nv) contributions to extra-mechanical term coefficients Z_{kl} to the corresponding mechanical term coefficient Y_{kl}, divided by the ratio of electronic to atomic mass.[a]

		Z^r_{kl}		Z^v_{kl}	
k	l	nr	nv	ad	nv
Li in ^{7}Li^{1}H					
0	1	0.77	...	−0.87	...
0	2	−0.95	0.67	−1.75	−0.67
1	0	...	...	−0.42	0.33
H in ^{7}Li^{1}H					
0	1	−0.75	...	−0.81	...
0	2	−2.72	−0.30	−1.36	0.30
0	3	−7.30	−1.09	−2.41	1.09
0	4	−13.2	−2.04	−3.27	2.04
1	0	...	...	−0.52	−0.15
1	1	−0.04	0.11	−0.49	−0.14
1	2	2.26	0.39	...	−1.13
2	0	...	...	...	−0.49
2	1	...	0.18	...	...
Si in ^{28}Si^{32}S					
0	1	−1.17	...	...	...
1	0	...	...	...	0.88
S in ^{28}Si^{32}S					
0	1	−1.55	...	...	...
1	0	...	...	...	−0.16

[a]The mark ... indicates that the pertinent quantities are indeterminate.

(that have a 'shape', 'volume' or other property of an isolated, electrically neutral atom in the gaseous phase).

APPLICATION OF THE MODEL OF A ROTATING DIPOLE

According to this model two stationary point masses, M_a and M_b, carry charges of equal magnitude but opposite sign, $+q$ and $-q$ respectively, and are separated a distance R_e. This stationary dipole possesses electric dipolar moment μ^e and exerts an electric field but no magnetic field relative to axes fixed in space. If such a dipole rotates about its centre of mass, the rotation of each separate pole is equivalent to an electric current in a loop of a conductor; each such loop generates a magnetic field. To the extent that masses M_a and M_b are not equal, the radii of the two loops differ and the magnetic fields thus fail to cancel one another. This net magnetic field implies the existence of a magnetic dipole, of which the magnetic dipolar moment is proportional to the rotational angular momentum of the rotating masses; the factor of proportionality involves the nuclear magneton, to carry units, and a rotational g factor that thereby lacks dimensions. For a rotating diatomic molecule as a rotating electric dipole, molecular rotation is supposed to induce an interaction with electronically excited states; if the electronic ground state lacks net electronic spin or orbital angular momentum, belonging to symmetry class $^1\Sigma^+$ or O^+, the rotational motion induces interaction with excited states of symmetry class $^1\Pi$ or 1; we ignore the effects of net intrinsic nuclear angular momentum. According to the relation[16]

$$g_J = g_J^{nr} + m_p|\mu^e|(1/M_a - 1/M_b)/(eR_e) \tag{12}$$

the rotational g factor contains two contributions, the former from inertia of electrons with respect to rotating nuclei or nonadiabatic rotational effects, and the latter due to a rotating electric dipole of moment having magnitude $|\mu^e|$ with a positive pole of mass M_a separated from the negative pole of mass M_b by a distance R_e; e is the charge on the proton and m_p its mass. When a rotating diatomic molecule interacts with an externally applied magnetic field appropriately oriented, spectral lines associated with pure rotational transitions become split; the extent of splitting depends on the rotational g factor, but direct determination of the sign may require a circularly polarised magnetic field. Although the vibrational dependence of g_J, formally an expectation value of a particular vibration-rotational state, is detectable in conventional experiments involving the Zeeman effect, rotational dependence presses present limits of sensitivity.

The relation between the radially dependent rotational g factor, of which the expectation value is measurable, and radial functions defined above is expressed as

$$g_J(R)(m_e/m_p) = m_e \left[\sum_{j=0} t_j^a z^j / M_a + \sum_{j=0} t_j^b z^j / M_b \right] \tag{13}$$

Making use of an approximation that rotational and vibrational dependences of the rotational g factor are commonly small for not too highly excited rotational and vibrational states, we partition expectation value g_J between atomic centres A and B according to the convention of polarity (A^+ and B^-):

$$t_0^a = \mu[g_J/m_p + 2|\mu^e|/(eR_eM_b)] \tag{14}$$

$$t_0^b = \mu[g_J/m_p - 2|\mu^e|/(eR_eM_a)] \tag{15}$$

Applicable directly to only electrically neutral diatomic molecules, these relations provide a practical means to interpret radial coefficients t_0^a and t_0^b evaluated by fitting frequencies of pure rotational and vibration-rotational transitions in terms of fundamental molecular properties g_J (dependent on reduced mass of a particular isotopic variant) and μ^e (independent of molecular mass for a net electrically neutral molecule). (To obtain nonadiabatic rotational contributions to vibration-rotational terms of LiH employed to generate table 1, we used calculated values[17] of g_J and μ^e as a function of R and fitted radial coefficients t_0^a and t_0^b to combinations of these functions according to these equations 14 and 15 in more general form.)

We applied this approach to estimate the electric dipolar moment and rotational g factor of GeS by fitting available spectral data by means of radial coefficients or equivalent parameters. Because only frequency data even for multiple isotopic variants are inadequate to estimate all adiabatic and nonadiabatic rotational and vibrational effects, we neglect adiabatic effects according to the justification already presented, as both Ge and S have relatively massive nuclides. Then by fitting 727 assorted spectral data (Ogilvie, to be published) we evaluated ten parameters specified in the following table.

All these parameters are independent of nuclear mass; the first seven define the function for potential energy within the range of internuclear distance $1.84 < R/10^{-10}$ m < 2.26, as the force coefficient k_e (related[5] to $U_{1,0}$ or c_0) indicates the curvature of this function at the distance R_e (related[5] to $U_{0,1}$) of minimum potential energy; the five coefficients c_j define the shape of the function. Three other parameters pertain, in s_0^{Ge}, to nonadiabatic vibrational effects of Ge and, in t_0 of Ge and S, to nonadiabatic rotational effects of each separate atomic centre. When we insert the fitted values of t_0^{Ge} and t_0^S into relations 14 and 15 above, solution of two simultaneous equations yields values $\mu^e/10^{-30}$ Cm $= 7.13 \pm 0.86$ (independent of isotopic variant) and $g_J = -0.07699 \pm 0.0064$ for specifically $^{72}Ge^{32}S$. As both t_0^{Ge} and t_0^S are signed quantities, they are consistent with only a relative polarity $^+GeS^-$, with the permanent electric dipolar moment having the indicated magnitude. Likewise according to this

Table 2. Parameters of GeS derived from spectral data

property	fitted value
$k_e/\text{N m}^{-1}$	433.65847 ± 0.00136
$R_e/10^{-10}$ m	2.0120442 ± 0.0000025
c_1	-2.059743 ± 0.000062
c_2	1.93805 ± 0.00031
c_3	-1.0611 ± 0.0152
c_4	-0.329 ± 0.119
c_5	3.89 ± 0.73
s_0^{Ge}	1.39 ± 0.41
t_0^{Ge}	-1.385 ± 0.130
t_0^{S}	-1.828 ± 0.081

novel interpretation of available spectral data that consist of only frequencies of pure rotational and vibration-rotational transitions, the rotational g factor has both a negative sign and the indicated magnitude.

For ^{74}Ge ^{32}S measurements of the Stark effect in vibrational state $v = 0$ on pure rotational transitions $J = 1 \leftarrow 0$ and $J = 3 \leftarrow 2$ yielded a mean expectation value of electric dipolar moment; the magnitude is $(6.671 \pm 0.20) \times 10^{-30}$ C m[18]. Separate measurement of the Zeeman effect on pure rotational transition $J = 4 \leftarrow 3$ of ^{72}Ge ^{32}S yielded an expectation value of the rotational g factor in the vibrational ground state, specifically the magnitude of g_J and, indirectly, its sign; the value is -0.06828 ± 0.00011[19]. Magnitudes of μ^e and g_J, and the sign of g_J, deduced according to only frequency data of spectra measured for samples without externally applied electric or magnetic field as discussed above, agree with these values from application of Stark and Zeeman effects, although the latter are much more precise. Uncertainties of these values deduced from t_0^{Ge} and t_0^{S} have orders of magnitude expected from the known precision of spectral measurements and the relative magnitudes of mechanical and extra-mechanical effects, as explained above for LiH. As the electric dipolar moment is derived from the difference of two quantities t_0^{Ge} and t_0^{S} that both have negative signs, its uncertainty is particularly sensitive to the uncertainties of these quantities, whereas, being derived from a weighted sum of t_0^{Ge} and t_0^{S}, g_J is more immune from accumulated uncertainties. If precise experimental values (or at least magnitudes) of μ^e and g_J are available (such as from application of Stark and Zeeman effects), a preferable procedure in spectral analysis is naturally to constrain t_0^{Ge} and t_0^{S} to their values consistent with the further data; other spectral parameters thereby assume greater 'physical' significance: such a preferred set of spectral parameters is to be presented elsewhere. In cases of either freely fitted or constrained values of t_0^{Ge} and t_0^{S}, all spectral data of GeS are equally well reproduced within the accuracy of their measurement: the normalised standard deviation of both fits is 0.95.

As this procedure becomes empirically justified through this successful application to GeS, we applied it to predict unmeasured values of μ^e and g_J of other molecules. In the case of GaH for which only vibration-rotational spectral data were available, the difference of values of t_0^{Ga} and t_0^{H} were so small, relative to their combined estimated standard errors, that the magnitude of electric dipolar moment derived from their difference lacked significance. However the value of g_J was well defined, -3.223 ± 0.011 for ^{69}Ga^{1}H; this value agrees satisfactorily with values in the range $[-3.44, -3.24]$ depending on the level of the quantum-chemical calculation (S.P.A. Sauer, to be published). The first such prediction of g_J was made for AlH on the basis of relatively imprecise spectral measurements on ^{27}Al^{1}H and ^{27}Al^{2}H; that value -2.2 ± 0.25[20] for ^{27}Al^{1}H became revised to -2.7 ± 0.5[21] when more precise spectral data were included in the analysis. In both cases lack of isotopic variant of Al made results relatively insensitive to the data, thus somewhat unreliable. Nevertheless it proved practicable to apply the latter value to estimate the paramagnetic contribution to the perpendicular component of the molecular magnetisability. The total magnetisability includes the diamagnetic contribution that was calculated for lack of experimental data; in this case the total perpendicular component of magnetisability 0.4×10^{-28} J T^2 has a small positive value, thus indicating a (marginally) net paramagnetic value as predicted theoretically, although the overall magnetic susceptibility is negative, consistent with a net diamagnetic property. This net paramagnetic component for AlH is the first experimental evidence, albeit indirect, for the overall paramagnetic susceptibility of BH predicted on the basis of quantum-chemical computations[22]. All three hydrides BH, AlH and GaH of elements of group 13 have electronic ground states of symmetry class formally $^1\Sigma^+$ or O^+, which are customarily associated with neither net unpaired electrons nor net orbital angular momentum, thus with diamagnetic susceptibility, apart from effects of net nuclear angular momenta. Preliminary analysis of spectral data of a hydride of a further element in this group indicates that g_J of InH is much less negative than -3, although the quality and consistency of data and the relatively decreased influence of mass effects between ^{113}In and ^{115}In preclude at present a definitive evaluation. For another member of diatomic hydrides in the family of elements in group 13, TlH, spectral data of insufficient quality and quantity preclude at present a corresponding analysis.

Spectral data of BrCl in its four isotopic variants enabled the deduction of both electric and magnetic properties, according to the procedure applied to data of GeS. Here the relatively small and imprecise difference between t_0^{Br} and t_0^{Cl} yielded a correspondingly imprecise value of electric dipolar moment, $\mu^e/10^{-30}$ C m $= 2.38 \pm 0.78^{23}$, but that value is consistent with the magnitude 1.732 ± 0.007^{24} from measurement of the Stark effect. For comparison with the more significant value $g_J = -0.02509 \pm 0.00070$ no measurement of the Zeeman effect is reported. The relative electric polarity $^+$BrCl$^-$ from these spectral data conforms to traditional chemical ideas.

In a systematic analysis[25] of available pure rotational and vibration-rotational spectra of fluorides of all elements of group 13 – BF, AlF, GaF, InF and TlF – data from Stark and Zeeman effects serve to constrain values of t_0^a and t_0^b of the latter four molecular species, whereas in the lack of such experimental data for BF values of g_J and μ^e derived from quantum-chemical computations were applied. According to such calculations[26], the sense of electric dipolar moment of AlF is $^+$AlF$^-$; analogous polarity is expected for succeeding members of this family. Although adiabatic effects are significant for BF, and although lack of isotopic variants of F other than ^{19}F precludes application of a procedure used successfully for GeS and BrCl, test fits of spectral data of BF indicated the polarity $^-$BF$^+$, perhaps contrary to chemical intuition. This analysis of spectral data uncovered the first experimental evidence in favour of this polarity, long predicted according to quantum-chemical computations[27].

Although spectral data of gaseous carbon dioxide in its several isotopic variants were analysed according to a more numerical approach (maintaining a firm theoretical basis), a value of the rotational g factor was derived[28] from only vibration-rotational spectral data of samples without externally applied electric or magnetic field that agreed with the value from experiments on molecular beams[29]. In the absence of a permanent electric dipolar moment, g_J of ^{12}C^{16}O$_2$ reflects purely nonadiabatic rotational effects. The assumption in the analysis[28] that adiabatic effects were negligible is justified by the agreement between the two values of g_J from distinct experimental data.

DISCUSSION

In this work we examine two models, one of great general importance and the other of narrow application. The results in table 1 confirm what chemists and physicists have long taken for granted, namely that the classical idea of a molecule to consist essentially of recognisable atomic centres is a practical model stable compounds under mild conditions. This assumption is least valid for elements of small atomic number; for instance the lengths of bonds between other elements, even those with molar masses as small as that of carbon, are found to be anomalously small according to xray data from diffraction of single crystals because the electronic density near the hydrogen nucleus is comparable with that in the binding region toward an adjacent atomic centre. From a more technical point of view, these results justify the approximation that Born sought to evaluate, first unsuccessfully with Heisenberg, then with increasing refinements published with Oppenheimer and Huang, followed in turn by other workers[13]. Even before the successful treatment, spectroscopists and practitioners of quantum-chemical calculations imposed this separate treatment of nuclear motions on their methods of considering molecular systems drawing their attention. What will prove of increased interest are relative magnitudes of mechanical and extra-mechanical effects for states of a diatomic molecule approaching the dissociation limit. van Vleck[7] expected adiabatic and nonadiabatic vibrational effects to increase with increasing vibrational quantum number. There is in table 1 scant evidence to support this supposition, but the available spectral data of LiH of usable quality pertain to energies only a small proportion of the energy at the first dissociation limit. Precise data of spectral frequencies of vibrational states of much increased quantum numbers and precise expectation values of g_J or accurate computations of g_J and μ^e over extended ranges of internuclear separation are required to enable tests of relative magnitudes of mechanical and extra-mechanical effects at much greater energies, but LiH remains an excellent diatomic species for such tests.

Methods of evaluating the sense of electric polarity of small molecules from experimental data are poorly developed. The original method involved the isotopic dependence of the rotational g factor, deduced from application of the Zeeman effect to molecules either in a conventional microwave spectrometer or in a molecular beam with mass-spectrometric or other detection. During the period between years 1980 and 1995 experiments on the Zeeman effect in microwave spectroscopy seem to have become less common than during the preceding decade or two, contrary to the trend of increasingly numerous scientists in general and microwave spectral experiments in particular. In fact the development of instruments to produce microwave spectra from Fourier transforms of temporal signals has so far generally precluded use of even the Stark effect. Quantum-chemical computations yield directly such polarities. In principle distributions of electronic and nuclear charge determined in experiments with diffraction of xrays also provide this information, but the most commonly practised experiment of this kind requires single crystals of generally pure substances; strong interactions within the lattice thus formally preclude attribution of deduced properties to individual molecules. In these circumstances our development of a further method to evaluate electric dipolar moments and polarities from precise data of spectral frequencies, such as those measured with new microwave spectrometers, is timely and valuable despite that the approach appears to be limited to diatomic and linear triatomic molecular species having sufficient isotopic variants for which precise and abundant pure rotational and vibration-rotational spectral data are obtainable.

Not only is the electric dipolar moment a chemically meaningful molecular property, but also the intensity of the pure rotational spectrum in absorption or emission is proportional to the square of this quantity. Thus nonadiabatic rotational effects, specifically $t_0^{a,b}$ deduced from frequency measurements, provide indirect information about spectral intensities in the pure rotational spectrum, with prospective analytical applications for samples in exotic environments. Further radial coefficients t_1, t_2 etc. are related to successive derivatives of the electric dipolar moment with respect to internuclear distance, generally evaluated at R_e. In principle such information defines intensities of vibration- rotational spectra, $d\mu^e/dR$ for the fundamental band, $d\mu^e/dR$ and $d^2\mu^e/dR^2$ for the first overtone band etc. In practice precise evaluation of radial coefficients t_j beyond $j = 0$ becomes increasingly difficult and susceptible to error propagated ultimately from uncertainty of measurements of transition frequencies for spectral data of finite extent of a particular compound, even with abundant isotopic variants; separation of adiabatic, nonadiabatic rotational and nonadiabatic vibrational effects also becomes increasingly difficult. As table 1 demonstrates, nonadiabatic vibrational effects vanish identically by cancellation for term coefficients $Z_{0,l}$ with $l > 0$; there is thus no interference from nonadiabatic vibrational effects in the evaluation of radial coefficients $t_0^{a,b}$ that are derived primarily from $Z_{0,1}^r$. Radial coefficients $s_0^{a,b}$ are also expected to provide information about the derivative of electric dipolar moment with respect to internuclear distance; models for pertinent nonadiabatic vibrational effects and their quantal justification are in course of development.

The model according to which a diatomic molecule consists of two atomic centres, which become neutral atoms or atomic ions in the event of dissociation, has a sound physical foundation established on data from experiments on diffraction of electrons and xrays. What is the physical foundation of a model of a diatomic molecule as a rotating electric dipole? Certainly diatomic molecules with nuclei having unlike protonic numbers generally possess electric dipolar moments (defined with respect to the nuclear frame). A static spatial distribution of positive and negative electric charges, or the mean field of electrons moving relative to atomic nuclei, can be expressed as a multipole expansion. After the net charge of the monopole, which vanishes for a net electrically neutral molecule, the dipolar moment is the next term in the expansion. If the spatial distribution is not static but varies temporally in an oscillatory manner, the dipole becomes a vibrating dipole; if the entire distribution of charges rotates about an internal origin, the dipole becomes analogously a rotating electric dipole, possibly generating a magnetic dipolar moment in the process. In either case interaction of a vibrating or rotating dipole with an electromagnetic wave provides classically a mechanism for absorption or emission of energy by the molecule. Thus although a diatomic molecule is not merely a rotating or vibrating electric dipole, electric dipolar moment is an important property of such a molecule; the model of a diatomic molecule as a rotating electric dipole that we employ here to evaluate electric and magnetic properties has thus a rational physical basis. A quantal treatment that justifies this model is currently under development.

In conclusion the models of a diatomic molecule to consist of two atomic centres strongly interacting and of a rotating electric dipole are shown not only to yield valuable information about structural, dynamic, electric and magnetic properties from analysis of molecular spectra but also to have a firm physical foundation consistent with traditional ideas about molecular structure. Structural (R_e and coefficients of the expansion for potential energy) and dynamic (force coefficient k_e, reflecting the resistance of the molecule to a small displacement from the equilibrium internuclear distance) information may be evaluated highly precisely, to the extent of parts in 10^6 (including uncertainties in fundamental physical constants h and N_A), whereas electric and magnetic properties deduced from extra-mechanical effects are necessarily much less precise because of their origin in secondary effects present to a minor extent (for vibrational states at energies far from the relevant limit of dissociation). Signs of electric dipolar moment and rotational g factor deduced directly according to this approach provide information not readily obtained otherwise.

REFERENCES

1. F.M. Fernandez and J.F. Ogilvie, *Chin. J. Phys.* 30:177 and 30:499 (1992)

2. P.R. Bunker and R.E. Moss, *Mol. Phys.* 33:417 (1977)

3. J.F. Ogilvie, *Proc. Roy. Soc. London* A378:287 (1981) and A381:479 (1982)

4. J.F. Ogilvie, *J. Chem. Phys.* 88:2804 (1988)

5. J.F. Ogilvie, *J. Phys. At. Mol. Opt. Phys.* B27:47 (1994)

6. J.L. Dunham, *Phys. Rev.* 41:721 (1932)

7. J.H. van Vleck, *J. Chem. Phys.* 4:327 (1936)

8. A.H.M. Ross, R.S. Eng and H. Kildal, *Opt. Commun.* 12:433 (1974)

9. W. Gordy and R.L. Cook, "Microwave Molecular Spectra", Wiley, New York, third edition (1984)

10. R.R. Freeman, A.R. Jacobson, D.W. Johnson and N.F. Ramsey, *J. Chem. Phys.* 63:2597 (1975)

11. T.R. Lawrence, C.H. Anderson and N.F. Ramsey, *Phys. Rev.* 130:1865 (1963)

12. J.F. Ogilvie, *J. Mol. Spectrosc.* 148:243 (1991) and 154:453 (1992)

13. F.M. Fernandez, *Phys. Rev.* 50:2953 (1994) and references therein

14. J.F. Ogilvie, *Ber. Bunsenges. Phys. Chem.* 100 (1996) accepted for publication

15. J.F. Ogilvie and M.C.C. Ho, *Chin. J. Phys.* 31:721 (1993)

16. E. Tiemann and J.F. Ogilvie, *J. Mol. Spectrosc.* 165:377 (1994)

17. J.F. Ogilvie, J. Oddershede and S.P.A. Sauer, *Chem. Phys. Lett.* 228:183 (1994)

18. J. Hoeft, F.J. Lovas, E. Tiemann, R. Tischer and T. Torring, *Z. Naturforsch.* 24a:1217 (1969)

19. R. Honerjager and R. Tischer, *Z. Naturforsch.* 32a:1 (1977)

20. J.F. Ogilvie, *J. Chin. Chem. Soc.* 39:381 (1992)

21. S.P.A. Sauer and J.F. Ogilvie, *J. Phys. Chem.* 98:8617 (1994)

22. R.M. Stevens and W.N. Lipscomb, *J. Chem. Phys.* 42:3666 (1965)

23. J.F. Ogilvie, *J. Chem. Soc. Faraday Trans.* 91:3005 (1995)

24. K.P.R. Nair, J. Hoeft and E. Tiemann, *Chem. Phys. Lett.* 58:153 (1978)

25. J.F. Ogilvie, H. Uehara and K. Horiai, *J. Chem. Soc. Faraday Trans.* 91:3007 (1995)

26. S. Huzinaga, E. Miyoshi and M. Sekiya, *J. Comput. Chem.* 14:1440 (1993)

27. R.A. Hegstrom and W.N. Lipscomb, *J. Chem. Phys.* 48:809 (1968)

28. J.L. Teffo and J.F. Ogilvie, *Mol. Phys.* 80:1507 (1993)

29. J.W. Cederberg, C.H. Anderson and N.F. Ramsey, *Phys. Rev.* 136:A960 (1964)

CAN QUANTUM MECHANICS ACCOUNT FOR CHEMICAL STRUCTURES?

Anton Amann

Laboratorium für Physikalische Chemie
ETH Zentrum
CH-8092 Zürich, Switzerland

Abstract

In quantum mechanics, there is no fundamental understanding of chemical structure: though chemical structures can be *imposed* on quantum-chemical calculations, there are many "strange" quantum states which do not fit at all into the world of chemical structure formulas. Such "strange" quantum states are usually notated as superpositions of states having chemical structure, e. g., as superpositions of two different isomers.

Most molecular species can be expected to possess a nuclear structure, whereas for some, e. g. ammonia or monodeuteroanlinine, the issue is not so obvious. The ammonia-MASER transition, for example, is thought to be a transition between "strange" molecular (pure) states *not* admitting a nuclear structure. Since one cannot sharply distinguish between molecular species "admitting strange (pure) states" and such which do not, it is plausible that all molecules can in principle be prepared in such strange (pure) states, but that the latter do not play an important role for most molecular species (with exceptions such as ammonia, for example). This suggests that chemical classical concepts–e. g. chirality, knot type and nuclear structure of molecules–are not *strictly* classical in the sense that superpositions between different structures are *strictly* forbidden. Such superpositions may still arise but are unstable under external perturbations and therefore usually not observed. Instability of "strange" states is thought to increase with increasing nuclear masses or with decreasing level splitting between the low lying eigenstates of the molecular hamiltonian, as, for example, in the sequence of species {monodeuteroaniline → ammonia → naphthazarin → ... → sugar/amino acid}. In other words: chemical classical structures are *fuzzy*, because "strange" superpositions–not compatible with classicality– are still admitted. But the fuzzyness of such classical structures decreases with increasing nuclear masses (because strange states "die out").

It is quite hard to substantiate these heuristic ideas by rigorous *conceptual* and mathematical reasoning. The most important conceptual problem in this respect is that a density operator (non-pure state) cannot *uniquely* be decomposed into pure states. Hence in *traditional statistical* quantum mechanics different decompositions of, say, a thermal density operator D_β are considered as being equivalent. Nevertheless different decompositions of a given thermal density operator can refer to *entirely different* physical situations, as, for example, to molecules with or without nuclear structure.

This implies that there is no fundamental understanding of chemical classical concepts. In particular, the fascinating idea of a chemical molecular structure, invented in traditional chemistry and taken over in the *Born–Oppenheimer scheme* of quantum chemistry, is not fully understood.

Here an attempt is made to adapt the quantum-mechanical formalism for these problems: A *canonical* decomposition of thermal density operators–stable under external perturbations–is introduced opening a way

- to understand the concept of chemical structure,
- to derive and study *deviations* from chemical structures in the sense that "strange" quantum-mechanical superpositions of different chemical structures appear for *certain* molecules like ammonia,
- to get a deeper understanding of the Born–Oppenheimer approximation,
- to make first steps towards a *fully quantum-mechanical* explanation of the stochastic aspects in single-molecule spectroscopy.

It is enormously interesting to understand the traditional chemical concepts in the light of quantum mechanics. Such an understanding inevitably leads to studying the *"fuzzy" quantum deviations from classical chemical behavior*. These quantum deviations are described here by *large-deviation techniques* and by stochastic dynamics on the set of pure states of a quantum system. A relatively simple example for such a stochastic dynamics is discussed in detail. Furthermore a large-deviations entropy is presented, describing the quantum fluctuations of a Curie–Weiss magnet around classical behavior. For molecules it is not yet possible to calculate large-deviation entropies describing quantum fluctuations around the classical behavior (the latter arising in the limit of infinite nuclear masses), but the Curie–Weiss example discussed here gives a good idea about the similar conceptual problems with molecules.

QUANTUM THEORY GOES BEYOND THE CHEMICAL STRUCTURES OF TRADITIONAL CHEMISTRY

The superposition principle of quantum mechanics gives immediate rise to what I call "strange" states.

Consider, for example, a *hydrogen bond* in some molecule[1]. The hydrogen atom in question can sit at different *fixed* positions within the molecule, described by wave functions Ψ_1 and Ψ_2. But *if* one considers an appropriate superposition of these two chosen wave functions

$$\Psi \overset{\mathrm{def}}{=} c_1\Psi_1 + c_2\Psi_2, \tag{1}$$

then the hydrogen atom has *no* fixed position any more with respect to the new wave function Ψ.

Such situations, where the nuclei of a molecule have *no* determined fixed position could arise everywhere in chemistry: in analogy to the above example one could always superpose different isomers and end up with such "strange" states, "strange" not from the point of view of quantum mechanics but "strange" from the point of view of traditional chemistry, where the structural formulas imply unequivocal positions for the nuclei in a molecule.

An interesting example in this respect is ammonia. Consider fig. 1: The double-minimum function there is (a sketch of) the Born–Oppenheimer potential for the electronic ground state of ammonia. It is an energy vs. an internal inversion coordinate diagram. Every value of the inversion coordinate corresponds to a particular nuclear frame of ammonia. The two minima, for example, correspond to pyramidal structures, whereas the maximum corresponds to a planar structure. At first sight, ammonia molecules roughly keep their traditional chemical structures even in a quantum description. The pyramidal forms, for example, are described by the rather peaked (non-stationary) wave functions Ψ_L and Ψ_R, and therefore admit at least an approximate nuclear frame. But as soon as we consider superpositions of Ψ_L and Ψ_R, things change dramatically. Look at the superposition

$$\Psi_+ := \frac{1}{\sqrt{2}}(\Psi_L + \Psi_R), \tag{2}$$

which describes the proper ground state of ammonia, and at the superposition

$$\Psi_- := \frac{1}{\sqrt{2}}(\Psi_L - \Psi_R), \tag{3}$$

which describes the first excited state of ammonia. Both, Ψ_+ and Ψ_-, are eigenstates of the underlying molecular hamiltonian and therefore *do not change under the time-evolution given by the time-dependent Schrödinger equation* apart from phase factors $\exp\{-itE_\pm/\hbar\}$, where E_+ and E_- are the respective eigenvalues. This implies that all expectation values $\langle\Psi_\pm|\hat{A}\Psi_\pm\rangle$ of observables $\hat{A}$ are invariant under the Schrödinger time evolution, and not just the expectation values of, say, the nuclear position operators. Since expectation values do not change, the states Ψ_+ and Ψ_- are *stationary*.

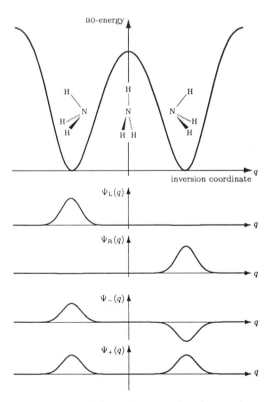

Figure 1. Illustration of the superposition principle using ammonia and ammonia-type molecules.

Fig. 2 is a sketch of the ground state Ψ_+ of ammonia. Though still drawn, the nitrogen-hydrogen *bonds* lose their meaning in this context. With respect to Ψ_+ and Ψ_-, a single ammonia molecule does not have a nuclear frame any more: The nitrogen and hydrogen nuclei do not sit at fixed positions; only probability distributions for their positions can be given, just as is usual with electrons. Hence ammonia does not admit a nuclear frame with respect to the states Ψ_+ and Ψ_-. It is this fact which makes these states appear *"strange" from the point of view of a chemist.*

The transition between the two stationary states Ψ_+ and Ψ_- is the ammonia-MASER transition with transition frequency $\nu = (E_- - E_+)/\hbar = 23\,870\,110\,000\,\mathrm{s}^{-1}$. The very existence of the ammonia-MASER transition suggests that the states Ψ_+ and Ψ_- of ammonia do indeed exist in reality and not just in the quantum-mechanical formalism.

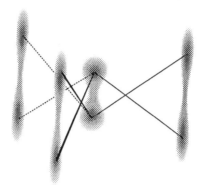

Figure 2. Sketched distribution of the nuclei in the ground state of an ammonia-type molecule.

Incidentally, the states Ψ_L and Ψ_R can be written as superpositions

$$\Psi_L := \frac{1}{\sqrt{2}}(\Psi_+ + \Psi_-) \tag{4}$$

$$\Psi_R := \frac{1}{\sqrt{2}}(\Psi_+ - \Psi_-) \tag{5}$$

of the ground and excited states of ammonia. They are non-stationary states with Schrödinger time evolution

$$\frac{1}{\sqrt{2}}[\Psi_+ \pm \Psi_-] \quad \rightarrow \quad \frac{1}{\sqrt{2}}[\Psi_+ \pm \exp\{-it(E_- - E_+)/\hbar\}\Psi_-]. \tag{6}$$

Recall that the time-evolved state on the right-hand side of eq. (6) does not just differ by a phase factor from the initial state $\frac{1}{\sqrt{2}}[\Psi_+ \pm \Psi_-]$. The Schrödinger time evolution (6) describes a tunneling process which transforms the state Ψ_L into Ψ_R and vice versa.

As far as the quantum-mechanical formalism is concerned, ammonia is by far no exception. *Similar situations can be constructed for all molecules:* Consider a molecule with internal hamiltonian H_0 (the kinetic energy of the center of mass being subtracted). Let, furthermore, I denote the unitary operator which implements space inversion

$$I\Psi(q_1, \dots, q_L; Q_1, Q_2, \dots, Q_M) \stackrel{\text{def}}{=} \Psi(-q_1, \dots, -q_L; -Q_1, -Q_2, \dots, -Q_M). \tag{7}$$

Here the q's are electronic coordinates and the Q's are nuclear coordinates. If the hamiltonian H_0 does not contain the weak neutral current terms[2,3], it is invariant under space-inversion, i. e.,

$$IH_0I^{-1} = H_0. \tag{8}$$

Then the eigenstates Ψ_n of H_0 can be chosen to be symmetry-adapted,

$$I\Psi_n = \pm\Psi_n, \quad n = 1,2,3,\dots . \tag{9}$$

and therefore will always fulfill

$$\langle\Psi_n|Q_j\Psi_n\rangle = \langle I\Psi_n|Q_jI\Psi_n\rangle = \langle\Psi_n|I^{-1}Q_jI\Psi_n\rangle = \langle\Psi_n|-Q_j\Psi_n\rangle = 0,$$
$$j = 1,2,\dots,M. \tag{10}$$

This is quite strange, since then

- a nuclear molecular frame does not exist,

- different isomeric molecular forms do not arise,

- and a *sequence* of monomers in a macromolecule does not make sense,

since nuclear frame, isomerism and sequential structure in a macromolecule need (at least approximately) localized nuclei, which is incompatible with the result of eq. (10).

The symmetry-adapted eigenstates in eq. (9) correspond to the states Ψ_+ and Ψ_- of ammonia. They do not fit into the molecular-structure scheme of traditional chemistry and are therefore considered as "strange" states from a chemical point of view. *Neither chemical structures nor chemical bonds do exist any more.*

CHEMICAL CONCEPTS ARE FUZZY CLASSICAL CONCEPTS

Chemical concepts like handedness and isomerism of molecules or the knot-type of a circular DNA-molecule[4,5] are often thought to be *strictly* classical concepts, i. e. to have strictly dispersion-free expectation values:

- Handedness of a molecule, for example, is thought to be *either* left *or* right, without any intermediate forms such as superpositions of the left- and right-handed states.

- The isomer type of a molecule is thought to be unequivocally determined, without any intermediate forms such as superpositions of different isomers.

- The knot type of a circular DNA-molecule is thought to be fixed without any intermediate forms such as superpositions of differently knotted circular DNA-molecules (having the same monomer sequence).

In case of NHDT, a potentially chiral derivate of ammonia, this assumption is *not* fulfilled. The stationary ground and excited states Ψ_+ and Ψ_- (analogous to those of ammonia) are actually states which do not fit into the left/right classification scheme, since they are symmetric under space inversion. Nevertheless these stationary states seem to exist in reality, as suggested by the associated MASER-transition between them (with a frequency which is smaller than the corresponding one of NH_3[6]).

For other ammonia-type molecules, such as naphthazarin[6,7,8,9] or properly chiral molecules (amino acids/sugars/sulfoxides[10,11] etc.) the situation is not so clear, see fig. 1. These molecular species admit an inversion coordinate and hence have the same structure of states as ammonia, i. e., stationary states Ψ_+ and Ψ_-, and nonstationary states Ψ_L and Ψ_R (in a two-level approximation); the main difference between them is the different level splitting $E_- - E_+$. Handwaving arguments based on quantum-mechanical perturbation theory[12] lead to the conclusion that the stationary states Ψ_+ and Ψ_- become unstable when the level

Table 1. Ammonia-type molecules

	Accessible states	Level Splitting
Monodeuteroaniline[26]	all pure states	$600\,\mathrm{J\,mol^{-1}}$
Ammonia	all pure states	$9.5\,\mathrm{J\,mol^{-1}}$
Naphthazarin[9]	Ψ_L and Ψ_R and ?	$\approx 0.02\,\mathrm{J\,mol^{-1}}$
Aspartic acid	Ψ_L and Ψ_R "only"	$\approx 10^{-60}\,\mathrm{J\,mol^{-1}}$

splitting is sufficiently small. This would mean that the states Ψ_+ and Ψ_-, though being stationary under the Schrödinger time evolution, would decay into more stable states *under the influence of small external perturbations*. Since the level splitting of properly chiral molecules such as amino acids or sugars can be estimated to be very small indeed[13], it seems to be no surprise that the states Ψ_+ and Ψ_- cannot easily be prepared experimentally[14,15,16,17].

In other words: Though the "strange" states Ψ_+ and Ψ_- might exist for *all* isolated or reasonably well screened ammonia-type molecules, they usually don't arise in case of small level splitting due to lability under external perturbations.

Seen from this point of view, *no phase transition is expected to take place in a sequence of molecular species with decreasing level splitting such as in the sequence {monodeuteroaniline → ammonia → ... → asparagic acid} of table 1. The heuristic idea is that the lability of the "strange" states Ψ_+ and Ψ_- increases *continuously* with decreasing level splitting. The most unstable situation is expected to take place for properly chiral molecules with very low level splitting (below $10^{-50}\mathrm{J\,mol^{-1}}$).

Therefore even handedness need not be a *strict* classical concept, but could be a *fuzzy* classical one: I expect the superposition principle to hold universally for *all* molecules in table 1 such that "strange" states may arise or be prepared experimentally; nevertheless I expect that such "strange" states arise only with small, but nonzero probability in thermal equilibrium (for species with low level splitting). This would mean that the fuzzyness of the classical concept "chirality" decreases *continuously* with decreasing level splitting.

These considerations will be substantiated in section "Fuzzy classical observables and large-deviation theory" below. For the moment, it might be helpful to sketch the *alternative* approach admitting a phase-transition in a sequence like {monodeuteroaniline → ammonia → ... → asparagic acid} of table 1. Such a phase transition is *defined* to happen if for large level splitting there exists only one thermal state with density operator D_β (for a given inverse temperature β), whereas for small level splitting two different thermal states $D_{\beta,L}$ and $D_{\beta,R}$ exist for left- and right-handed molecular species. Change of behaviour between the two different regimes would happen at some critical level splitting. Here I do *not* want to follow this line of reasoning, since it is not appropriate for thermal situations, as we shall see below. For considerations of ground states, the situation is different[13,18,19,20,21,22,23,24,25].

Another interesting example of a classical chemical concept is the *nuclear frame of a molecule*. The usual Born–Oppenheimer approximation for molecules shows that it is *not* a strict classical structure: with respect to the states Ψ_L and Ψ_R (see fig. 1), for example, the nuclei's positions in ammonia show a small dispersion and not a zero one (with respect to the states Ψ_+ and Ψ_- the dispersions of the nuclear positions are substantially increased, anyway). Hence again there is no reason to expect the molecular nuclear frame to be a *strict* classical concept, but much more a *fuzzy* classical one. From this point of view all molecular species admit "strange" states without nuclear frame such as superpositions of two (or more) different isomers; nevertheless such "strange" states can be expected to survive only in isolated molecules and to decay under the influence of external perturbations.

The parameter "level splitting" in case of the two-level ammonia-type molecules discussed above can now be replaced by the molecules' nuclear masses. I expect that with increasing nuclear masses M_j, $j = 1, 2, \ldots, K$, one gets a smaller probability to find "strange" states such as superpositions of different isomers (in a thermal situation). In the limit $M_j \to \infty$ I expect the nuclei to behave entirely classically. Again there is no reason to believe that some phase transition takes place at particular values for the nuclear masses.

Summarizing, there are several important points to be remembered:

- Chemists' molecules are usually not in strict eigenstates, since the latter do not show chemical structure (handedness, isomerism, knot type). From a quantum-mechanical point of view, it is not clear why states showing chemical structure should arise[13,24,25,27,28,29,30,31,32,33,34,35,36,37,38,39,40,41].

- Ammonia is the paradigm of a molecule without chemical structure. Its ground state, in particular, does not admit a nuclear molecular frame. But for *all* molecules with at least two possible isomeric structures the same "strange" conclusions could be drawn in theory.

- An *isolated* molecule can be prepared in any stationary or non-stationary pure state. There is no reason to restrict oneself to eigenstates of the hamiltonian, i. e., solutions of the time-independent Schrödinger equation. Hence *isolated molecules do not show chemical structure*. Deriving chemical structure from the Schrödinger equation for an isolated molecule can only work by tricks or approximations hidden in complicated mathematics.

- Without consideration of the molecular environment, it is impossible to *derive and understand* chemical structure. Possible environments to be considered are the radiation or gravitation field, collisions with other molecules etc.

- The Born–Oppenheimer idea of chemical structure goes back to traditional chemistry of the pre-quantum era and is-as I am convinced-essentially correct. The problem is to understand chemical structure in quantum mechanical terms and to work out and discuss borderline cases not conform to traditional chemical thinking.

- Chemical concepts usually are considered to be *strict* classical concepts. Here arguments will be given in favor of *fuzzy* classical chemical concepts, where the fuzzyness decreases with increasing nuclear masses etc.

STATISTICAL VS. INDIVIDUAL FORMALISM OF QUANTUM MECHANICS

In quantum theory, non-pure states have at least two different meanings. They can arise as *mixtures of pure states* or as a *restriction of a pure state onto a subsystem*.

In the latter case, the respective non-pure state is called an "improper mixture"[42,43] and a decomposition into pure states does not necessarily make sense. To get a feeling for the matter, one can play with 4×4-matrices $\mathcal{M}_4$ and restrict a pure state on $\mathcal{M}_4$ (i. e., a vector in 4-dimensional Hilbert space) to the two-level subsystem described by the block-matrix observables

$$\begin{pmatrix} a & b & 0 & 0 \\ c & d & 0 & 0 \\ 0 & 0 & 1 & 0 \\ 0 & 0 & 0 & 1 \end{pmatrix}. \tag{11}$$

Actually *every* non-pure state of a two-level system can be viewed as the restriction of some pure state on a four-level system: For an arbitrarily chosen density operator

$$\begin{pmatrix} d_1 & d_2 \\ d_3 & d_4 \end{pmatrix} \tag{12}$$

of a two-level system, there exists a (normalized) vector ξ in 4-dimensional Hilbert space such that "the respective expectation values coincide" on observables of the two-level system, i. e., for all $a,b,c,d \in \mathbf{C}$, one has

$$\mathrm{Trace}\left(\begin{pmatrix} d_1 & d_2 \\ d_3 & d_4 \end{pmatrix} \begin{pmatrix} a & b \\ c & d \end{pmatrix} \right) = (\bar{\xi}_1, \bar{\xi}_2, \bar{\xi}_3, \bar{\xi}_4) \begin{pmatrix} a & b & 0 & 0 \\ c & d & 0 & 0 \\ 0 & 0 & 1 & 0 \\ 0 & 0 & 0 & 1 \end{pmatrix} \begin{pmatrix} \xi_1 \\ \xi_2 \\ \xi_3 \\ \xi_4 \end{pmatrix}. \tag{13}$$

If the non-pure state described by the density operator in eq. (12) is actually the restriction of the pure state (of the four-level system) described by the vector ξ in eq. (13) *in physical reality and not just by mathematical convenience*, there is no sense of decomposing this non-pure state on the two-level system into pure states (of the two-level system) again. Here "physical reality" could mean, for example, that the 4-level system describes a two-level system coupled to a two-level environment.

Let us focus here on the other situation, where a non-pure state (described by some density operator D) is *not* a restriction of a pure state on a bigger system, but a real mixture or reflecting our ignorance of the actual pure state of the system in question. For illustration, consider again an ammonia-type molecule in two-level description (as before) and introduce 2×2-density operators D_+, D_-, D_L, D_R corresponding to the pure state vectors $\Psi_+, \Psi_-, \Psi_L, \Psi_R$ in the sense that

$$\mathrm{Trace}(D_+ \hat{A}) = \langle \Psi_+ | \hat{A} \Psi_+ \rangle \tag{14}$$

holds true for all 2×2-matrices $\hat{A}$, and similarly for D_-, D_L, D_R. The density operators D_+, D_-, D_L, D_R describe pure states, of course, but we can look at proper mixed states, such as

$$D_{\mathrm{rac}} \overset{\mathrm{def}}{=} \frac{1}{2} D_L + \frac{1}{2} D_R, \tag{15}$$

describing a mixture of 50% left- and 50% right-handed molecules. Or consider a thermal state at inverse temperature β as being a mixture

$$D_\beta \overset{\mathrm{def}}{=} \frac{e^{-\beta E_+} D_+ + e^{-\beta E_-} D_-}{\mathrm{Trace}(e^{-\beta E_+} D_+ + e^{-\beta E_-} D_-)} \tag{16}$$

of the eigenstates Ψ_+ and Ψ_- with appropriate Boltzmann weights.

Summarizing, a non-pure state (and the corresponding density operator) can have two different meanings, and it is often not easy to figure out which one is appropriate. It is, for example, not at all obvious that a thermal state D_β can be considered as a mixture of pure states as insinuated in eq. (16) (apart from mathematical convenience).

These difficulties in interpretation of non-pure states give rise to *different versions and interpretations of quantum mechanics,* namely statistical and individual ones. Quantum theory in its usual *statistical* version claims that:

- A quantum system is always intrinsically coupled to its quantum environment and therefore cannot be expected to be in a pure state. Subsystems of a quantum system are not objects in their own right.

- Consequently a decomposition of a non-pure state into pure states does not make physical sense.

- Decomposition of a non-pure state (into pure ones) can be used as a mathematical tool, but different decompositions are entirely equivalent.

Quantum theory in its *individual* version, on the other hand, claims that

- Quantum systems are often intrinsically coupled to their quantum environment and cannot generally be expected to be in a pure state. Nevertheless in many cases the coupling is "weak" and a description in terms of pure states is appropriate.

- In the latter case a decomposition of a non-pure state into pure states can make sense.

- Different decompositions of a non-pure state are not necessarily physically equivalent[44,45,46].

The respective third items concerning *different* decompositions of non-pure states into pure ones will be discussed in the next section. For the moment, we shall concentrate on the relevance of pure and the meaning of non-pure states as such.

As can be seen from the vague formulation of individual quantum theory, things are not so clear. Molecules in a vessel at a pressure of 1 mbar are usually considered as being objects in their own right and therefore in individual pure states. An electron of an ammonia molecule, on the other hand, is definitely not in a pure state (as can be computed, e. g., starting with the overall ground state Ψ_+ of the ammonia molecule). A proper discussion of these problems is difficult and needs *dressing procedures*[47]. Here we shall simply *assume* that the assumptions of individual quantum theory are fulfilled, if the quantum system considered is large enough (with large parts of the environment incorporated, if necessary) or *assume* that an appropriate dressing has already been made.

Treating a two-level system in an individual way is, of course, a delicate matter and quite misleading, as will be seen later on in a discussion of the Curie–Weiss model. It is only simplicity which makes me use two-level systems for purposes of illustration.

THE DECOMPOSITION OF A NON-PURE STATE INTO PURE STATES IS NOT UNIQUE

The problem discussed here is the *non-uniqueness* of decompositions of non-pure thermal states into pure ones. This non-uniqueness is nicely illustrated in case of a *two-level system,* because there the state space is isomorphic to the sphere in three-dimensional Euclidean space, the so-called *Bloch sphere.*

Remark: To every non-pure state of a two-level system with associated density operator D there corresponds a uniquely determined vector $\vec{b}$ in three-dimensional Euclidean space defined by

$$\begin{pmatrix} b_1 \\ b_2 \\ b_3 \end{pmatrix} \overset{\text{def}}{=} \begin{pmatrix} \text{Trace}(D\sigma_1) \\ \text{Trace}(D\sigma_2) \\ \text{Trace}(D\sigma_3) \end{pmatrix} , \tag{17}$$

with $\sigma_j, j = 1, 2, 3$, being the Pauli matrices. Under this correspondence, *pure states* are mapped to the *surface* S_2 of the Bloch sphere, whereas non-pure states are mapped to its inner points. *Mixing of pure states corresponds to mixing of vectors* $\vec{b}$: If the density operators D_1 and D_2 are mapped to the vectors $\vec{b}_1$ and $\vec{b}_2$, respectively, then the mixture

$$D := \lambda_1 D_1 + \lambda_2 D_2 \tag{18}$$

is mapped to the vector

$$\vec{b} := \lambda_1 \vec{b}_1 + \lambda_2 \vec{b}_2 . \tag{19}$$

Representing the states Ψ_+ and Ψ_- as

$$\Psi_- = \begin{pmatrix} 1 \\ 0 \end{pmatrix}, \quad \Psi_+ = \begin{pmatrix} 0 \\ 1 \end{pmatrix} , \tag{20}$$

one gets as corresponding points on the Bloch sphere the north- and southpole, respectively. A pure state

$$\Psi \overset{\text{def}}{=} c_+ \Psi_+ + c_- \Psi_- = \begin{pmatrix} c_- \\ c_+ \end{pmatrix} \tag{21}$$

corresponds to the point

$$\begin{pmatrix} 2\,\text{Re}(c_-^* c_+) \\ 2\,\text{Im}(c_-^* c_+) \\ c_-^* c_- - c_+^* c_+ \end{pmatrix} \tag{22}$$

Recall again, that it is quite delicate to use two-level systems (instead of, say, an ammonia molecule) in individual quantum theory. Nevertheless, two-level systems can be quite instructive, precisely because simple visualization is possible by use of the Bloch sphere.

Let us devise a *Gedankenexperiment* for a gas of ammonia-type molecules in a vessel at the pressure $p = 1\,\text{mbar}$ and a given inverse temperature β by assuming that:

- all molecules can be treated as two-level systems;

- all the molecules are in a pure state;

- the vessel has a small aperture, through which molecules escape (at a rate of, say, 1 molecule per second). The pure state of each of these escaping molecules is measured by a "protective measurement", e. g., in the sense of Aharonov and Anandan[48];

- the mixture of the molecules' pure states is given by the thermal density operator

$$D_\beta \approx \frac{1}{2} \begin{pmatrix} 1 & 0 \\ 0 & 1 \end{pmatrix} . \tag{23}$$

This is the case if either the level splitting is low enough (as with chiral molecules, see table 1) or the temperature is high enough. The last assumption is made only for simplicity.

The question then is: *What is the distribution of pure states found in such an experiment, when the ammonia molecules in the vessel are in thermal equilibrium?*

Does one find

- a mixture of eigenstates Ψ_+ and Ψ_-, each with 50% probability? This would correspond to a decomposition of the thermal density operator D_β into pure states as $D_\beta = \frac{1}{2}D_+ + \frac{1}{2}D_-$.

- a mixture of chiral states Ψ_L and Ψ_R, each with 50% probability? This would correspond to a decomposition of the thermal density operator D_β as
$D_\beta = \frac{1}{2}D_L + \frac{1}{2}D_R$.

- a mixture of "alternative" chiral states

$$\tilde{\Psi}_L \overset{\text{def}}{=} \frac{1}{\sqrt{2}}(\Psi_+ + i\Psi_-), \tag{24}$$

$$\tilde{\Psi}_R \overset{\text{def}}{=} \frac{1}{\sqrt{2}}(\Psi_+ - i\Psi_-), \tag{25}$$

which again are transformed into one another under space-inversion? This would correspond to a decomposition of the thermal density operator D_β as
$D_\beta = \frac{1}{2}\tilde{D}_L + \frac{1}{2}\tilde{D}_R$.

- an ensemble of 25% left-handed, 25% right-handed, 25% ground and 25% excited states? This would correspond to a decomposition of D_β as
$D_\beta = \frac{1}{4}D_+ + \frac{1}{4}D_- + \frac{1}{4}D_L + \frac{1}{4}D_R$.

In fig. 3 some of these different decompositions are visualized using the Bloch sphere. The thermal density operator $D_\beta = \frac{1}{2}\mathbf{1}$ from eq. (23) corresponds to the zero vector in three-dimensional space, i. e., to the center of the Bloch sphere. The statements about decompositions of D_β made before can then be verified by computing the vectors $\vec{b}_+, \vec{b}_-, \vec{b}_L, \vec{b}_R, \tilde{\vec{b}}_L$ and $\tilde{\vec{b}}_R$ corresponding to the respective pure state vectors $\Psi_+, \Psi_-, \Psi_L, \Psi_R, \tilde{\Psi}_L$ and $\tilde{\Psi}_R$. This can be done using eq. (22). The chiral states, for example, correspond to points on the Bloch sphere having zero y- and z-coordinates. Then a mixture of 25% $\vec{b}_L$, 25% $\vec{b}_R$, 25% $\vec{b}_+$ and 25% $\vec{b}_-$ obviously gives the vector $\vec{0}$ corresponding to our thermal density operator $D_\beta = \frac{1}{2}\mathbf{1}$, which verifies one of the above claims (cf. eq. (19)).

Summarizing, one has infinitely many different ways of decomposing a thermal density operator into pure states. *Different decompositions can refer to entirely different physical or chemical points of view:*

- The decomposition of $D_\beta = \frac{1}{2}\mathbf{1}$ into chiral states Ψ_L and Ψ_R gives rise to an approximate nuclear molecular structure [simply because we assume that the molecule is in either of these states with equal probability and both, Ψ_L and Ψ_R, admit an approximate nuclear structure, see fig. 1].

- The decomposition of $D_\beta = \frac{1}{2}\mathbf{1}$ into eigenstates states Ψ_+ and Ψ_- does *not* admit an approximate nuclear molecular structure [simply because we assume that the molecule is in either of these states with equal probability and both, Ψ_+ and Ψ_-, do *not* admit an approximate nuclear structure, see fig. 1].

- The decomposition of $D_\beta = \frac{1}{2}\mathbf{1}$ into alternative chiral states $\tilde{\Psi}_L$ and $\tilde{\Psi}_R$ does *not* admit an approximate nuclear molecular structure [simply because we assume that the molecule is in either of these states with equal probability and both, $\tilde{\Psi}_L$ and $\tilde{\Psi}_R$, do *not* admit an approximate nuclear structure].

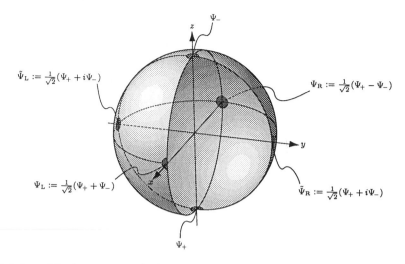

Figure 3. A thermal density operator can be decomposed into pure states in infinitely many different ways. Mixing the vectors corresponding to eigenstates, *or* chiral states *or* alternative chiral states with 50% probability always leads to the zero vector, i. e., the center of the Bloch sphere, corresponding to the density operator $D_\beta = \frac{1}{2}\mathbf{1}$.

It is, of course, not compulsory to decompose a thermal density operator into only two or finitely many pure states. We could equally well try do decompose a thermal density operator into a continuum of pure states, or even into all the pure states of the system in question. In this case probabilities $\mu(\tilde{B})$ must be given for finding the system in question in a certain subset $\tilde{B}$ of the set of all pure states. In case of a two-level system, the set of all pure states is isomorphic to the three-dimensional sphere S_2; hence a subset $\tilde{B}$ of the set of all pure states can be specified by giving the corresponding subset B of S_2 (see fig. 4).

As a typical example for such a probability distribution consider the *equidistribution of pure states*, with probabilities

$$\mu_{\text{eqp}}(\tilde{B}) \overset{\text{def}}{=} \int_{B \subseteq S_2} \frac{\sin\vartheta\, d\vartheta\, d\varphi}{4\pi}, \tag{26}$$

with spherical coordinates ϑ and φ used. Here the subset $\tilde{B}$ of all the pure states of the two-level system in question corresponds to the subset B of the surface of the threedimensional sphere. The probability distribution in eq. (26) of pure states is called *equidistribution* of pure states since it is invariant under all symmetries of the two-level system, i. e.,

$$\mu(V\tilde{B}V^*) = \mu(\tilde{B}) \tag{27}$$

holds for all unitary 2×2-matrices V. Mixing all the pure states of the two-level system with probability μ_{eqp} corresponds (see eq. (19)) to mixing of all the vectors $\vec{b}$ according to

$$\int_{S_2} \vec{b}(\vartheta, \varphi) \frac{\sin\vartheta\, d\vartheta\, d\varphi}{4\pi} \tag{28}$$

which is clearly the zero vector $\vec{0}$ corresponding to the density operator $D_\beta = \frac{1}{2}\mathbf{1}$. Hence the mixture of all the pure states of the two-level system with probability μ_{eqp} results in the non-pure state with associated density operator $D_\beta = \frac{1}{2}\mathbf{1}$. Hence this gives a new decomposition of $D_\beta = \frac{1}{2}\mathbf{1}$, specified by the equipartition probabilities in eq. (26). In terms of density operators, this decomposition would read

$$\text{Tr}((\tfrac{1}{2}\mathbf{1})A) = \int_{S_2} \text{Tr}(D_{\vartheta,\varphi}A)\, \frac{\sin\vartheta\, d\vartheta\, d\varphi}{4\pi}, \tag{29}$$

where A is an arbitrary observable of the two-level system (i. e., an arbitrary 2×2-matrix) and where $D_{\vartheta,\varphi}$ is the density operator corresponding to that (pure) state, which in the Bloch-sphere description has spherical coordinates ϑ and φ. In more abstract notation, this decomposition would be denoted by

$$\text{Tr}((\tfrac{1}{2}\mathbf{1})A) = \int_{S_2} \text{Tr}(D_{\vartheta,\varphi}A)\, \mu_{\text{eqp}}(d\vartheta, d\varphi), \tag{30}$$

with μ_{eqp} being the equipartition measure on the pure states of the two-level system (with two-level systems only being used because of their simplicity).

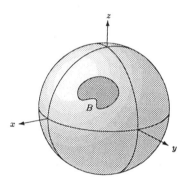

Figure 4. A subset $\tilde{B}$ of all the pure states of a two-level system can be specified by giving the corresponding subset of the surface S_2 of the Bloch sphere.

In fig. 5 again different decompositions of some thermal state D into pure states are sketched: This time the density operator D is *not* chosen to be equal to $\frac{1}{2}\mathbf{1}$. Since every thermal density operator D can be written as a mixture of the eigenstates of the underlying hamiltonian, the corresponding vector $\vec{b}_D$ must be some mixture of the north- and southpole vectors corresponding to the eigenstates Ψ_+ and Ψ_-; hence $\vec{b}_D$ must be on the z-axis, as indicated in fig. 5.

In the left upper part of fig. 5 the spectral decomposition of D into orthogonal eigenstates is illustrated. Since it is not compulsory to decompose a density operator into *orthogonal* states, also another decomposition into two non-orthogonal pure states could be chosen, as indicated in the upper right part of fig. 5. Since it is neither compulsory to decompose into *two* pure states, decompositions into continuously many pure states are also possible; e. g., into all the pure states lying on the same parallel of latitude as D, indicated in the lower left part of the figure; or, e. g., into all the pure states of the two-level system under discussion as illustrated in the lower right part of fig. 5, where the shading indicates the probability to find a pure state in the respective region of the set of all pure states.

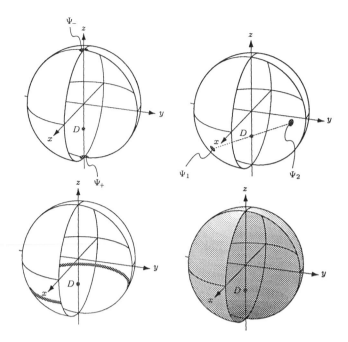

Figure 5. A non-pure state (with density operator D) can be decomposed in many different ways into pure states. Here the situation is sketched for 2×2-matrices. The upper left part of the figure shows a decomposition of a density operator D into its eigenstates. The upper right part of the figure shows a decomposition of D into two non-orthogonal pure states. The lower left part of the figure shows a decomposition into pure states all lying on the same parallel of latitude. The lower right part of the figure illustrates the decomposition of D into *all* possible pure states (of a two-level system). The probability density of pure states arising in this decomposition is indicated by the shading.

A decomposition of a thermal state D_β into pure states is nothing else than the specification of probabilities $\mu(\tilde{B})$ that the system under discussion is in one of the pure states of the subset $\tilde{B}$ of all pure states possible. A typical example is given as

$$\mu_f(\tilde{B}) \stackrel{\text{def}}{=} \int_{B \subseteq S_2} f(\vartheta, \varphi) \frac{\sin \vartheta \, d\vartheta \, d\varphi}{4\pi}, \tag{31}$$

where B is the subset of the Bloch sphere corresponding to the chosen subset $\tilde{B}$ of pure states and where $f : S_2 \to [0, \infty[$ is now some fixed function from the Bloch sphere into the positive real numbers. The respective decomposition would then read as

$$\text{Tr}(D_\beta A) = \int_{S_2} \text{Tr}(D_{\vartheta, \varphi} A) \, f(\vartheta, \varphi) \frac{\sin \vartheta \, d\vartheta \, d\varphi}{4\pi}, \tag{32}$$

or in more abstract notation as

$$\text{Tr}(D_\beta A) = \int_{S_2} \text{Tr}(D_{\vartheta, \varphi} A) \, \mu(d\vartheta, d\varphi), \tag{33}$$

The probability to find the system in any pure state must be equal to 1, of course,

$$\int_{S_2} f(\vartheta, \varphi) \frac{\sin \vartheta \, d\vartheta \, d\varphi}{4\pi} = 1. \tag{34}$$

The function f is a probability density, indicated by the shading of the lower right part of fig. 5. To get a decomposition of D_β into *eigenstates*, one must choose the "function" f as a linear combination of Dirac delta functions (i. e., as an appropriate distribution).

CHEMICAL VS. QUANTUM-MECHANICAL POINTS OF VIEW

A given thermal density operator can be decomposed into *stationary* or *non-stationary* pure states. Depending on this choice, two different points of view are adopted, which-for simplicity-I shall call the *spectroscopist's* and the *chemist's* point of view, respectively.

Let us again look at ammonia molecules:

- From a *spectroscopist's point of view*, ammonia molecules are in the stationary eigenstates Ψ_+ and Ψ_- with stochastically occurring "quantum jumps" between them (corresponding to the MASER-transition). Note that the expectation value of the dipole moment operator $\hat{\vec{\mu}}$ is zero with respect to these stationary states.

- From a *traditional* chemist's point of view, ammonia molecules have a (dispersion-free) nuclear structure, oscillating back and forth between the two pyramidal forms, preserving a nuclear structure during the whole oscillation process and with a planar transition state.

- Matching the traditional chemist's point of view with quantum mechanics, one arrives at what I call now the *chemist's point of view*, omitting the epithet "traditional": ammonia molecules are in non-stationary (pyramidal) states $\Psi_L = \frac{1}{\sqrt{2}}[\Psi_+ + \Psi_-]$ and $\Psi_R = \frac{1}{\sqrt{2}}[\Psi_+ - \Psi_-]$ with Schrödinger time-evolution

$$\frac{1}{\sqrt{2}}[\Psi_+ \pm \Psi_-] \quad \rightarrow \quad \frac{1}{\sqrt{2}}[\Psi_+ \pm \exp\{-it(E_- - E_+)/\hbar\}\Psi_-]. \tag{35}$$

During this tunneling process, the nuclear molecular frame is not conserved; in between the "alternative chiral" states $\frac{1}{\sqrt{2}}[\Psi_+ \pm i\Psi_-]$ arise, which do not admit a nuclear structure. Incidentally, for small level splitting $(E_- - E_+)$, the tunneling process is very slow and the question comes up which of the available chiral states (on the equator of the Bloch sphere) actually arise in a properly chiral molecule.

Ammonia is, of course, no isolated example. An analogous situation appears when one considers *hydrogen bonds* between two oxygen atoms (one in carbonyl form, the other in hydroxy form, as in naphthazarin, citrinin and di-carboxylic acids). The questions then are again: does the hydrogen "sit" near *one* of the oxygens, does it oscillate back and forth or should one consider a stationary state (analogous to the ground state of ammonia) without a nuclear structure, i. e., without a dispersion-free position of the hydrogen atom (and other nuclei)?

In such situations the different descriptions mentioned can be used:

- A *"quantum"* description by some stationary state Ψ_o or by various stationary states with an additional stochastic dynamics (imposed on the usual Schrödinger dynamics),

- or a *"chemical"* description by non-stationary states whose superposition is Ψ_o.

This dichotomy reflects itself in different choices for decompositions of the thermal density operator D_β into pure states:

- A decomposition of the thermal density operator D_β into (symmetry-adapted) eigenstates of the hamiltonian. If superpositions of these eigenstates are *not* considered, one gets a *classical energy observable*.

- A decomposition of the thermal density operator D_β into pure handed states. If superpositions of these handed states are *not* considered, one gets a *classical chirality observable*.

- A decomposition μ of the thermal density operator D_β into pure states Ψ such that the average dispersion

$$\langle \Psi | \hat{Q}_j^2 \Psi \rangle - \langle \Psi | \hat{Q}_j \Psi \rangle^2 \tag{36}$$

for the expectation values of the nuclear position operators $\hat{Q}_j, j = 1, 2, \ldots, K$, is minimal. Using the nomenclature of two-level systems (compare eq. (30)), this average dispersion is defined as

$$\int_{S_2} \left\{ \mathrm{Tr}(D_{\vartheta,\varphi} \hat{Q}_j^2) - \mathrm{Tr}(D_{\vartheta,\varphi} \hat{Q}_j)^2 \right\} \mu(d\vartheta \, d\varphi), \tag{37}$$

If superpositions of these minimal-dispersion states are *not* considered, one gets an approximate *classical nuclear structure,* which is the best possible nuclear structure compatible with the thermal density operator D_β.

This way of *imposing classical structures "by hand"* by simply omitting superpositions between certain chosen states is a very common procedure. Often, the same phenomenon (e. g., the spectrum of ammonia) can be explained in different ways (using, e. g., pyramidal non-stationary states instead of eigenstates), giving a "chemical" or a "quantum-mechanical" explanation in our sense. Nevertheless such different explanations are not compatible (and sometimes even complementary), since *either* energy *or* handedness is a classical observable but not both together.

Here another approach will be advocated, assuming that a quantum object *is* indeed in some pure state (without claiming that this pure state can always be easily determined). Hence pure states in quantum mechanics are not considered as toys which can be exchanged ad libitum. Consequently we shall look for a *canonical* decomposition of thermal density operators D_β into pure states. This canonical decomposition should depend on the particular situation: For ammonia, the eigenstates can be expected to play an important role (i. e., have high probability with respect to the canonical decomposition) whereas for a sugar molecule the handed states can be expected to be important (i. e., have high probability with respect to the canonical decomposition).

With a *canonical* decomposition at hand, also the fuzzyness of some observable $\hat{A}$ can be unambiguously described, giving, e. g., the probability to find eigenstates of $\hat{A}$ with respect to the canonical decomposition of the thermal density operator D_β. If different decompositions are admitted, one cannot say anything about the fuzziness of some observable: the energy, for example, is dispersion-free with respect to the decomposition into eigenstates but not with respect to other decompositions of D_β into pure states. Consequently the distribution of the expectation values of the energy or the distribution of the dispersions of the molecular energy is very different depending on the particular decomposition of D_β into pure states.

EFFECTIVE THERMAL STATES

Consider *two or more different isomers* of some molecular species. Since different isomers have the same underlying hamiltonian $H_\circ$, the respective thermal density operator

$$D_\beta \overset{\text{def}}{=} \frac{\exp(-\beta H_\circ)}{\mathrm{Tr}(\exp(-\beta H_\circ))} \tag{38}$$

is *identical* for different isomers. Consequently, the expectation values

$$\mathrm{Tr}(D_\beta \hat{Q}_j), \quad j = 1, 2, \ldots, K, \tag{39}$$

of the nuclear position operators $\hat{Q}_j$ are *identical* for different isomers. But this result is in obvious contradiction with chemical experience, saying that for different isomers the nuclei sit at *different* (relative) positions.

The conclusion is that the recipes of statistical quantum mechanics, e. g., the way a thermal state is defined in eq. (38), cannot explain chemical phenomena without taking over concepts from traditional chemistry in an *ad hoc* manner. These recipes

- neither give rise to molecular isomers,

- nor to handed molecules,

- nor to monomer sequences in a macromolecule,

- nor to differently knotted macromolecules.

For all these chemically well known concepts *different* expectation values of the nuclear position operators are necessary.

The main reason for this bewildering observation is that thermal density operators D_β describe a *strictly* stationary situation. To make this point clear, consider a "thermal state" $D_{\beta,L}$ describing an ensemble of left-handed molecules of some given species. Since handed molecules racemize–slowly, but nevertheless–one may expect that $D_{\beta,L}$ evolves into $\frac{1}{2}(D_{\beta,L} + D_{\beta,R})$ for large times. Consequently the density operator $D_{\beta,L}$ is *not* strictly stationary and therefore cannot be a thermal state in the strict sense.

"Thermal state in the strict sense" is to mean that the state fulfils the KUBO-MARTIN-SCHWINGER-boundary condition, which is a stability requirement[49,50] under a specified class of external perturbations. In a quantum system having finitely many degrees of freedom there exists precisely one KMS-state for every positive temperature, namely the state defined in eq. (38).

Summarizing: The usual recipes of statistical quantum mechanics cannot explain chemical phenomena such as isomerism and handedness of molecules. The question is how to introduce effective *thermal states for different isomers or differently handed molecules etc.*

With the Born–Oppenheimer idea in mind, one could try to restrict oneself to a particular minimum of the BO-potential: In fig. 1, for example, one could only consider state vectors living in the left *or* the right minimum and construct corresponding effective thermal states by some ad-hoc procedure (and decompose the overall thermal state in eq. (38) accordingly, as $D_\beta = \frac{1}{2}(D_{\beta,L} + D_{\beta,R})$, for example). But then the old question comes again which asks *why should other state vectors, like the ground state vector* Ψ_+ *or like the alternative handed states* $\tilde{\Psi}_L$ *and* $\tilde{\Psi}_R$ *of an ammonia-type molecule, be excluded?* Hence any ad-hoc procedure introducing effective thermal states is based on a particular choice of state vectors or on a particular choice for a decomposition of the overall thermal state D_β of eq. (38)).

Therefore we end up again with the problem to find a *canonical* decomposition of the "overall" thermal density operator D_β of some molecular species into pure states. Based on such a canonical decomposition of D_β (and dynamical arguments, see below) one can introduce effective thermal states in some ad-hoc manner .

Another option has been proposed in algebraic quantum mechanics[40,49,50,51,52], where it has been emphasized that

- a quantum system with *finitely* many degrees of freedom allows *one* thermal state D_β for every temperature

- whereas a quantum system with *infinitely* many degrees of freedom may admit several different thermal states $D_{\beta,j}, j = 1, 2, \ldots$ for a given temperature.

The idea behind all this is, of course, to find several different (even strict, i. e., KMS-) thermal states for

- different coexisting phases, such as water and steam at 100 centigrades,

- different molecular isomers,

- differently handed molecules, etc.

Let us check this option for another, perhaps more transparent example, namely a Curie–Weiss magnet below the Curie temperature, i. e., with the possibility of having two different phases with positive and negative permanent magnetization (compare the references[53,54,55] for mean-field models). Then

- considering *finitely* many spins leads always to *one* thermal state D_β^N only (N being the number of spins),

- whereas the Curie–Weiss model consisting of *infinitely* many spins admits *two* different (strict, i. e., KMS-) thermal states $D_{\beta,+}$ and $D_{\beta,-}$ with positive and negative magnetization. In this case the "overall" thermal state D_β as defined in eq. (38) is a mixture $D_\beta = \frac{1}{2}(D_{\beta,+} + D_{\beta,-})$.

This result is somewhat irritating: Why is it necessary to consider a *limit* of infinitely many spins to derive two thermal states $D_{\beta,+}$ and $D_{\beta,-}$ for a Curie–Weiss magnet?

Interestingly enough, in the infinite limit, one gets a superselection rule excluding superpositions between the mentioned two thermal states $D_{\beta,+}$ and $D_{\beta,-}$. For *any* finite number of spins, on the other hand, one has always full validity of the superposition principle! Is it legitimate to ask *how fast superpositions of states with positive and negative magnetization disappear with increasing number of spins?*

The particular example of a Curie–Weiss magnet will be studied in below in more detail. The interesting point in this example is

- that in the limit of *infinitely* many spins the "overall" thermal state decomposes (as can be shown by theorem) into the (non-pure) states $D_{\beta,+}$ and $D_{\beta,-}$
  [because the specific magnetization operator becomes a strict classical observable, taking only two different values of specific magnetization, which are called here $+m_\beta$ and $-m_\beta$; pure states Ψ arising in any further decomposition of $D_{\beta,+}$ and $D_{\beta,-}$ into pure states have the same expectation values $+m_\beta$ and $-m_\beta$, respectively],

- whereas in the case of *finitely* many spins it is not clear at all how to decompose the unique thermal state D_β^N into some states $D_{\beta,+}^N$ and $D_{\beta,-}^N$ corresponding to the states $D_{\beta,+}$ and $D_{\beta,-}$ in the infinite limit. It is not at all compulsory to decompose D_β^N into such two effective thermal states
  [it is, in particular, not at all compulsory that pure states Ψ in a decomposition of D_β^N into pure states lead to the expectation values $+m_\beta$ and $-m_\beta$ of the specific magnetization operator].
  Actually a *canonical* decomposition of D_β^N is expected to do the job, i. e., to lead to a definition of $D_{\beta,+}^N$ and $D_{\beta,-}^N$.

When introducing a canonical decomposition of thermal states into pure ones, it is therefore important to check compatibility with these limit considerations. Hence expectation values of the specific magnetization operator [with respect to pure states Ψ in a *canonical* decomposition] other than $+m_\beta$ and $-m_\beta$ must disappear with an increasing number of spins in the

Curie–Weiss model. In this way the specific magnetization operator gets "more and more classical," because the probability to find other expectation values than $+m_\beta$ and $-m_\beta$ [in the canonical decomposition] gets smaller and smaller with increasing number of spins (see section "Fuzzy classical observables and large-deviation theory"). In this and similar situations we shall speak of a *fuzzy classical observable* or *fuzzy classical structure*. Getting "more and more classical" means that the fuzziness disappears, finally ending up with a strict classical observable in the infinite limit having dispersion-free expectation values.

Summarizing: In statistical quantum mechanics one can derive *strict* classical observables (such as a classical specific magnetization operator) by appropriate limit considerations (such as a limit of infinitely many spins in case of the Curie–Weiss model). Nevertheless statistical quantum mechanics cannot cope with *fuzzy* classical observables (for finitely many degrees of freedom) since different decompositions of a thermal state D_β are considered to be equivalent. Introducing a canonical decomposition of D_β into pure states will give rise to an *individual* formalism of quantum mechanics in which fuzzy classical observables can be treated in a natural way.

STOCHASTIC DYNAMICS ON THE LEVEL OF PURE STATES

We have argued heuristically that certain molecular states are unstable under external perturbations. The symmetric ground state Ψ_+ of an ammonia-type molecule, for example, seems to be unstable under external perturbations *if* the level splitting $(E_- - E_+)$ is very small (see table 1). Or superpositions of states describing different molecular isomers seem to be unstable etc. This is *not* to say that such "strange" states, as I called them, cannot be prepared experimentally. To the contrary, it is reasonable to argue that a molecule can be prepared in *every* pure state available in the quantum mechanical formalism as long as it is screened well enough from external influence.

External perturbations can be of various sorts:

- Quantum radiation and gravitation field,

- collisions with neighbor particles,

- coupling to some heat bath etc.

Such external perturbations cannot easily be treated theoretically in a rigorous manner: the main problem is that pure states of the joint quantum system {molecule & quantum environment} are usually not product states and therefore their restriction to the molecular part is not a pure state any more (see eq. (13)). Nevertheless I shall accept here the idea that appropriate dressing transformations[56] can cure this situation (almost at least: some effects due to non-product states will survive even the best possible dressing[56]). Hence molecule *and* perturbing environment are both thought to be in pure states.

The pure state of the molecular environment is never precisely known. Since the dynamics of the molecular pure state depends on this unknown environment's (pure) state, one gets a *stochastic* dynamics for the molecular pure states. For a two-level system–with its pure states being describable by the Bloch sphere–the situation is illustrated in fig. 6:

- The dynamics of some given molecular initial state is not only governed by the Schrödinger equation, but also by external influence.

- Depending on the (pure) state of the environment, different final molecular states can be reached.

- Usually only probabilistic predictions can be given (and no information about the precise trajectory of pure states, i. e., the trajectory on the Bloch sphere).

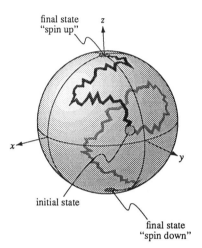

Figure 6. Sketch of a stochastic dynamics for pure states of a two-level molecule: Starting from an arbitrary initial state one ends up with certain states such as, for example, eigenstates of some operator. Which final state is reached depends on the (pure) state of the molecular environment.

Stochastic dynamics for pure states of a quantum system have been proposed or derived in various different contexts[57,58,59,60,61,62]. Also the transition rates of Fermi's Golden Rule[63]

$$W(n \rightarrow m) = \frac{\pi}{2\hbar^2} |\langle \Psi_n | B \Psi_m \rangle|^2 \delta(\omega - |\omega_{nm}|), \qquad (40)$$

between eigenstates of a molecular hamiltonian give rise to a stochastic dynamics for pure states (with B being the coupling operator between molecule and radiation field, as, e. g., the dipole moment operator).

Unfortunately, a rigorous derivation of stochastic pure-state dynamics is still missing. Nevertheless it is rather convincing that such stochastic dynamics are important and in fact the clue of a quantum theory of *individual* (quantum) objects. One hint in this direction comes from **S**ingle-**M**olecule **S**pectroscopy[64,65,66,67,68,69,70,71,72,73,74,75,76,77], where "single molecule" is always to be understood as a single molecule embedded into a polymorphic matrix or a crystal. The example used in fig. 7 is a single terrylene defect molecule embedded in a hexadecane matrix, the actual individual quantum object therefore being the joint quantum system {terrylene defect & hexadecane matrix}. Another well investigated example is a pentacene defect in a p-terphenyl crystal[64]. Investigation is done in these examples by fluorescence excitation spectroscopy, i. e., by exciting the system under discussion with a laser measuring at the same time the overall emission of radiation (excluding the frequency of the laser itself). The interesting point with these SMS-experiments is the *stochastically migrating behavior* of some of the spectral lines. This migrating behavior can ·

- either be investigated by changing the laser frequency over some range (say 2000 MHz with one sweep per second) and checking *at which frequency* absorption (and reemission) takes place (in function of time),

- or by leaving the laser frequency fixed and recording if absorption and reemission takes place at all (in function of time).

In fig. 7 the second method has been used, showing very different stochastic behavior for different terrylene molecules in a hexadecane matrix. In the corresponding *ensemble* spectroscopy–investigating a large number of terrylene defects at the same time–all the possible spectral lines appear at the same time and hence a migrating behavior cannot be observed.

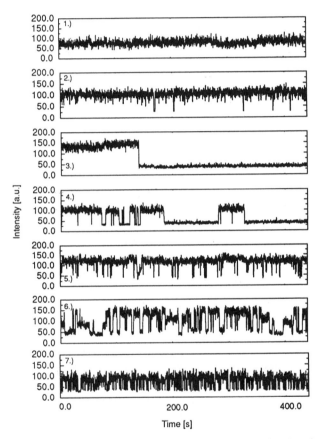

Figure 7. Migrating absorption frequencies attributed to single terrylene defects in a hexadecane matrix. The examples shown here change between roughly two possible absorption frequencies. Depending on the particular situation, one observes a more or less pronounced stochastic behavior of spectral migration. This particular experiment has been performed in the group of Prof. Urs Wild at ETH-Zürich. Interested readers are invited to consult http://www.chem.ethz.ch/sms/ at internet.

Summarizing, one has a regular dynamics on the level of density operators (not differing much from the thermal state), whereas the dynamics on the level of pure states is stochastic. Averaging over all possible stochastic "paths" of pure states results in the regular density-operator dynamics. To get an understanding of single-molecule spectroscopy, for example, it is vital to investigate stochastic dynamics on the pure-state space of large quantum systems (and not only two-level systems, of course).

AN EXAMPLE OF A STOCHASTIC DYNAMICS ON THE SET OF PURE STATES OF A TWO-LEVEL SYSTEM

In this section (which can be skipped at a first reading), a "simple" example for a stochastic dynamics on the pure-state space of a system is discussed. In particular, transition probabilities will be computed for transitions between certain regions of the set of pure states (cf. fig. 4). These transitions are

- *not* necessarily transitions between eigenstates of some operator (e. g., the hamiltonian)

- but transitions between arbitrary regions of the pure-state space (e. g., starting from an arbitrary pure initial state to an arbitrary part B of the surface of the sphere)

Again, a two-level system is only taken for simplicity. Interesting systems in this respect could be single molecules or a magnet (for the latter, cf. section "Fuzzy classical observables and large-deviation theory"). The system in question is thought to be perturbed from "outside", i. e., its dynamics also depends on the (pure) state ω of the environment. Hence the state ω becomes a stochastic variable in the description of the system.

Remark: In this context, not only the system but also its *environment is assumed to be in a pure state*, though this pure state ω is usually not known precisely and can only be estimated. Depending on ω, the initial state ϕ_o of the original system will develop dynamically in a different way. The result is a stochastic dynamics on the pure states ϕ of the system, cf. ref.[46,57,58,59,60,61,78].

Let us now come back to the example of a two-level system. Starting with some arbitrary initial state, one could, for example, be interested to know the probability that the dynamics leads to the north- or to the south pole. In the Bloch-sphere parametrization used here, north- and south pole correspond to eigenstates of the hamiltonian, but they could equally well correspond to eigenstates of, say, the dipole moment operator (which couples a molecule to the radiation field). Or one could be interested in the *time* necessary to get to the north- or south pole. A realistic situation could, of course, be much more complex. Also, given the philosophy of the present paper, eigenstates would not necessarily play a predominant role (neither eigenstates of the hamiltonian nor eigenstates of the dipole moment operator). It would, in particular, be *interesting to compute lifetimes of the ground state of a chiral molecule* (see first section), i. e., to know how long it takes for the proper ground state of a chiral molecule to decay into the handed states.

Here, no particular physical point of view is taken. The only goal is to give some idea how stochastic dynamics on the set of pure states of a quantum system can, in principle, be discussed.

Starting from some initial state, we deal with trajectories depending on the state ω of the environment (which arise with a certain probability P, i. e., P is a probability distribution on the pure-state space of the environment. These trajectories are *trajectories in the set of pure states*, and not trajectories in the sense of position and momentum values (p,q) varying with time (see fig. 6). The pure state ϕ_t depends on the state ω of the environment

$$\phi_t = \phi_t(\omega) \,. \tag{41}$$

This nomenclature could be misleading for a mathematician. It should simply indicate that ϕ_t depends an an additional variable ω. If one chooses a fixed initial state ϕ_o for the system (as in fig. 6), one has

$$\phi_o(\omega) = \phi_o \tag{42}$$

for all pure states ω of the environment. Let us now draw little circles around the north- and south poles, respectively, and ask when and where a particular trajectory ω passes these circles* (little circles are used, since an actual trajectory will usually not reach a pole precisely). The respective time will be denoted by $\tau(\omega)$ and called a *stopping time*[79]. Some of the trajectories will first pass the circle near the north pole, and some of the trajectories will first pass the circle near the south pole (*here*, we are not interested in what happens later on, though that would be very interesting to discuss). It will be tried below to compute the transition probabilities, i. e., the probabilities to pass first the circle near the north- or the south pole, respectively.

Only a relatively simple example will be discussed, the technique being applicable to any stochastic dynamics on the pure-state space. The example is given by the following class of (Stratonovich) stochastic differential equations

$$d\vartheta_t = -2\kappa \sin(\vartheta_t) \cos(\vartheta_t) \, dt + 2\nu \sin(\vartheta_t) \circ dW_t , \tag{43}$$

with κ and ν being fixed positive constants. Here ϑ is one of the spherical coordinates (for simplicity, it is assumed that the problem does not depend on the azimutal angle φ), and W_t is a one-dimensional Wiener process[79,80].

Remark: The Wiener process W depends on two variables: The time t and the stochastic variable ω, which specifies the particular path and which describes the (pure) state of the environment in our context. A path ω is explicitly given by

$$[0,\infty] \ni t \rightarrow W_t(\omega) \in \mathbf{R} . \tag{44}$$

For all paths ω, one has $W_{t=0}(\omega) = 0$. The Wiener process is characterized by the following properties:

- if $t_0 < t_1 < \ldots t_n$, then the increments W_{t_0}, $(W(t_1) - W(t_0))$, $\ldots (W(t_n) - W(t_{n-1}))$ are independent,
- the probability that an increment $(W_{s+t} - W_s)$ lies in an interval $[a,b]$ is given by

$$\int_a^b (2\pi t)^{-1/2} e^{-x^2/2t} \, dt , \tag{45}$$

- with probability 1 the paths (44) are continuous.

More general stochastic dynamics could be formulated (adding, e. g., the experimenter's input to the so-called drift term $-2\kappa \sin(\vartheta_t) \, dt$). Actually, all the variables also depend on the stochastic variable ω, which describes the state of the environment ($\equiv$ the path of the Wiener process in the mathematical description). Hence eq. (43) could also be written as

$$d\vartheta_t(\omega) = -2\kappa \sin\{\vartheta_t(\omega)\} \cos\{\vartheta_t(\omega)\} \, dt + 2\nu \sin\{\vartheta_t(\omega)\} \circ dW_t(\omega) . \tag{46}$$

An equivalent integral version of eq. (46) is

$$\vartheta_t(\omega) = \vartheta_\circ + \int_0^t (-2\kappa) \sin\{\vartheta_s(\omega)\} \cos\{\vartheta_s(\omega)\} \, ds + \int_0^t 2\nu \sin\{\vartheta_s(\omega)\} \circ dW_s(\omega) . \tag{47}$$

Roughly spoken, these integrals are computed for every path ω. The "difficult" integral is the second one (with the Wiener process W_t), which for a *Stratonovich* stochastic differential equation is defined as the limit

$$\lim_{\delta \to 0} 2\nu \sum_{k=1}^K \frac{\sin\{\vartheta_{s_k}(\omega)\} + \sin\{\vartheta_{s_{k+1}}(\omega)\}}{2} \left(W_{s_{k+1}}(\omega) - W_{s_k}(\omega)\right) , \tag{48}$$

---

*Pure states of the environment and pure-state trajectories of the system are in one-one correspondence and hence denoted by the same variable ω.

where $s_k, k = 1, 2, 3, \ldots, K$ is a partition of the interval $[0, t]$ with a maximal distance δ between the partition points[80]. In an *Itô* stochastic differential equation[80], on the other hand, this integral would be defined as

$$\lim_{\delta \to 0} 2\nu \sum_{k=1}^{K} \sin\{\vartheta_{s_k}(\omega)\} \, (W_{s_{k+1}}(\omega) - W_{s_k}(\omega)). \tag{49}$$

It is important to realize that the definitions in eqs. (48, 49) give rise to different results (since the Wiener paths are not differentiable). *Physical* reasoning leads to *Stratonovich* stochastic differential equations, which can be reformulated as an *Itô* equation with different drift term[80]. The Stratonovich eq. (43), for example, corresponds to the Itô equation

$$d\vartheta_t = \left\{ -2\kappa \sin(\vartheta_t) \cos(\vartheta_t) + \frac{1}{2} \frac{\partial \sin(\vartheta_t)}{\partial \vartheta} \sin(\vartheta_t) \right\} dt + 2\nu \sin(\vartheta_t) \, dW_t, \tag{50}$$

Reformulation as *Itô* equation has mathematical advantages: An *Itô* integral (as in eq. (49)), depending on t, is a martingale[†], whereas the corresponding *Stratonovich* integral is not. The concept of a *martingale* is explained in the appendix.

For the particular values $\kappa = \nu = \frac{1}{2}$, the stochastic process (43) is the Gisin process[61], which has the correct transition probabilities $\frac{1}{2}(1 \pm \cos \vartheta)$ for transitions to the poles, i. e., reproduces the transition probabilities of the *projection postulate*. Hence the class (43) of stochastic dynamics is not entirely artificial. Note that the transition probabilities discussed here are not transition probabilities between eigenstates of the hamiltonian, but probabilities for an arbitrary pure initial state to get to some chosen region of the pure-state space (some region on the surface of the sphere in case of a two-level system).

Incidentally, the equation (43) can be transformed by introducing a new variable $x_t := -\log \tan(\frac{1}{2}\vartheta_t)$ into

$$dx_t = 2\kappa \tanh(x_t) \, dt - 2\nu \, dW_t, \tag{51}$$

which will be used for the mathematical discussion below (the heuristic physical discussion will still be lead in terms of fig. 6). Note that for this stochastic integral equation the Stratonovich and Itô versions coincide (since the diffusion term does only contain the Wiener differential dW_s, but not contain a function of W_s or a function of the solution x_s of the stochastic equation). The mentioned transformation changes the domain of definition from $0 \leq \vartheta \leq \pi$ to $-\infty \leq x \leq +\infty$. Hence small circles around north- and south pole do now correspond to scalars x_- and x_+, where x_- is very negative ("near" $-\infty$) and x_+ is very positive ("near" $+\infty$). The initial value of the trajectories considered (corresponding to the initial state on the sphere S_2) will be denoted by x_o.

Consider again the stopping time $\tau(\omega)$ introduced above. At time $\tau(\omega)$, the trajectory

$$t \mapsto x_t(\omega) \tag{52}$$

with initial value x_o reaches either x_- or $_+$, i. e.,

$$x_{\tau(\omega)}(\omega) = x_- \quad \text{or} \quad x_{\tau(\omega)}(\omega) = x_+. \tag{53}$$

Assume now that there exists a function

$$h : \mathbf{R} \to \mathbf{R} \tag{54}$$

[†]Here martingales are always taken with respect to the time t and not with respect to particle number N.

such that

$$h(x_t) = h(x_t(\omega)), \quad t \in \mathbf{R} \tag{55}$$

is a *martingale* (cf. the appendix). Recall that a martingale

$$t \to Z_t$$
$$Z_t = Z_t(\omega) \tag{56}$$

is a *typical stochastic process in the usual sense*, i. e., given the values $Z_{s_1}(\omega)$ for some time s_1, the best prediction for Z_{s_2}; $s_2 > s_1$ is Z_{s_1} itself. A Wiener process, for example, is a martingale. For a martingale $h(x_t)$, Doob's optional stopping theorem[79] implies that

$$h(x_\circ) = \int_{\text{states of the environment}} h(x_{\tau(\omega)}(\omega)) \, dP(\omega) \tag{57}$$
$$= h(x_-) P\{\omega | x_{\tau(\omega)}(\omega) = x_-\} + h(x_+) P\{\omega | x_{\tau(\omega)}(\omega) = x_+\}. \tag{58}$$

The equality in eq. (58) is an immediate consequence of eq. (53), with $P\{\omega | x_{\tau(\omega)}(\omega) = x_-\}$ and $P\{\omega | x_{\tau(\omega)}(\omega) = x_+\}$ being the transition probabilities to (small circles around) the south- and the north pole, respectively, which sum up to one,

$$P\{\omega | x_{\tau(\omega)}(\omega) = x_-\} + P\{\omega | x_{\tau(\omega)}(\omega) = x_+\} = 1. \tag{59}$$

Denoting these transition probabilities by P_- and P_+, one arrives at the simple result

$$P_+ = \frac{h(x_-) - h(x_\circ)}{h(x_-) - h(x_+)}. \tag{60}$$

$$P_- = \frac{h(x_\circ) - h(x_+)}{h(x_-) - h(x_+)} \tag{61}$$

Therefore, slightly generalizing, *if one has a real-valued function h on the pure-state space such that* $h(\phi_t) = h(\phi_t(\omega))^{\ddagger}$ *is a martingale* (where ϕ is an element of the pure-state space), eqs. (60, 61) allow to compute the transition probabilities looked for. If the set of pure states is, as in our example, the surface S_2 of the 3-sphere, one would have to find an appropriate function

$$h : S_2 \to \mathbf{R} \tag{62}$$

such that

$$h(\phi_t) = h(\phi_t(\omega)) \tag{63}$$

is a martingale. For the computation of transition probabilities, the function h has to be chosen in such a way that it is constant on the little circles around north- and south-pole (or conversely, these little "circles" have to be replaced by little curves around the poles, on which the function h is constant).

Hence the problem is finally to *find appropriate real-valued functions h on the pure-state space such that* $h(\phi_t) = h(\phi_t(\omega))$ *is a martingale*. Martingales are time-conserved quantities in the sense of eqs. (98, 99), or integrals of motion, which has already been used in eq. (57). *In the individual formalism of quantum mechanics (with a stochastic dynamics on the pure-state space) martingales replace the usual "integrals of motion" such as, e. g., the energy or the angular momentum*. Hence, for a complete description of the quantum processes arising

---

‡This notation should only indicate that ϕ_t and $h(\phi_t)$ are functions of the state ω of the environment.

during a spectroscopic investigation, one would need not only a stochastic dynamics (on the pure states) but also enough martingales to get a clear impression of the stochastic process and as a means for the computation of transition probabilities.

Martingales can be found using Itô's change-of-variable formula (see appendix). Consider our example of a stochastic differential equation (eq. (51)),

$$dx_t = 2\kappa \tanh(x_t)\,dt - 2\nu\,dW_t\,. \tag{64}$$

Then by Itô's formula (108), any differentiable function $h : \mathbf{R} \to \mathbf{R}$ fulfills

$$
\begin{aligned}
h(x_t) - h(x_0) &= \\
&= \int_0^t h'(x_s)\left\{2\kappa \tanh(x_s)\,ds - 2\nu\,dW_s\right\} + \frac{1}{2}\int_0^t h''(x_s)\,4\nu^2\,ds \\
&= \int_0^t \left\{2\kappa \tanh(x_s)\,h'(x_s) + 2\nu^2\,h''(x_s)\right\} + \int_0^t (-2\nu)\,h'(x_s)\,dW_s\,.
\end{aligned} \tag{65}
$$

Since

$$\int_0^t (-2\nu)\,h'(x_s)\,dW_s \tag{66}$$

is a martingale, it is clear that the $h(x_t) = h(x_t(\omega))$ is a martingale *if*

$$h''(x) + \frac{\kappa}{\nu^2}\tanh(x)\,h'(x) = 0 \quad x \in \mathbf{R}, \tag{67}$$

holds. Therefore the function

$$
\begin{aligned}
h(x) &= \int_0^x \exp\left\{\int_0^y -\frac{\kappa}{\nu^2}\tanh(z)\,dz\right\}dy \tag{68}\\
&= \int_0^x (\cosh y)^{-\frac{\kappa}{\nu^2}}\,dy \tag{69}
\end{aligned}
$$

is a solution leading to a martingale $h(x_t) = h(x_t(\omega))$. In case of the Gisin process (with $\kappa = \nu = \frac{1}{2}$) one arrives at

$$h(x) = \tanh x, \tag{70}$$

and therefore gets transition probabilities (see eqs. (60, 61)

$$
\begin{aligned}
P_+ &= \frac{\tanh(x_-) - \tanh(x_0)}{\tanh(x_-) - \tanh(x_+)}\,. \tag{71}\\
P_- &= \frac{\tanh(x_0) - \tanh(x_+)}{\tanh(x_-) - \tanh(x_+)} \tag{72}
\end{aligned}
$$

Since the function tanh converges quickly to ± 1 for $x \to \pm\infty$, these transition probabilities (to reach small circles around the poles) converge quickly to

$$\frac{1 \pm \tanh x_0}{2} = \frac{1 \pm \cos\vartheta}{2} \tag{73}$$

and hence reproduce the projection postulate. This is not the case for other values of $\frac{\kappa}{\nu^2}$. For $\frac{\kappa}{\nu^2} = 4$, e. g., one arrives at

$$h(x) = -\frac{1}{3}\tanh^3 x + \tanh x, \tag{74}$$

which gives rise to transition probabilities

$$\frac{2 \pm 3\cos\vartheta \pm \cos^3\vartheta}{4}, \tag{75}$$

differing from the "correct" transition probabilities in eq. (73) up to 10% (depending on the initial state).

To get results about the *time necessary for a transition*, similar techniques can be applied. One may, for example, consider functions $h : \mathbf{R}^2 \to \mathbf{R}$ such that

$$h(x_t, [x,x]_t) = h(x_t(\omega), [x,x]_t(\omega)) \tag{76}$$

is a martingale (where $[x,x]$ is the Doob–Meyer bracket, introduced in the appendix).

The main problem with this and other, more involved, computations is that one needs numerical capacity to solve (usual, i. e., *non-* stochastic) partial differential equations.

Summarizing: It is desirable to have a dynamical theory on the level of pure states, not only encorporating the hamiltonian dynamics of the system considered, but encorporating also external stochastic perturbations. Transition probabilities and transition times can be computed using martingale theory (and numerical procedures to solve partial differential equations). Martingales are conserved quantities, replacing the usual conserved quantities such as energy, angular momentum, etc. Given such a dynamical approach on the pure-state level, an interesting question to solve would be: How long does it take for the symmetric ground state of a chiral molecule to decay into left- and right-handed (or other) states?

A CANONICAL DECOMPOSITION OF THERMAL STATES INTO PURE STATES

The question to be posed here is the following: Can one give a canonical decomposition of thermal states which

- is not necessarily concentrated on eigenstates of the underlying hamiltonian and therefore can also give rise to a "chemical" description (in the sense of section "Chemical vs. quantum-mechanical points of view")

- is concentrated on eigenstates of the hamiltonian in case of ammonia and concentrated on handed states for properly chiral molecules (compare table 1),

- is compatible with limits (number of spins going to ∞; or nuclear masses in a molecule going to ∞; etc.) in the sense of section "Effective thermal states",

- is dynamically *stable* under small external perturbations.

Let us look at the last *requirement of stability* and assume that a decomposition $\mu_\circ$ of a thermal state D_β has been chosen arbitrarily (cf. eq. (33)). This decomposition $\mu_\circ$ is considered as giving the *probability that a certain state ϕ is an initial state* [in the last section, *one* fixed initial state was chosen for simplicity, whereas now it is an initial distribution]. Assume now that the pure states arising in this decomposition $\mu_\circ$ are exposed to some stochastic dynamics (stemming from an external perturbation). Then the decomposition $\mu_\circ$ develops in time, $t \to \mu_t$, and usually converges to some limit μ_∞. To every probability distribution μ_t of pure states there corresponds a density operator D_t (cf. eq. (33)) defined by

$$\mathrm{Tr}(D_t A) = \int_{S_2} \mathrm{Tr}(D_{\vartheta,\varphi} A)\, \mu_t(d\vartheta, d\varphi). \tag{77}$$

The density operator $D_\circ$ corresponding to $\mu_\circ$ is the thermal state D_β. If no energy is transferred to the system from outside (on an average, and after large times t), the inverse temperature is again β after long time and we have $D_\infty = D_\beta$. Nevertheless the respective probability distribution μ_∞ of pure states might be different from the original one $\mu_\circ$ (if the latter was unstable under small external perturbations). In contrast to $\mu_\circ$, the limit probability distribution μ_∞ does *not* change under the particular perturbation (i. e., the particular stochastic dynamics) considered, and hence is a suitable *candidate for a stable distribution of pure states*.

Let us use as an example a properly chiral molecule, like a sugar or an amino acid: Take $\mu_\circ$ be the decomposition of the thermal state D_β into its eigenstates Ψ_+ and Ψ_-. These eigenstates are unstable under external perturbations and could decay, for example, into the chiral states Ψ_L and Ψ_R, the final decomposition μ_∞ being concentrated on these chiral states.

Stochastic dynamics on the pure states of some quantum system cannot easily be discussed. Hence it would be desirable to have some criterion to *estimate* the "most stable" decomposition of a given thermal state D_β.

A very useful criterion in this respect is given by the *maximum entropy principle* in the sense of Jaynes[81]. The ingredients of the maximum entropy principle are

- some reference probability distribution on the pure states,

- and a way to estimate the quality of some given probability distribution μ on the pure states with respect to the reference distribution.

As reference probability distribution I shall take the equidistribution μ_{eqp}, defined in eq. (26) for a two-level system (this definition of equipartition can be generalized to arbitrary $d \times d$-matrices, being the canonical measure on the d-dimensional complex projective plane[82,83]). The relative entropy of some probability distribution μ_f [see eq. (31)] with respect to μ_{eqp} is defined as

$$S(\mu_f | \mu_{eqp}) = - \int_{S_2} f(\vartheta, \phi) \ln[f(\vartheta, \phi)] \frac{\sin\vartheta\, d\vartheta\, d\phi}{4\pi}. \tag{78}$$

Note that the entropy as defined in eq. (78) is not necessarily related to the von Neumann entropy $\mathrm{Tr}(D_\beta \ln D_\beta)$ of the thermal state. The entropy in eq. (78) estimates only the information contained in the particular decomposition μ.

The maximum-entropy principle can then be formulated as follows: *Assume that a given quantum system is in a pure state but that this pure state is unknown. Assume furthermore that the only knowledge about the system is some non-pure state with density operator D_β. Then the probability of finding the pure state in some subset of all pure states is described by the probability distribution μ_{max} of pure states having maximum entropy with respect to equipartition. The probability distribution μ_{max} is chosen among all the probability distributions μ of pure states yielding the given density operator D_β via mixing in the sense of eq. (33).*

Let me stress again that different decompositions of a molecular thermal state D_β into pure states refer to entirely different physical or chemical descriptions:

1. A thermal state D_β can be decomposed into eigenstates of the underlying hamiltonian. This so-called *spectral decomposition* may be unstable under external perturbations, in particular, for high energies or low level splitting. Furthermore (strict) eigenstates such as the states Ψ_+ and Ψ_- for properly chiral molecules go not conform with chemical intuition.

2. A molecular thermal state D_β can be decomposed into pure states in such a way that the (average) dispersions of the nuclear position operators Q_j, $j = 1, 2, \ldots, K$, are minimal (see eq. (37)). In this case the question is: why should one single out the nuclear *position* operators from the very beginning?

3. A thermal state D_β can be decomposed according to the maximum entropy principle. The advantages are

- that this decomposition goes conform with dynamical stability under external perturbations[84],

- that it may lead to "quantum" *or* "chemical" descriptions (in the sense of section "Chemical vs. quantum-mechanical points of view"), depending on the particular situation (i. e., the level splitting or the number of spins considered etc.)

- that no particular operators such as the nuclear position operators are distinguished,

- that the maximum entropy decomposition μ_{max} is uniquely determined.

In the following, the maximum entropy decomposition μ_{max} of a given thermal state D_β will be defined to be its *canonical* decomposition.

At a later stage of development, one could perhaps consider very particular mechanisms of external perturbation (i. e., particularly chosen stochastic dynamics on the pure states of the system in question). This might lead to a refined definition for the canonical decomposition of thermal states. There is certainly *no* reason to insist dogmatically on the maximum entropy decomposition. The maximum entropy decomposition should only be considered as a serious candidate for a canonical decomposition, the only one at hand for the moment.

FUZZY CLASSICAL OBSERVABLES AND LARGE-DEVIATION THEORY

The overall picture of an individual quantum object developed here is the following:

- The individual quantum object is in a pure state, which may be unstable under external influence. The dynamics is not only given by the Schrödinger equation, but specified by additional stochastic terms (cf. ref.[84]). The probability distribution of pure states in a thermal situation is given by the maximum entropy decomposition μ_{max} of the respective thermal density operator D_β.

- All statistical results in terms of thermal density operators D_β are preserved. Consideration of the canonical decomposition μ_{max} allows to discuss molecular structure or other fuzzy classical concepts, which is not yet possible by considering thermal density operators alone.

It is, unfortunately, not simple to compute the maximum entropy decomposition in molecular situations. We shall therefore consider again the simpler example of the (quantum-mechanical) Curie–Weiss model with hamiltonian

$$H_N = -\frac{J}{2N} \sum_{i,j=1}^{N} \sigma_{z,i}\sigma_{z,j}. \tag{79}$$

Here $\sigma_{z,j}$ are the Pauli matrices (spin $=\frac{1}{2}$) in z-direction for the particles $1, 2, \ldots, N$. The corresponding thermal state $D_{\beta,N}$ is defined as in eq. (38). The hamiltonian H_N is invariant under the symmetry transformation, defined by

$$\sigma_{x,i} \rightarrow \sigma_{x,i} \tag{80}$$

$$\sigma_{y,i} \rightarrow -\sigma_{y,i} \tag{81}$$

$$\sigma_{z,i} \rightarrow -\sigma_{z,i} \tag{82}$$

for all spins i, $i = 1, 2, \ldots, N$, in the model. The specific magnetization operator $\hat{m}_N$ is defined as

$$\hat{m}_N \overset{\text{def}}{=} \frac{1}{N} \sum_{j=1}^{N} \sigma_{z,j}. \tag{83}$$

All we want to understand in the following is: Does the specific magnetization become a "more and more classical observable" and "how fast" does the fuzzyness of the specific magnetization operator go to zero *with increasing number of spins in the magnet,* ending up with a strictly classical observable in the limit of infinitely many spins. This question is analogous to the one asking "how fast" the fuzzyness of the nuclear molecular frame disappears *with increasing masses of the nuclei,* leading to a strictly classical nuclear frame in the limit of infinite nuclear masses.

The discussion to follow is based on the respective remarks in section "Effective thermal states". Recall again that the thermal state can be decomposed in many different ways into pure states:

1. One could, for example, decompose $D_{\beta,N}$ into symmetry-adapted eigenstates Ψ_n, $n = 1, 2, \ldots, 2^N$ of the hamiltonian H_N. Similarly as in eq. (10) it can be shown that the expectation values of $\hat{m}_N$ with respect to these eigenstates Ψ_n are zero, $\langle \Psi_n | \hat{m}_N \Psi_n \rangle = 0$. Hence the probability $P_{\beta,N,\text{spectral}}[m_1, m_2]$ to find an expectation value $\langle \Psi | \hat{m}_N \Psi \rangle$ in an interval $[m_1, m_2]$ (with respect to the spectral decomposition) is

$$P_{\beta,N,\text{spectral}}[m_1, m_2] = 1, \quad \text{if} \quad 0 \in [m_1, m_2] \tag{84}$$

$$P_{\beta,N,\text{spectral}}[m_1, m_2] = 0, \quad \text{otherwise}. \tag{85}$$

Consequently, the respective probability density $p_{\beta,N,\text{spectral}}$ for the distribution of the expectation values $\langle \Psi | \hat{m}_N \Psi \rangle$ defined by

$$P_{\beta,N,\text{spectral}}[m_1, m_2] =: \int_{m_1}^{m_2} p_{\beta,N,\text{spectral}}(m) \tag{86}$$

is a Dirac δ-function at $m = 0$, i. e., $p_{\beta,N,\text{spectral}}(m) = \delta_0(m)$.

If the number N of spins in the model is increased, this probability density $p_{\beta,N,\text{spectral}}$ does *not* change. In particular, the expectation values $\langle \Psi | \hat{m}_N \Psi \rangle$ with respect to pure states Ψ in the spectral decomposition do *not* converge to the values $+m_\beta$ and $-m_\beta$ expected in the infinite limit (cf. section "Effective thermal states").

2. Or one could decompose $D_{\beta,N}$ according to the maximum entropy principle. This results in a different picture (cf. ref.[84] and the large-deviation discussion below): The respective probability density $p_{\beta,N,\text{max}}$ has two peaks (for high enough N, see the sketch in fig. 8), i. e., the expectation values $\langle \Psi | \hat{m}_N \Psi \rangle$ concentrate around two different values. The interesting point is that the densities $p_{\beta,N,\text{max}}$ get more sharply peaked with increasing number N of spins. Therefore the expectation values $\langle \Psi | \hat{m}_N \Psi \rangle$ converge to the values $+m_\beta$ and $-m_\beta$ in the limit of infinitely many spins (cf. section "Effective thermal states" and ref.[84]).

With respect to the maximum entropy decomposition, the specific magnetization operator $\hat{m}_N$ gets more and more classical, till–in the limit of infinitely many spins–one has a strictly classical observable (see section "Effective thermal states") with only two possible expectation values of $\hat{m}_N$, namely $+m_\beta$ and $-m_\beta$. Nevertheless the superposition principle remains universally valid for every finite N. The only new aspect here is that certain pure states, such

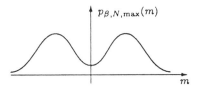

Figure 8. Sketch of the probability density $p_{\beta,N,\max}$ for the expectation values of the specific magnetization operator $\hat{m}_N$ with respect to the maximum entropy decomposition of the thermal state D_β of the Curie–Weiss model. This figure refers to some fixed inverse temperature. The expectation values of $\hat{m}_N$ concentrate around two possible values for positive and negative magnetization, respectively. For increasing N, the density $p_{\beta,N,\max}$ becomes more sharply peaked, converging to a sum of two Dirac δ-functions in the limit $N \to \infty$.

as symmetry-adapted eigenstates, get more and more unstable with increasing number N of spins in the magnet.

Recall again our main question: "How fast" do such symmetry-adapted eigenstates (and more generally, pure states with expectation values of $\hat{m}_N$ equal to zero) die out *with increasing N?*

The answer is: *exponentially fast.* The probabilities $P_{\beta,N,\max}[m_1,m_2]$ go either exponentially fast to zero or exponentially fast to one with increasing N, depending on the particular interval $[m_1,m_2]$ of expectation values considered. Furthermore the decay constant is given by an *entropy in the sense of large-deviation theory* [as general introductions see refs.[85,86,87]; the concept of a large-deviation entropy is different from an entropy in the sense of Jaynes, as used in eq. (78)]:

$$\int_{[m_1,m_2]} p_{\beta,N,\max}(m) \sim \exp\left\{-N \inf_{m\in[m_1,m_2]} (s_{\beta,\mathrm{mean}}(m))\right\}. \tag{87}$$

To understand this formula, look at fig. 9, showing the large-deviation entropy $s_{\beta,\mathrm{mean}}$. The figure is based on a computation (using some approximations) in ref.[84], the temperature now being chosen as one third of the Curie temperature. For a given interval $[m_1,m_2]$ one has to find the respective infimum of $s_{\beta,\mathrm{mean}}$ on $[m_1,m_2]$ and this is the decay constant used in eq. (87). Two different situations can occur:

- either the interval $[m_1,m_2]$ does *not* contain one of the special values $+m_\beta$ or $-m_\beta$ (at which the function $s_{\beta,\mathrm{mean}}$ takes its minimum zero). In this case one gets a strictly positive decay constant for eq. (87) and hence an exponential decay of the probabilities $P_{\beta,N,\max}[m_1,m_2]$ to 0,

- or the interval $[m_1,m_2]$ contains one of the special values $+m_\beta$ or $-m_\beta$ (at which the function $s_{\beta,\mathrm{mean}}$ takes its minimum zero). In this case one gets a zero decay constant for eq. (87) and hence an exponential decay of the probabilities $P_{\beta,N,\max}[m_1,m_2]$ to 1.

The large-deviation entropy $s_{\beta,\mathrm{mean}}$ describes "how fast" the fuzzyness of the fuzzy classical observable $\hat{m}_N$ goes to zero with increasing number N of spins in a Curie–Weiss magnet:

- For small and fixed chosen N the system behaves quantum-mechanically, i. e., pure states Ψ with zero expectation value $\langle\Psi|\hat{m}_N\Psi\rangle = 0$ arise with relatively high probability in the thermal ensemble $\mu_{\beta,N,\max}$ of pure states.

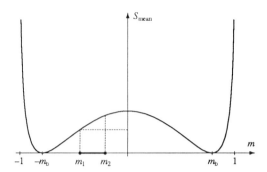

Figure 9. An entropy function s_{mean} in the sense of fluctuation (i. e., large-deviation) theory, describing how fast the mean magnetization of a spin system gets classical with an increasing number of spins. The figure is based on an approximate calculation for the Curie–Weiss model. The temperature is fixed and has been taken here as one third of the critical (Curie) temperature. Above the Curie temperature the respective entropy s_{mean} would only have one minimum, namely at $m = 0$.

- With increasing number N of spins such pure states die out quickly, determined by the value $s_{\beta,\text{mean}}(m = 0)$ of the large-deviation entropy at $m = 0$ in eq. (87). In the infinite limit $N \to \infty$, only the expectation values $+m_\beta$ and $-m_\beta$ survive. This goes conform with the result of algebraic quantum mechanics, that

$$\hat{m} \stackrel{\text{def}}{=} \lim_{N \to \infty} \hat{m}_N \qquad (88)$$

is a *strict* classical observable, i. e., superpositions of *pure* states Ψ_1 and Ψ_2 having *different* expectation values

$$\langle \Psi_1 | \hat{m}_N \Psi_1 \rangle = +m_\beta \qquad (89)$$
$$\langle \Psi_2 | \hat{m}_N \Psi_2 \rangle = -m_\beta \qquad (90)$$

are forbidden.

Summarizing: the specific magnetization operators $\hat{m}_N$ are *fuzzy* classical observables, and the fuzzyness decreases continuously with increasing number N of spins in the Curie–Weiss magnet. This decrease of fuzzyness is described by the large-deviation entropy $s_{\beta,\text{mean}}$ in eq. (87). In statistical mechanics, large deviations are "large" *fluctuations around some mean-value*, around the internal energy, for example. In the present context, large deviations are *fluctuations around "classical values" of certain observables*. In the infinite limit (e. g., number of spins going to infinity) only classical values are admitted, whereas "before the limit" (i. e., for $N < \infty$) also other expectation values (with respect to pure states) arise.

THE STRUCTURE OF SINGLE MOLECULES

Let us now again consider single ammonia-type molecules and first review the problems of the *usual statistical formalism of quantum mechanics* (which uses density operators and *no* particular decompositions of thermal density operators into pure states:

- Thermal states as defined by the usual recipe of statistical quantum mechanics (see eq. (38)) are in contradiction with traditional chemistry: neither isomers, nor handed states nor any other structural chemical concepts can be described or explained thereby.

- Fuzzy classical observables *cannot* be described; only strict classical observables can be derived, but only in some appropriate limit (as, e. g., with the number of spins or molecular nuclear masses going to infinity).

- These limits are somewhat dubious, because the superposition principle holds universally before the limit (e. g., for *every* finite number N of spins) and is broken afterwards.

- The infinite limit should be the zeroth term in some expansion in the variable $\frac{1}{N}$. The interesting correction terms for large but finite N cannot be written down in the language of usual statistical quantum theory.

A typical question which makes *no obvious sense* in the usual statistical formalism of quantum mechanics asks *if there is some change in the behavior* in the sequence {monodeuteroaniline $\rightarrow$ ammonia $\rightarrow \ldots \rightarrow$ asparagic acid} of table 1 such as a phase transition. As exposed in section "Chemical concepts are fuzzy classical concepts", a possibility for such a phase transition to happen would be that there exists only one thermal state with density operator D_β for large level splitting $(E_- - E_+)$ (for a given inverse temperature β), whereas for small level splitting two different thermal states $D_{\beta,L}$ and $D_{\beta,R}$ exist for left- and right-handed molecular species, respectively. Change of behaviour between the two different regimes would happen at some critical level splitting. Two arguments–one "experimental" and the other theoretical–show that a phase transition in this strict sense *cannot* exist:

- Racemization implies that strict thermal states $D_{\beta,L}$ and $D_{\beta,R}$ (in the sense of KMS-states) cannot exist (cf. section "Effective thermal states").

- Rigorous Araki perturbation theory[49] shows that a change in level splitting $(E_- - E_+)$ [mathematically speaking, this is a perturbation of finite norm] cannot lead to a change in the structure of the β-KMS-states (for a given inverse temperature β). Hence *if* ammonia admits only one KMS-state then also all the other species in table 1 admit only one KMS-state, and conversely. This remark holds, in particular, for systems with infinitely many degress of freedom such as the joint system {molecule & radiation field}. For finitely many degrees of freedom there exists only one thermal state (for every temperature), anyway. [This result does not apply to a Curie–Weiss magnet consisting of infinitely many spins, because a change in the coupling constant J is not a perturbation of finite norm].

If therefore the overall strict thermal state D_β, as defined in eq. (38), is decomposed into thermal states $D_{\beta,L}$ and $D_{\beta,R}$,

$$D_\beta = \tfrac{1}{2}(D_{\beta,L} + D_{\beta,R}), \tag{91}$$

then these states $D_{\beta,L}$ and $D_{\beta,R}$ can only be *effective* and not strict thermal states, and similarly for different isomers of some molecular species. Moreover–since decompositions such as in eq. (91) are not accepted in usual statistical quantum mechanics–such effective thermal states do not make sense there (and are, by the way, not uniquely defined, since different decompositions, if ever accepted, are considered as being equivalent).

Hence starting from usual statistical quantum mechanics it is impossible to understand the traditional chemical theories about single molecules.

Let us now turn to the *individual* formalism of quantum mechanics again, where thermal states D_β are decomposed in a canonical way according to the maximum entropy principle. I am not yet capable of computing maximum entropy decompositions and large deviation entropies for molecular situations. Nevertheless the (simpler) example of the Curie–Weiss magnet suggests that also in molecular situations

- it is sufficient to consider (large but) *finite* nuclear molecular masses M_j, $j = 1, 2, \ldots, K$, and that it is not necessary to go to the limit $M_j \to \infty$,

- fuzzy classical structures like handedness or the nuclear structure of molecules can arise,

- distributions of expectation values are well defined in thermal situations.

An example of the latter can be given by some internal coordinate operator, such as the inversion coordinate operator $\hat{q}$ for an ammonia-type molecule (cf. fig. 1):

- Decomposing the thermal molecular state D_β into (strict) symmetry-adapted eigenstates, such as Ψ_+ and Ψ_-, leads to zero expectation values of $\hat{q}$,

$$\langle \Psi_\pm | \hat{q} \Psi_\pm \rangle = 0, \tag{92}$$

cf. eq. (10). Consequently the probability density $p_{\beta,\mathrm{spectral}} = p_{\beta,\mathrm{spectral}}(q)$ of the expectation values q of $\hat{q}$ [defined analogously to those for the specific magnetization in eq. (86)] is a Dirac δ-function $\delta_\circ(q)$.

- Decomposing the thermal molecular state D_β according to the maximum entropy principle is expected to lead to a probability density $p_{\beta,\mathrm{max}} = p_{\beta,\mathrm{max}}(q)$ with two or more peaks (see fig. 10), analogously to what was shown for the specific magnetization of the Curie–Weiss model.

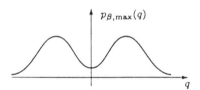

Figure 10. Sketch of the expected probability density $p_{\beta,\mathrm{max}} = p_{\beta,\mathrm{max}}(q)$ for the expectation values of the inversion coordinate operator $\hat{q}$ with respect to the maximum entropy decomposition of the molecular thermal state D_β of an ammonia-type molecule. This figure refers to some fixed inverse temperature. The expectation values of $\hat{q}$ concentrate around two possible values for positive and negative inversion coordinate, respectively. For increasing nuclear masses M_j, $j = 1, 2, \ldots, K$, the density $p_{\beta,\mathrm{max}}$ is expected to become more sharply peaked, converging to a sum of two Dirac δ-functions in the limit $M_j \to \infty$.

Similarly to the large-deviation considerations for the Curie–Weiss model, one could try to characterize molecules by some large-deviation entropy which describes "how fast" a nuclear molecular structure appears *with increasing molecular nuclear masses*. Such a large-deviation entropy would describe the decrease in fuzzyness of the molecular nuclear structure when the nuclear masses increase. In the limit of infinite nuclear masses one expects a strictly classical nuclear frame, not being fuzzy any more at all. Such a large-deviation entropy would also nicely describe the *quantum fluctuations* around the strictly classical nuclear structure.

Assume now that the probability density $p_{\beta,\mathrm{max}} = p_{\beta,\mathrm{max}}(q)$ does indeed show a peaked structure (see fig. 10) similar to the probability density $p_{\beta,N,\mathrm{max}}$ of the Curie–Weiss magnet in fig. 8. Then *effective* thermal states can be introduced in an ad-hoc way.

To this end, let

$$\mathrm{Tr}(D_\beta T) = \int\limits_{\substack{\text{all pure} \\ \text{states with} \\ \text{state vector } \Psi}} \langle\Psi|T\Psi\rangle\, \mu_{\beta,\mathrm{max}}(d\Psi). \tag{93}$$

be the maximum entropy decomposition of a thermal molecular state D_β [the only difference to eq. (33) is that now the molecule is not written in two-level approximation, and that therefore the probability distribution $\mu_{\beta,\mathrm{max}}$ lives on all the pure states with state vectors Ψ and not on S_2; the notation $\mu_{\beta,\mathrm{max}}(d\Psi)$ is not entirely correct, since the probability distribution $\mu_{\beta,\mathrm{max}}$ lives on pure states and not on the state vectors.] Then $D_{\beta,\mathrm{L}}$ and $D_{\beta,\mathrm{R}}$ are defined as

$$\mathrm{Tr}(D_{\beta,\mathrm{L}}T) = \frac{\int\limits_{\{\Psi|\langle\Psi|\hat{q}\Psi\rangle\leq 0\}} \langle\Psi|T\Psi\rangle\, \mu_{\beta,\mathrm{max}}(d\Psi)}{\int\limits_{\{\Psi|\langle\Psi|\hat{q}\Psi\rangle\leq 0\}} \mu_{\beta,\mathrm{max}}(d\Psi)}, \tag{94}$$

$$\mathrm{Tr}(D_{\beta,\mathrm{R}}T) = \frac{\int\limits_{\{\Psi|\langle\Psi|\hat{q}\Psi\rangle\geq 0\}} \langle\Psi|T\Psi\rangle\, \mu_{\beta,\mathrm{max}}(d\Psi)}{\int\limits_{\{\Psi|\langle\Psi|\hat{q}\Psi\rangle\geq 0\}} \mu_{\beta,\mathrm{max}}(d\Psi)}. \tag{95}$$

The denominators in these equations (being equal to $\frac{1}{2}$) are introduced to get *normalized* density operators, i. e., density operators D with $\mathrm{Tr}(D) = 1$.

Hence we simply "cut apart" the set of pure molecular states into the

- set of pure states $S_\mathrm{L} \stackrel{\mathrm{def}}{=} \{\Psi|\langle\Psi|\hat{q}\Psi\rangle \leq 0\}$ with negative expectation value of $\hat{q}$

- and the set of pure states $S_\mathrm{R} \stackrel{\mathrm{def}}{=} \{\Psi|\langle\Psi|\hat{q}\Psi\rangle \geq 0\}$ with positive expectation value of $\hat{q}$.

This procedure is *ad-hoc* because *effective* thermal states can never really be defined in a *completely* unique way: The question, for example, if some state Ψ fulfilling $\langle\Psi|\hat{q}\Psi\rangle = 0$ belongs to the set of left- or right-handed states, can never be conclusively answered or defined in an unambiguous way. A similar problem arises when one partitions the set of all pure states into sets S_j, $j = 1, 2, \ldots$, for different isomers of some molecular species. The deeper reason for the ad-hoc character of such partitioning procedures is that

- handed states and isomers are *only rigorously defined in some classical limit*, as, e. g., in the limit of infinite nuclear molecular masses,

- before the limit many "quantum-mechanical" pure states, called "strange pure states" in the first section, arise which simply are not conform at all to a particular classical classification,

- these "strange pure states" are *forced into some classical classification*.

For "strange" states–such as the ground state Ψ_+ of ammonia–it is, of course, senseless to attribute them to one of the sets S_L or S_R, because Ψ_+ is definitely neither left- nor right-handed. Hence the main motivation for defining the effective thermal states $D_{\beta,\mathrm{L}}$ and $D_{\beta,\mathrm{R}}$ in eq. (94, 95) is that the probability *with respect to the maximum entropy decomposition* to find "strange" states is expected to get smaller and smaller with increasing nuclear molecular masses, or with decreasing level splitting for ammonia-type molecules.

Consequently, the definitions of effective thermal states $D_{\beta,\mathrm{L}}$ and $D_{\beta,\mathrm{R}}$ in eq. (94, 95) make sense *if* the respective probability density $p_{\beta,\mathrm{max}}$ sketched in fig. 10 is peaked sharply enough around "non-strange" pure states conform to the classical classification in the infinite limit (e. g., the classification into left- and right-handed states). Similar remarks apply to the problem of introducing different isomers of some molecular species.

Note again that a decomposition of the thermal state D_β into *symmetry-adapted eigenstates* leads to a situation where one has zero expectation values of the inversion coordinate operator $\hat{q}$, $\langle \Psi | \hat{q} \Psi \rangle = 0$, with probability 1. Using this decomposition it is not possible to introduce effective thermal states $D_{\beta,L}$ and $D_{\beta,R}$ as in eq. (94, 95).

CONCLUDING REMARKS

I do not claim to have presented a fully worked-out theory here. All material is in a preliminary form. In particular, a convincing derivation of nonlinear stochastic dynamics on the pure states of a quantum system is still missing. Also many of the old unsolved problems of quantum mechanics come into play, and there is no hope to get rid of these without a major effort.

The questions posed refer to many different fields:

- Can quantum mechanics give a full description of the stochastic behaviour in single-molecule spectroscopy?

- Is it possible to explain handedness, nuclear structure and isomerism of molecular species?[84]

- Can *fuzzy* classical observables be properly defined?

- Can *effective* thermal states (for handed molecules of chemical isomers) be defined?

- Can one understand fuzzy classical behavior in chemistry by specifying some *strictly* classical behaviour in an appropriate limit (corresponding to *crisp* sets in fuzzy set theory) and giving quantum deviations therefrom for finitely many degrees of freedom or finite nuclear molecular masses (corresponding to proper fuzzy sets in mathematical fuzzy-set theory)?

- Is it possible to compute the decay rates of "strange" states?

The molecular picture devised here is dynamical rather than static: Under the influence of external perturbations, the pure states of a molecule develop in a stochastic way. The stochastic dynamics on the pure-state space is called *ergodic* if, starting from some initial state ϕ_0, all other pure states can be reached by the stochastic dynamics (for mathematical definitions of ergodicity, see ref.[86]). *For systems with a partial classical structure (chirality, isomerism, magnetization), ergodicity effectively breaks down (with respect to the entire state space):* Starting with a left-handed initial state ϕ_0 of a chiral molecule, for example, the stochastic dynamics will not leave the subensemble of left handed states (for a long time). Hence, the maximum-entropy ensemble corresponding to the "global" thermal density operator D_β will split up into subensembles, which are structurally stable for long times. On the subensembles, the dynamics still acts ergodically. The structurally stable subensembles (for, e. g., left- and right-handed states) behave as attractors, whereas "forbidden" superpositions (of states taken from different structurally stable subensembles) are transient states, finally ending up in one of the structurally stable subensembles.

In this way, density operators like $D_{\beta,L}$ and $D_{\beta,R}$ can also be introduced by dynamical arguments and the respective approximate classical structure gets a dynamical touch. These density operators correspond to (effectively) stationary distributions of pure states (i. e., to stationary measures on the attractors).

90

In such a discussion, quantum systems are very much treated like a classical dynamical system:

- the respective phase space consists of all the pure states of the quantum system in question (e. g., the surface S_2 of a sphere in case of a two-level system),

- the dynamics on this phase space is stochastic (cf. fig. 6),

- handed molecules or isomers of some molecular species correspond to attractors of the stochastic dynamics,

- "symmetry breaking" (i. e., generation of non-symmetric pure states such as Ψ_L and Ψ_R) corresponds to generation of attractors in the phase space,

- "strange" pure states (such as the symmetric ground state Ψ_+ of a chiral molecule) are transient states under the stochastic dynamics,

- chaos, non-vanishing Lyapunov coefficients and other interesting phenomena in the theory of classical dynamical systems[88] can, in principle, also arise in the discussion of genuine individual quantum systems; it would be a challenge to discuss single-molecule spectroscopy from that point of view.

It was my goal here to give the main ideas of a new *individual* approach in quantum mechanics. There is no question that these ideas are only in a preliminary form. Nevertheless a proper understanding of chemical concepts needs such an *individual* quantum-mechanical approach. The traditional statistical approach is too restricted to give a rigorous understanding of chemistry.

APPENDIX: RULES FOR MARTINGALES

The appendix is designed to help when studying section "An example of a stochastic dynamics on the set of pure states of a two-level system". This section as well as the appendix can be skipped at a first reading. An introduction to probability theory can be found in ref.[89]. An introduction into the theory of stochastic dynamical systems can be found in ref.[80]. Ref.[79] gives a more advanced account of diffusion equations from a martingale point of view.

Def.: Consider a probability space (Ω, P) possessing an increasing family of σ-algebras

$$\mathcal{F}_t, t \geq 0. \tag{96}$$

Then a *martingale* is defined as a progressively measurable, P-integrable and continuous stochastic process§ $Z_t = Z_t(\omega)$, $\omega \in \Omega, t \geq 0$, such that the conditional expectations $E^P[Z_{t_2}|\mathcal{F}_{t_1}]$ fulfill

$$E^P[Z_{t_2}|\mathcal{F}_{t_1}] = Z_{t_1}, \quad 0 \leq t_1 < t_2. \tag{97}$$

For many purposes, it is not necessary to understand this mathematical definition in detail. The important consequences are:

- The integrals over the stochastic variables Z_t's are independent of time¶,

$$\int_\Omega Z_t(\omega)\, P(d\omega) = \int_\Omega Z_\circ(\omega)\, P(d\omega), \quad t \geq 0, . \tag{98}$$

---

§ Again the notation used is not entirely correct. See the corresponding footnote in section "An example of a stochastic dynamics on the set of pure states of a two-level system".

¶ Here martingales are always taken with respect to the time t and not with respect to particle number N.

- If τ is a stopping time, then

$$\int_\Omega Z_{\tau(\omega)}(\omega)\,P(d\omega) = \int_\Omega Z_\circ(\omega)\,P(d\omega).$$ (99)

This is a consequence of Doob's optional stopping theorem[79].

Hence martingales are *conserved quantities*, even when they are time-dependent (as functions). It is important to be able to *construct* martingales. This can be done by use of Itô's formula (see below).

Some few rules[90,91] and notations are necessary to deal with a set of d martingales $Z_t := (Z_t^1, Z_t^2, \ldots, Z_t^d)$, e. g., Wiener processes $(W_t^1, W_t^2, \ldots, W_t^d)$ for different coordinate directions $1, 2, \ldots, d$. Note that martingales depend on the "time" t and the stochastic variable ω (the "path"). In the following, all the stochastic integrals are Itô-integrals:

- The joint quadratic variation $[X,Y]$ of two stochastic processes $X_t = X_t(\omega)$ and $Y_t = Y_t(\omega)$ is defined as the stochastic process

$$[X,Y]_t \overset{\text{def}}{=} X_t Y_t - X_\circ Y_\circ - \int_0^t X_s\,dY_s - \int_0^t Y_s\,dX_s.$$ (100)

The bracket $[X,Y]$ is called the Doob–Meyer bracket of the stochastic processes X and Y. It is symmetric

$$[X,Y] = [Y,X].$$ (101)

- A stochastic process X is said to be of finite variation if the paths $s \mapsto X_s(\omega)$ are of finite variation for every ω on every finite time interval $[0,t]$. In this case the Doob–Meyer bracket with any other stochastic process Y vanishes, $[X,Y] = 0$.

- For arbitrary stochastic processes, the Doob–Meyer bracket $[X,Y] = [X,Y]_t$ is of bounded variation (and hence not a martingale). The Doob–Meyer bracket for independent Wiener processes $(W_t^1, W_t^2, \ldots, W_t^d)$ can be computed to give

$$[W^i, W^j]_t = t\,\delta_{ij},$$ (102)

with δ_{ij} being the Kronecker-Delta.

- The Doob–Meyer bracket can be dealt with by observing that

$$[\textstyle\int H_s\,dX_s,\ \int K_s\,dY_s]_t = \int_0^t H_s K_s\,d[X,Y]_s.$$ (103)

- If X and Y are martingales, then

$$X_t Y_t - [X,Y]_t$$ (104)

is again a martingale. The process $(W_t^j)^2 - t$, for example, is a martingale.

- If $F = F(\omega)$ is a real-valued function and if $Z = Z_t$ is a martingale, then the stochastic process

$$\int_0^t F(\omega)\,dZ_s, \quad j = 1, 2, \ldots, d,$$ (105)

is a martingale (with respect to t).

- A *semimartingale* Y is defined to be the sum of a martingale Z and a process of finite variation A,

$$Y_t = Z_t + A_t. \tag{106}$$

Usual functions are of finite variation. The drift part in eq. (43), for example, is of finite variation, whereas the diffusion part is a martingale. Hence the solution of eq. (43) is a semimartingale.

- *Itô's change of variable formula:* Given a differentiable function

$$f : \mathbf{R}^d \to \mathbf{R} \tag{107}$$

and d continuous semimartingales $X_t^1, X_t^2, \dots, X_t^d$, then

$$f(X_t) - f(X_\circ) =$$
$$\sum_{i=1}^{d} \int_0^t D_i f(X_s) \, dX_s^j + \frac{1}{2} \sum_{i=1}^{d} \sum_{j=1}^{d} \int_0^t D_{ij} f(X_s) \, d[X^i, X^j]. \tag{108}$$

Here $D_i f$ is the first partial derivative of the function f with respect to the i-th variable, and $D_{ij} f$ is the second partial derivative of f with respect to the ith and jth variables. Note that Itô's formula is quite general. If, for example, a function h depends on two variables, and if W_t is a Wiener process, then eq. (108) implies that

$$h(W_t, t) - h(W_\circ, 0) = \tag{109}$$
$$\int_0^t D_1 h(W_s, s) \, dW_s + \int_0^t D_2 h(W_s, s) \, ds + \int_0^t D_{11} h(W_s, s) \, ds,$$

because $[W_s, s] = 0$ and $[W_s, W_s] = s$ (see above). Since the integral

$$\int_0^t D_1 h(W_s, s) \, dW_s \tag{110}$$

is a martingale, it follows that $h(W_t, t)$ is a martingale, if

$$D_2 h + D_{11} h = 0 \tag{111}$$

holds.

REFERENCES

1. Gilli, P., Ferretti, V. and Gilli, G., Hydrogen bonding models: their relevance to molecular modeling, *in*: present volume (1996).

2. Quack, M., Molecular quantum dynamics from high resolution spectroscopy and laser chemistry, *J. Mol. Struct.* 292: 171 (1993).

3. Quack, M., On the measurement of CP-violating energy differences in matter-antimatter enantiomers, *Chem. Phys. Lett.* 231: 421 (1994).

4. Stasiak, A. and Koller, T., Analysis of DNA knots and catenanes allows to deduce the mechanism of action of enzymes which cut and join DNA strands, *in*: "Fractals, Quasicrystals, Chaos, Knots and Algebraic Quantum Mechanics." Proceedings of the NATO ASI "New Theoretical Concepts in Physical Chemistry", October 1987, Maratea (Italy), NATO ASI Series C: Vol. 235," A. Amann, L. Cederbaum and W. Gans, ed., Kluwer, Dordrecht (1988).

5. Sumners, D. W., Using knot theory to analyze DNA experiments, *in*: "Fractals, Quasicrystals, Chaos, Knots and Algebraic Quantum Mechanics." Proceedings of the NATO ASI "New Theoretical Concepts in Physical Chemistry", October 1987, Maratea (Italy), NATO ASI Series C: Vol. 235," A. Amann, L. Cederbaum and W. Gans, ed., Kluwer, Dordrecht (1988).

6. Papousek, D. and Spirko, V., A new theoretical look at the inversion problem in molecules, *in*: "Topics in Current Chemistry Vol. 68," ed., Springer, Berlin (1976).

7. Herbstein, F. H., Kapon, M., Reisner, G. M., Lehman, M. S., Kress, R. B., Wilson, R. B., Shiau, W.-I., Duesler, E. N., Paul, I. C. and Curtin, D. Y., Polymorphism of naphthazarin and its relation to solid-state proton transfer. Neutron and X-ray diffraction studies on naphthazarin C, *Proc. R. Soc. Lond. A* 399: 295 (1985).

8. Rentzepis, P. M. and Bondybey, V. E., Large molecule relaxation: spectroscopy, structure, and vibrational energy redistribution, *J. Chem. Phys.* 80: 4727 (1984).

9. Vega, J. R. d. l., Busch, J. H., Schauble, J. H., Kunze, K. L. and Haggert, B. E., Symmetry and tunneling in the intramolecular proton exchange in naphthazarin, methylnaphthazarin, and dimethylnaphthazarins, *J. Am. Chem. Soc.* 104: 3295 (1982).

10. Mislow, K., Axelrod, M., Rayner, D. R., Gotthardt, H., Coyne, L. M. and Hammond, G. S., *J. Am. Chem. Soc.* 87: 4958 (1965).

11. Hammond, G. S., Gotthardt, H., Coyne, L. M., Axelrod, M., Rayner, D. R. and Mislow, K., *J. Am. Chem. Soc.* 87: 4959 (1965).

12. Reed, M. and Simon, B., "Methods of Modern Mathematical Physics. Volume IV: Analysis of Operators," Academic Press, New York (1978).

13. Pfeifer, P., "Chiral Molecules - a Superselection Rule Induced by the Radiation Field," Thesis ETH-Zürich No. 6551, ok Gotthard S+D AG, Zürich (1980).

14. Quack, M., Structure and dynamics of chiral molecules, *Angew. Chem. Int. Ed. Engl.* 28: 571 (1989).

15. Quack, M., On the measurement of the parity violating energy difference between enantiomers, *Chem. Phys. Lett.* 132: 147 (1986).

16. Cina, J. A. and Harris, R. A., On the preparation and measurement of superpositions of chiral amplitudes, *J. Chem. Phys.* 100: 2531 (1994).

17. Cina, J. A. and Harris, R. A., Superpositions of handed wave functions, *Science* 267: 832 (1995).

18. Spohn, H., Ground state(s) of the spin-boson hamiltonian, *Commun. Math. Phys.* 123: 277 (1989).

19. Spohn, H. and Dümcke, R., Quantum tunneling with dissipation and the Ising model over R, *J. Stat. Phys.* 41: 389 (1985).

20. Amann, A., Molecules coupled to their environment, *in*: "Large-Scale Molecular Systems: Quantum and Stochastic Aspects - Beyond the Simple Molecular Picture", NATO ASI Series B 258, W. Gans, A. Blumen and A. Amann, ed., Plenum, London (1991).

21. Amann, A., Theories of molecular chirality: A short review, *in*: "Large-Scale Molecular Systems: Quantum and Stochastic Aspects - Beyond the Simple Molecular Picture," NATO ASI Series B 258, W. Gans, A. Blumen and A. Amann, ed., Plenum, London (1991).

22. Amann, A., Ground states of a spin-boson model, *Ann. Phys.* 208: 414 (1991).

23. Amann, A., Applying the variational principle to a spin-boson hamiltonian, *J. Chem. Phys.* 96: 1317 (1992).

24. Amann, A., Must a molecule have a shape?, *South African J. Chem.* 45: 29 (1992).

25. Amann, A., The Gestalt problem in quantum theory: generation of molecular shape by the environment, *Synthese* 97: 125 (1993).

26. Quack, M., Die Symmetrie von Zeit und Raum und ihre Verletzung in molekularen Prozessen, *Jahrbuch der Akademie der Wissenschaften zu Berlin 1990 - 1992* (1993).

27. Claverie, P. and Diner, S., The concept of molecular structure in quantum theory: interpretation problems, *Israel J. Chem.* 19: 54 (1980).

28. Claverie, P. and Jona-Lasinio, G., Instability of tunneling and the concept of molecular structure in quantum mechanics: The case of pyramidal molecules and the enantiomer problem, *Phys. Rev. A* 33: 2245 (1986).

29. Sutcliffe, B. T., The concept of molecular structure, *in*: "Theoretical Models of Chemical Bonding, Part 1: Atomic Hypothesis and the Concept of Molecular Structure," Z. B. Maksiç, ed., Springer, Berlin (1990).

30. Sutcliffe, B. T., The chemical bond and molecular structure, *J. Mol. Struct. (Theochem)* 259: 29 (1992).

31. Davies, E. B., Nonlinear Schrödinger operators and molecular structure, *J. Phys. A* 28: 4025 (1995).

32. Davies, E. B., Symmetry breaking for molecular open systems, *Ann. Inst. Henri Poincaré* 35: 149 (1981).

33. Woolley, R. G., Quantum theory and molecular structure, *Advanc. Phys.* 25: 27 (1976).

34. Woolley, R. G., Must a molecule have a shape?, *J. Amer. Chem. Soc.* 100: 1073 (1978).

35. Woolley, R. G., Natural optical activity and the molecular hypothesis, *in*: "Structures versus Special Properties," ed., Springer, Berlin Heidelberg (1982).

36. Woolley, R. G., Molecular shapes and molecular structures, *Chem. Phys. Lett.* 125: 200 (1986).

37. Woolley, R. G., Must a molecule have a shape?, *New Scientist 22 October 1988*: 53 (1988).

38. Woolley, R. G., Quantum theory and the molecular hypothesis, *in*: "Molecules in Physics, Chemistry and Biology. Vol. 1," J. Maruani, ed., Kluwer, Dordrecht (1988).

39. Woolley, R. G., Quantum chemistry beyond the Born-Oppenheimer approximation, *J. Mol. Struct. (Theochem)* 230: 17 (1991).

40. Primas, H., "Chemistry, Quantum Mechanics, and Reductionism. Perspectives in Theoretical Chemistry," Springer, Berlin (1983).

41. Primas, H., Kann die Chemie auf die Physik reduziert werden?, *Chemie in unserer Zeit* 19: 109 (1985).

42. Espagnat, B. d., An elementary note about "mixtures", *in*: "Preludes in Theoretical Physics," A. De-Shalit, H. Feshbach and L. v. Hove, ed., North-Holland, Amsterdam (1966).

43. Amann, A. and Primas, H., What is the referent of a non-pure quantum state?, *to appear in*: "Experimental Metaphysics–Quantum Mechanical Studies in Honor of Abner Shimony," R. S. Cohen and J. Stachel, ed., (1996).

44. Beltrametti, E. and Bugajski, S., A classical extension of quantum mechanics, *J. Phys. A* 28: 3329 (1995).

45. Misra, B., On a new definition of quantal states, *in*: "Physical Reality and Mathematical Description," C. P. Enz and J. Mehra, ed., Reidel, Dordrecht (1974).

46. Amann, A., The quantum-mechanical measurement process in the thermodynamic formalism, *in*: "Symposium on the Foundations of Modern Physics 1993 - Quantum Measurement, Irreversibility, and the Physics of Information," P. Busch, P. Lahti and P. Mittelstaedt, ed., World Scientific, Singapore (1994).

47. Amann, A., An individual interpretation of quantum mechanics, *in*: *to appear in*: Proceedings of the "Fourth Winter School on Measure Theory", Liptovsky Jan, Slovakia, Jan. 29 - Feb. 3, 1995,(1995).

48. Aharonov, Y., Anandan, J. and Vaidman, L., Meaning of the wave function, *Phys. Rev. A* 47: 4616 (1993).

49. Bratteli, O. and Robinson, D. W., "Operator Algebras and Quantum Statistical Mechanics Vol. 2," Springer, New York (1981).

50. Sewell, G. L., "Quantum Theory of Collective Phenomena," Clarendon Press, Oxford (1986).

51. Strocchi, F., "Elements of Quantum Mechanics of Infinite Systems," World Scientific Publishing Co., Singapore (1985).

52. Bratteli, O. and Robinson, D. W., "Operator Algebras and Quantum Statistical Mechanics Vol. 1," Springer, 2nd revised edition, New York (1987).

53. Werner, R. F., Large deviations and mean-field quantum systems, *in*: "Quantum Probability and Related Topics Vol. VII," L. Accardi, ed., World Scientific, Singapore (1992).

54. Bóna, P., The dynamics of a class of quantum mean-field theories, *J. Math. Phys.* 29: 2223 (1988).

55. Bóna, P., Equilibrium states of a class of quantum mean-field theories, *J. Math. Phys.* 30: 2994 (1989).

56. Atmanspacher, H., Wiedenmann, G. and Amann, A., The endo-exo-distinction and its relevance for the study of complex systems, to appear in *Complexity* (1995).

57. Primas, H., Induced nonlinear time evolution of open quantum objects, *in*: "Sixty-two Years of Uncertainty: Historical, Philosophical, and Physical Inquiries into the Foundations of Quantum Mechanics," A. I. Miller, ed., Plenum Press, New York (1990).

58. Primas, H., The measurement process in the individual interpretation of quantum mechanics, *in*: "Quantum Theory without Reduction," M. Cini and J.-M. Lévy-Leblond, ed., IOP Publishing Ltd., Bristol (1990).

59. Ghirardi, G. C., Rimini, A. and Weber, T., Unified dynamics for microscopic and macroscopic systems, *Phys. Rev. D* 34: 479 (1986).

60. Ghirardi, G. C., Pearle, P. and Weber, T., Markov processes in Hilbert space and continuous spontaneous localization of systems of identical particles, *Phys. Rev. A* 42: 78 (1990).

61. Gisin, N., Quantum measurements and stochastic processes, *Phys. Rev. Lett.* 52: 1657 (1984).

62. Pearle, P., Reduction of the state vector by a nonlinear Schrödinger equation, *Phys. Rev. D* 13: 857 (1976).

63. Fermi, E., "Notes on Quantum Mechanics," The University of Chicago Press, Chicago (1961).

64. Ambrose, W. P. and Moerner, W. E., Fluorescence spectroscopy and spectral diffusion of single impurity molecules in a crystal, *Nature* 349: 225 (1991).

65. Moerner, W. E. and Basché, T., Optical spectroscopy of single impurity molecules in solids, *Angew. Chem. Int. Ed.* 32: 457 (1993).

66. Moerner, W. E., Examining nanoenvironments in solids on the scale of a single, isolated impurity molecule, *Science* 265: 46 (1994).

67. Orrit, M., Bernard, J. and Personov, R. L., High-resolution spectroscopy of organic molecules in solids: from fluorescence line narrowing and hole burning to single molecule spectroscopy, *J. Phys. Chem.* 97: 10256 (1993).

68. Croci, M., Müschenborn, H.-J., Güttler, F., Renn, A. and Wild, U. P., Single molecule spectroscopy: pressure effect on pentacene in p-terphenyl, *Chem. Phys. Lett.* 212: 71 (1993).

69. Pirotta, M., Güttler, F., Gygax, H.-R., Renn, A., Sepiol, J. and Wild, U. P., Single molecule spectroscopy. Fluorescence-lifetime measurements of pentacene in p-terphenyl, *Chem. Phys. Lett.* 208: 379 (1993).

70. Güttler, F., Sepiol, J., Plakhotnik, T., Mitterdorfer, A., Renn, A. and Wild, U. P., Single molecule spectroscopy: fluorescence excitation spectra with polarized light, *J. Luminescence* 56: 29 (1993).

71. Güttler, F., Irgnartinger, T., Plakhotnik, T., Renn, A. and Wild, U. P., Fluorescence microscopy of single molecules, *Chem. Phys. lett.* 217: 393 (1994).

72. Moerner, W. E., Plakhotnik, T., Irngartinger, T., Wild, U. P., Pohl, D. W. and Hecht, B., Near-field optical spectroscopy of individual molecules in solids, *Phys. Rev. Lett.* 73: 2764 (1994).

73. Moerner, W. E., Plakhotnik, T., Irngartinger, T., Croci, M., Palm, V. and Wild, U. P., Optical probing of single molecules of terrylene in a Shpol'skii matrix: a two-state single-molecule switch, *J. Phys. Chem.* 98: 7382 (1994).

74. Wild, U. P., Güttler, F., Pirotta, M. and Renn, A., Single molecule spectroscopy: Stark effect of pentacene in p-terphenyl, *Chem. Phys. Lett.* 193: 451 (1992).

75. Wild, U. P., Croci, M., Güttler, F., Pirotta, M. and Renn, A., Single molecule spectroscopy: Stark-, pressure-, polarization-effects and fluorescence lifetime measurements, *J. Luminescence* 60 & 61: 1003 (1994).

76. Plakhotnik, T., Moerner, W. E., Palm, V. and Wild, U. P., Single molecule spectroscopy: maximum emission rate and saturation intensity, *Opt. Comm.* 114: 83 (1995).

77. Wrachtrup, J., "Magnetische Resonanz an einzelnen Molekülen und kohärente ODMR-Spektroskopie an molekularen Aggregaten in Festkörpern," Thesis FU Berlin, Berlin (1994).

78. Amann, A., Modeling the quantum mechanical measurement process, *Int. J. Theor. Phys.* 34: 1187 (1995).

79. Stroock, D. W. and Varadhan, S. R. S., "Multidimensional diffusion processes," Springer, Berlin (1979).

80. Arnold, L., "Stochastic Differential Equations: Theory and Applications," Interscience, New York (1974).

81. Jaynes, E. T., Information theory and statistical mechanics, *Phys. Rev.* 106: 620 (1957).

82. Kobayashi, S. and Nomizu, K., "Foundations of Differential Geometry. Vol. I," Wiley, New York (1963).

83. Petz, D. and Sudár, C., Geometries of quantum states, preprint (1995).

84. Amann, A., Structure, Dynamics and Spectroscopy of Single Molecules: A Challenge to Quantum Mechanics, to appear in *J. Math. Chem.* (1996).

85. Ellis, R. S., "Entropy, Large Deviations, and Statistical Mechanics," Springer, New York (1985).

86. Deuschel, J.-D. and Stroock, D. W., "Large Deviations," Academic Press, San Diego (1989).

87. Lanford, O. E., Entropy and equilibrium states in classical statistical mechanics, *in*: "Statistical Mechanics and Mathematical Problems," A. Lenard, ed., Springer, Berlin (1973).

88. Peinke, J., Parisi, J., Rössler, O. E. and Stoop, R., "Encounter with Chaos. Self Organized Hierarchical Complexity in Semiconductor Experiments," Springer, Berlin (1992).

89. Cohn, D. L., "Measure Theory," Birkhäuser, Boston (1980).

90. Emery, M., "Stochastic Calculus in Manifolds," Springer, Berlin (1989).

91. Dellacherie, C. and Meyer, P. A., "Probabilités et Potentiels B," Hermann, Paris (1980).

ENVIRONMENTAL FACTORS IN MOLECULAR MODELLING

Jan C.A. Boeyens

Centre for Molecular Design
Department of Chemistry
University of the Witwatersrand
Johannesburg

INTRODUCTION

The most revolutionary result of quantum theory is also the one most commonly ignored by some theoreticians. It is quantum theory that shows how everything in the world is part of everything else, without boundaries and without isolated parts. However, in applications outside of quantum field theory, it is almost invariably an isolated system which is chosen for analysis. In chemistry it could be an isolated electron, an isolated atom, an isolated molecule, or an isolated crystal. It is easy to believe that this assumption would not introduce significant errors, and therefore difficult to initiate serious debate of the issue. Fact is that an isolated electron has no spin, an isolated atom has no size, and an isolated molecule has no shape. It follows that all of the important familiar properties of chemically significant entities arise through interaction with the environment.

Although the influence of the environment is rarely recognized in modelling studies, it is securely built into all empirical modelling schemes. In the modelling of molecular shape, for instance, the empirical force field is based on parameters measured in solid crystals, and not in isolated molecules. Although the method may therefore appear to simulate the shape of a single molecule, it is well to remember that the reference structure is as observed in a crystal, surrounded by, and interacting with many neighbours.

This means that the force-field parameters are conditioned by the crystal environment and any attempt to derive them from first principles should likewise take the influence of the environment into account.

The situation is the same for chemical bonds. All empirical facts about chemical bonds derive from measurements on material in the bulk. Chemical bonds are formed, either under crowded conditions or during high-energy collisions. In both instances there is a close encounter between activated species which implies previous interaction with external influences. The one thing not implied is that chemical bonds are generated spontaneously between isolated entities, either atoms or molecules. As before, this means that in order to simulate the process of chemical binding or to understand the characteristics of chemical bonds, it is necessary to recognize the driving effect of the environment.

The most fundamental issue in chemical binding is electron spin. Electrons without spin cannot stabilize bonds, which therefore cannot be understood without appreciating the origin of spin. At the next level of understanding is the atom, which fails to interact chemically unless it is first promoted by environmental compression into a valence state. Finally, there is the molecule, supposedly endowed with a characteristic three-dimensional shape, which is the objective of most modelling studies. No simulation, however successful can be complete if the link between the environment and molecular shape is ignored. It means that no theory, however fundamental, can account for all the details of molecular modelling without explicitly incorporating environmental factors. The quantum theory of isolated entities is one of these theories that cannot provide force-field parameters from first principles. Some procedures to make simple modifications to the theory, in order to compensate for first-order environmental effects are described next.

ELECTRON SPIN

The usual Schrödinger equation for a free electron has a form related to the classical kinetic energy expression

$$E = p^2/2m$$

through the operators $E \to i\hbar\partial/i\partial t$, $p \to i\hbar\nabla$, i.e. $2mi\hbar\partial\psi/\partial t - \hbar\nabla^2 = 0$, or suitably modified to $H \to (\hbar i\partial/\partial t - V)$ in a potential field. In its most familiar form, $V = -e^2/r$ for the central Coulomb field, as in the hydrogen atom. A complete solution is possible in terms of three quantum numbers, one of which quantizes the total angular momentum of the electron, but in a form that violates its conservation, in comparison with spectroscopic measurements[1]. The reason for this is that the Schrödinger equation, formulated as shown above, considers the electron as completely free of external influences. The electron (or the hydrogen atom, as the case may be) is in fact considered to behave as if independent of the symmetry of space. It means that in effect the influence of the environment, in this case of the vacuum, is ignored. It can be taken into account by stipulating that the wave equation correctly reflect the known Galilean invariance of three-dimensional space[2].

This is equivalent to demanding that the Schrödinger operator, $S \to E - p^2/2m$, should have a linear form, such that $(AE + Bp + C)^2 = 0$. This is possible only if the coefficients A, B, C are 4×4 anti-commuting matrices, and the wave function is a four-dimensional vector, made up of two-dimensional spinors, or wave functions with a spin component. This has been interpreted[3] to mean that spin is not a property of an electron, but a manifestation of the interaction between an electron and the vacuum. The conservation of angular momentum is restored when spin is taken into account and this indicates a special direction in space-time that breaks its rotational symmetry. An applied magnetic field projects this direction into three-dimensional space and reveals a two-level system for electron spin.

In the present context this shows that an empirical scheme which correctly models the known property of electron spin can never be reduced to the wave functions of isolated electrons. In practice the problem is considered as solved by the ad hoc stipulation of antisymmetric wave functions, without disclosing the origin of spin.

COMPRESSED ATOMS

Like electron spin, the valence state of an atom has no meaning in terms of free-atom wave functions. Like spin it could be added by an ad hoc procedure, but this has never been achieved beyond the qualitative level. All conventional methods of quantitative quantum chemistry endeavour to simulate atomic behaviour in terms of free-atom wave functions. Although it's an open secret that these calculations cannot compete with empirical simulations, there is a reluctance to admit that many known features of molecular behaviour simply have no meaning within the scheme. Rather than argue the point it is more rewarding to examine possible procedures whereby free-atom wave functions could be modified to reflect environmental influences.

In an environment of atoms in collision interatomic contacts consist of negative charge clouds interacting[4]. This environment for an atom is therefore well approximated by a uniform electrostatic field, which has a well-defined effect on the phases of wave functions for the electrons of the atom. It amounts to a complex phase or gauge transformation of the wave function

$$\psi(x) \rightarrow \psi'(x) = \exp(-i\xi(x)) \cdot \psi(x),$$

such that the probability density remains invariant. The effect of this has been shown[5] to be like a Faraday cage that confines the electron within a spherical surface. It is therefore no longer appropriate to formulate the wave functions for these atoms in terms of infinite boundary conditions, and $\psi(\infty) = 0$ should be replaced by $\psi(r_0) = 0$, for finite r_0.

Simulation of this condition which amounts to uniform compression of atoms was studied by numerical methods[6]. It shifts the electronic energy to uniformly higher levels, and eventually leads to ionization. It means that environmental pressure activates the atom, promotes it into the valence state and prepares it for chemical reaction. The activation consists therein that sufficient energy is transferred to a valence electron to decouple it from the atomic core. The wave function of such a freed electron in a spherical closure is a linear combination of Bessel functions, which in general is a slowly-varying function that indicates a state of low kinetic energy.

It should be clear that the valence electron of a promoted atom could readily interact with other activated species in its vicinity to form chemical bonds. The mechanism is the same for all atoms, since the valence state always consists of a monopositive core, loosely associated with a valence electron, free to form new liaisons. If the resulting bond is of the electron-pair covalent type its properties, like bond length and dissociation energy can be calculated directly by standard Heitler-London procedures[7], using valence-state wave functions.

The situation is simplified even further if the low kinetic energy of the valence electrons is ignored altogether. That leaves electrostatics as the only consideration in the calculation of bond properties, provided an appropriate atomic radius can be assigned in all cases. The radii are in fact known from first principles as the boundary parameters ro at which ionization occurs on uniform compression[6]. The calculation is now an entirely classical one, which has been used successfully to calculate bond dissociation energies[8] and molecular-shape factors[9].

It is important to note that molecular mechanics has the same classical basis[10]. Despite its success in accounting for molecular structures and thermodynamic properties, but because of its neglect of kinetic energy, it is rarely treated without suspicion. However, the reason why it works now becomes obvious. Because of environmental gauge transformation of electronic wave functions, the kinetic energy of valence electrons is of little practical importance and a classical calculation reproduces virtually all important properties. In the hands of the practising chemist molecular mechanics is therefore adequate. The classical form of the calculations is no cause for alarm. It is indeed the correct theoretical form, when recalling that it does not apply to the shape of isolated molecules, but only to conformations that occur in condensed phases.

EMERGENCE OF MOLECULAR SHAPE

There is no simple demonstration of how molecular shape, like spin or the atomic valence state, emerges as a consequence of environmental pressure, but there is the compelling argument that it never features in molecular physics, unless introduced by an ad hoc postulate.

Apart from subtle exceptions, an isolated molecule differs from a molecule in a crystal in that the isolated molecule has no shape, whereas in the crystal it acquires shape, but loses its identity as an independent entity.

To understand this paradoxical situation we can examine the transition between these two states of the molecule in terms of the famous Goldstone theorem, which for our purposes relates each phase transition, or broken symmetry, to a special interaction. We understand that the truly isolated molecule does not exist, but for argument's sake we can so describe a molecule in the physical vacuum, consisting of the gravitational and back-ground radiation fields only. This state of the molecule has been modelled in terms of the spin-boson model, which couples a spin system to an infinite number of harmonic oscillators[11]. It was found to differ from the hypothetical isolated state mainly by small changes in the molecular energy levels, of the order of Lamb shifts only. In this sense an isolated molecule can be viewed as a collection of atoms that move together because of randomly oriented cohesive interactions, also known as covalent interactions. Both nuclei and electrons behave quantum-mechanically and are not located more precisely than in terms of probability densities.

When introduced into an atmosphere consisting of other molecules of its own kind, a phase transition occurs as the molecule changes its ideal (gas) to non-ideal behaviour, and induced by Van der Waals-London dispersion interaction with its environment. In an applied electric field another phase transition mediated by electromagnetic interaction is possible.

Typically it leads to polarization of the the molecular charge density, which can cause some degree of nuclear alignment within the molecule. When the field is switched off the inverse transition occurs and the structure disappears.

Another common phase change, to the liquid state, is caused by increased concentration and can be promoted by increased pressure or reduced temperature. It is mainly rotational symmetry that breaks down. It effects the heavier nuclei more than the electrons and short-range intermolecular interactions become important. For lack of better terminology these can be called polarization interactions and their effect is the disappearance of molecular units, and although the covalent interaction still predominates, it is no longer completely undirected. The arrangement of nuclei in space changes more slowly and transient structures appear. Below this transition point the symmetry is too low to identify molecules by the same wave functions as before. The molecule is gaining structure, but losing identity. Interaction with the environment has now become too dominant to ignore in an effort to specify individual molecules.

The most dramatic phase transition is from the liquid to the solid state and this happens as the vibrational motion of individual molecules spreads through the bulk of the material to reappear as lattice modes. The molecules are now coupled into a periodic array, but not necessarily in the same orientation. As crystals are cooled down further it is common for a whole series of phase transitions to occur as additional orientational forces become dominant and cause a lowering of the symmetry at each transition. The symmetry is often found to decrease from high-symmetry cubic forms through orthorhombic and monoclinic modifications, as ordering increases, ending up as triclinic crystals on approaching the zero point.

In this state of matter the distribution of nuclei can no longer be considered as a probability distribution, with the possible exception of isolated units, like the perchlorate group discussed before, trapped in voids within the host lattice. In general however, it appears that crystals consist of an ordered array of nuclei, kept at equilibrium sites by delocalized interactions and hence, to represent a classical distribution. The electrons in the structure can still be treated quantum-mechanically, using the nuclear distribution as a classical framework to define the potential field. This is in line with the mathematical procedures to partition a comprehensive theory that embodies both classical and non-classical variables[12].

It is interesting to note that the interaction of the electron density with the potential field of an X-ray beam does not reveal any excitation to higher quantum levels, during diffraction experiments, except for momentum states, controlled by the lattice spacings in terms of Bragg's law. This periodicity of electron density to match the lattice structure defined by the nuclei, is caused by interaction with the X-ray beam. The results of X-ray crystallography are therefore routinely interpreted in terms of a continuous classical charge distribution.

This general scheme of events, although valid for all materials, does not predict the same effects under the same conditions for all species. It depends quite critically on the complexity of the molecule. Large biomolecules or polymers have no gas-phase existence and even in solution, or the liquid state, have a well-defined molecular shape, while small molecules like ammonia settle into a classical lattice structure only at very low temperatures.

REFERENCES

1. H. Primas and U. Müller-Herold, "Elementare Quantenchemie", Teubner, Stuttgart (1984)
2. J.-M. Lévy-Leblond, *Commun. Math. Phys.* 6:286 (1967)
3. J.C.A. Boeyens, *J. Chem. Ed.* 72:412 (1995)
4. J.C.A. Boeyens, *Electronic J. Theor. Chem.* (1995), in press.
5. J.C.A. Boeyens, Gauge chemistry, *in:* "Large-Scale Molecular Systems: Quantum and Stochastic Aspects", W. Gans, A. Blumen and A. Amann, (eds) Nato ASI Series B258, Plenum, New York (1991)
6. J.C.A. Boeyens, *J. Chem. Soc. Faraday Trans.* 90:3377 (1994)
7. J.C.A. Boeyens, *S. Afr. J. Chem.* 33:14 (1980)
8. J.C.A. Boeyens, *J. S. Afr. Chem. Inst.* 26:94 (1973)
9. A.Y. Meyer, *J. Mol. Struct. (Theochem)* 179:83 (1988)
10. J.C.A. Boeyens, *Structure and Bonding* 63:65 (1985)
11. A. Amann, *Ann. Phys.* 208:414 (1991)
12. H. Primas, "Chemistry, Quantum Mechanics, and Reductionism, Perspectives in Theoretical Chemistry," Springer-Verlag, Berlin (1983)

KNOWLEDGE ACQUISITION FROM CRYSTALLOGRAPHIC DATABASES: APPLICATIONS IN MOLECULAR MODELLING, CRYSTAL ENGINEERING AND STRUCTURAL CHEMISTRY

Frank H. Allen

Cambridge Crystallographic Data Centre
12 Union Road
Cambridge
UK

INTRODUCTION

The interplay between theory and experiment has always been crucial to the development of structural chemistry. This interplay has become more sharply focused in recent decades, as increasingly detailed calculations and increasingly precise experimental structure determinations are facilitated by rapid technological advances. It is these advances, particularly in raw computing power and in graphics capabilities, that now bring the most sophisticated computational methods to the chemists desk, and underpin the branch of chemistry that forms the subject of this book: molecular modelling. So persuasive are these models, however, that there is a tendency to imbue each graphic, each energy value, each atomic partial charge, with a veracity that is neither warranted, nor claimed by the originators of the software. There are many computational techniques, and many different levels of sophistication. Each technique makes different assumptions and approximations, and each must be understood and assessed in deciding its suitability in a given problem domain.

It is important, therefore, to assess the level of acceptability of computed molecular models against relevant experimental results whenever these data are available. Any positive link between theory and experiment, whether direct or indirect, adds credibility to the computations and generates increased confidence in proposed solutions to the problem in hand. This idealised scenario, of course, begs two important questions: What experimental results should we use? and How accessible is this information? The answer to the former is, perhaps, problem dependent. Since this Chapter is concerned primarily with modelling applied to rational molecular design (pharmaceuticals, agrochemicals, etc.), then experiments carried out on condensed phases, such as NMR or X-ray crystallography, are likely to be more relevant than in-vacuo data on isolated molecules. The answer to the second question may be somewhat more pragmatic in a problem-oriented environment: crystallographic data is readily accessible. It is organised into well established, fully retrospective and rapidly searchable databases, particularly for the small molecule and protein structures that are of fundamental importance in rational molecular design.

Fundamental Principles of Molecular Modeling
Edited by Werner Gans *et al.*, Plenum Press, New York, 1996

This Chapter, then, is concerned with the role of experimental crystallographic data in molecular modelling. More specifically, it is concerned with the derivation of systematic structural knowledge from the 3D results of more than 150,000 single crystal diffraction analyses, and with applications of that knowledge in a modelling environment. Here, the modeller frequently needs answers to some very basic questions:

(a) How big is this molecule — what are the molecular dimensions?

(b) What shape is this molecule — what is its overall conformation and what are the relative conformations of its functional substructures?

(c) How is this molecule likely to interact and bind at a receptor site?

At the molecular level (a,b), crystallographic information can both support and extend the computational results. Predicted dimensions and, particularly, the computed conformational energy minima can be compared with dimensions and conformational preferences observed in crystal structures. Further, the very broad spectrum of molecules that have been studied crystallographically means that database analyses can be extended into chemical areas where limitations, e.g. of force-field parameters or of suitable basis sets, would preclude the use of computational methods in any case. At the intermolecular level (c), crystallographic data is unique in providing direct experimental observations of the non-covalent interactions that are involved in molecular recognition processes.

CRYSTALLOGRAPHIC DATABASES: THE CAMBRIDGE STRUCTURAL DATABASE SYSTEM

Four major databases[1] now store complete results from all published crystallographic analyses. This Chapter is concerned with the largest of these, the Cambridge Structural Database (CSD)[2] of organo-carbon small molecules, and with its associated software system. Detailed descriptions of CSD information content and software functions are available elsewhere[3,4,5] and only a brief summary will be given here.

The CSD currently (October 1995) contains details of nearly 150,000 structures and the information content of each CSD structural entry is best summarised according to its 'dimensionality':

- Bibliographical and chemical text (1D)

- Chemical structural formula (2D): a compact and fully searchable representation of the chemical connectivity in terms of element types, bond types, hydrogen counts, formal charges, etc.

- Crystallographic structural data (3D): cell parameters, space group, and atomic coordinates that are mapped directly to the relevant atoms of the 2D chemical formula.

The CSD software system comprises three main programs, each having a fully interactive graphical interface for user applications:

- QUEST3D is the main search program, permitting searches of all available information fields. Most important among these are (a) the location of 2D substructures, which may then be further constrained using a wide variety of 3D geometrical criteria, and (b) the location of non-covalent intermolecular contacts in the extended crystal structure, again using geometrical criteria. The program will generate output files of user-defined geometrical parameters for any substructure (intramolecular or intermolecular) located in the search process. An example of such a tabulation is shown as Figure 1.

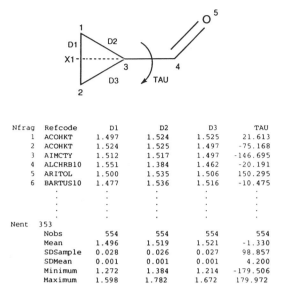

Nfrag	Refcode	D1	D2	D3	TAU
1	ACOHKT	1.497	1.524	1.525	21.613
2	ACOHKT	1.524	1.525	1.497	-75.168
3	AIMCTY	1.512	1.517	1.497	-146.695
4	ALCHRB10	1.551	1.384	1.462	-20.191
5	ARITOL	1.500	1.535	1.506	150.295
6	BARTUS10	1.477	1.536	1.516	-10.475
	.	.	.	.	.
	.	.	.	.	.
	.	.	.	.	.
Nent	353				
	Nobs	554	554	554	554
	Mean	1.496	1.519	1.521	-1.330
	SDSample	0.028	0.026	0.027	98.857
	SDMean	0.001	0.001	0.001	4.200
	Minimum	1.272	1.384	1.214	-179.506
	Maximum	1.598	1.782	1.672	179.972

Figure 1. Cyclopropyl-carbonyl substructure: CSD search fragment and section of the resultant geometry tabulation.

- VISTA reads the geometrical table(s) generated by QUEST3D and provides extensive facilities for the graphical representation (histograms, scattergrams, polar plots, etc) and statistical analysis (summary statistics, regression, principal component analysis, etc) of the numerical data. The polar histogram of Figure 2, and many of the other plots in this Chapter were generated by VISTA.

- PLUTO is used to visualise crystal and molecular structures in a variety of styles.

Other databases within the total CSD System now comprise:

- A compacted version of the Protein Data Bank, in which all non-coordinate information of the PDB, including sequence information, can be searched through the QUEST3D program.

- A database of literature references (currently over 500) to scientific applications of the CSD in a research environment.

MEAN MOLECULAR DIMENSIONS

The derivation of simple descriptive statistics for standard geometrical parameters, particularly bond lengths or valence angles, is a relatively straightforward application of the CSD System. These data are provided routinely as part of the standard QUEST3D tabulations (see Figure 1), and histograms of individual distributions can be rapidly generated using VISTA. As an aid to structural chemists and modellers, two major compilations of 'standard' bond lengths were produced in the late 1980's, derived for a wide range of organic[6] and metallo-organic[7] bond environments. Since this time, software has improved considerably and mean values for environments not covered by these compilations can be simply generated as required in just a few minutes of 'real' time.

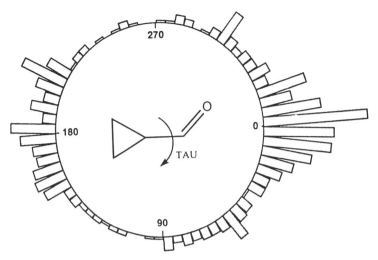

Figure 2. Cyclopropyl-carbonyl substructure: Polar histogram of the torsion angle O=C-C-X1 (TAU).

CONFORMATIONAL PREFERENCES

Conformational analysis, the detection of global and local energy minima in the potential energy hypersurface, is a common application of theoretical techniques. Such computations are not without their difficulties: incomplete parameterisation in molecular mechanics procedures, the inadequacies of semi-empirical Hamiltonians and the computational intensity of ab initio methods at higher basis sets. Further, the conformational space must be sampled at realistic intervals so as to ensure complete location of global and local minima.

In many cases, then, it may be useful to supplement the computational results by establishing the conformational preferences that are observed in crystal structures. In some cases, and for reasons described above, this may the only way of generating realistic conformational models. In this application of the CSD, the large volume of structural data that is often now available for a substructure of interest can be used to good effect. While it is unlikely that the substructure will adopt its minimum-energy form in all crystallographic examples, there is now a growing body of evidence to suggest that the distribution of observed conformations is correlated, in a qualitative way, with the low-energy features of the relevant potential energy hypersurface[8]. There are well-known problems[9], however, associated with the quantification of the relationship between potential energy and conformational frequencies deduced from distribution of experimental observations.

The cyclopropyl-carbonyl substructure of Figures 1,2 is a good example of a simple conformational application of the CSD. The single torsion angle, TAU = O1-C4-C1-X1 (where X1 is the midpoint of the C2-C3 bond), describes the orientation of the carbonyl group with respect to the three-membered ring. The polar histogram of TAU (Figure 2) shows a clear preference for values close to 0^0 (cis-bisected conformers) and, to a lesser extent, close to 180^0 (trans-bisected). These crystal structure observations[10] are in complete agreement with theory[11,12] and are of considerable importance in the modelling of pyrethroid insecticides. These compounds are synthetic modifications of chrysanthemic acid, the active constituent of a weak insecticide that occurs naturally in the chrysanthemum plant, and which incorporates the cyclopropyl-carbonyl function as a central substructural feature.

Figure 2, of course, represents a simple 1D problem. However, problems involving two or three torsions can also be handled quite simply, through the detection of clusters of observations in 2D or 3D scattergrams, objects that are both visually and graphically tractable. Problems of higher dimensionality, e.g. the conformational preferences of n-membered rings described by the n intra-annular torsions, can be addressed using the techniques of multivariate analysis[4], such as principal component analysis (PCA) or cluster analysis (CA).

PCA is a dimension reduction technique that attempts to re-express the original n-dimensional data with respect to a smaller number of mutually orthogonal axes — the principal components of the dataset. PC plots, i.e. with data points expressed as coordinates in PC-space, can often reveal clusterings, representing preferred conformations, and links between them, representing conformational interconversion pathways. 'Conformational' PCA, which has a formal relationship to Cremer-Pople analysis[13], has been applied to both cyclic[14,15,16] and acyclic[17] systems and is well illustrated in these references.

Cluster analysis is a purely mathematical technique that aims to locate unique dissections of the dataset in n-dimensions, i.e. to locate clusters of points having similar coordinates in an n-dimensional space. A variety of algorithms are available, and the technique has again been successfully applied to both rings[14,16,18] and to acyclic systems[18].

The above examples provide some evidence, at least, for a qualitative relationship between crystallographic conformer distributions and the low-energy features of the potential energy hypersurface. Recently[19], we have studied this relationship in a more systematic manner for a series of twelve 1D conformational problems. All of these substructures would be expected to show one symmetric (anti) torsional minimum and two symmetry-related asymmetric (gauche) minima. In each case the crystallographic torsional distribution was compared with the potential energy profile computed using the 6-31G* basis set via the program GAMESS-UK[20]. Results for three representative substructures are shown in Figure 3a,b,c. Optimized ab initio values of the asymmetric (gauche) torsion angle vary from below 55^0 to above 80^0 across the twelve substructures. A scatterplot of these values against the average crystallographic torsion for gauche conformers (Figure 3d) shows a high degree of linearity and a correlation coefficient of 0.831.

Two other results of the study were that: (a) torsion angles with higher strain energies (> 4.5 kJ mol^{-1}) are rarely observed in crystal structures (< 5 %), and (b) taken over many structures, the conformational distortion due to crystal packing appears to be the exception rather than the rule. We can only conclude that crystal structure observations are good guides to conformational preferences and are a valuable adjunct to computational methods, particularly in 'difficult' systems.

NON-COVALENT INTERACTIONS

Crystal structures can be viewed as the archetypal supramolecules: molecules and ions forming an extended structure according to the dictats of the electrostatic, charge transfer and other interactions that are responsible for molecular recognition processes. The unique experimental observation of the effects of these interactions in crystal structures makes their systematic study one of the most important applications of crystallographic databases. Such systematic information is of considerable importance in molecular modelling, since it is fundamental to obtaining realistic visualisations of possible modes of protein-ligand binding. However, such information has a more general importance, since it is invaluable knowledge in supramolecular chemistry[21] and in crystal engineering[22].

In all of these areas, we need information on three basic aspects of non-covalent interactions: (a) what sort of interactions are important in molecular recognition? (b) do these interactions show directional properties? and (c) how strong are the interactions? Crystal structure data is ideally suited to answer questions (a) and (b). Here, the high chemical diversity

Fragment 1 :

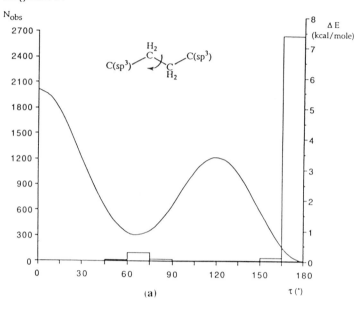

(a)

Fragment 2 :

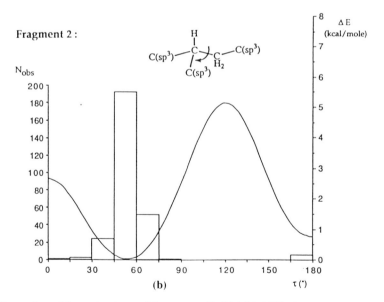

(b)

Figure 3. Comparison of the torsional potential energy profile (ab initio MO) with the torsional distribution observed in crystal structures for three representative fragments (a,b,c). Frame (d) shows the plot of the mean crystallographic torsion angle τ_{ave} vs. the ab initio optimum value τ_{opt} for the gauche conformers of all twelve fragments included in the study.

Fragment 10 :

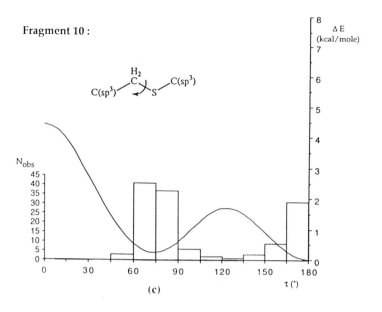

(c)

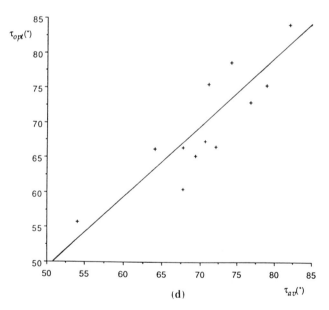

(d)

Figure 3. (*continued*)

of sources such as the CSD, coupled again with the fact that we are studying condensed phase information and at small computational costs, are benefits of the database approach. For (c), the best we can achieve is to record the lengths of non-covalent interactions relative to some yardstick (principally sums of van der Waals radii) to obtain some qualitative indication of strength. Quantitative estimates of interaction energies can, of course, be obtained by high-level molecular orbital calculations, such as the intermolecular perturbation theory (IMPT) of Hayes and Stone[23,24]. These calculations are cpu-intensive and are representative of an in-vacuo environment. However, a preliminary analysis of crystallographic data can suggest appropriate model systems, and can narrow the choices of suitable relative orientations of the component model molecules to be used in the calculations.

In a Chapter of this length, it is only possible to give representative examples of the types of study that can be carried out using the CSD: hydrogen bonding at 'univalent' sulphur (C=S), and dipolar carbonyl...carbonyl interactions are chosen for this purpose. However, leading references are given to studies of other types of non-covalent interactions as a lead into the relevant literature.

Hydrogen bonding at 'univalent' sulphur (C=S)[25]

The 'non-bonded contact search' mechanism of QUEST3D permits the location of supramolecular substructures, such as that in Figure 6a, and the user-definition of suitable geometrical parameters to describe the fragment. Certain special parameters, θ and ϕ in Figure 4a, are available to quantify the direction of approach of H to the acceptor S: values of θ, ϕ that are close to 0, 120^0 would indicate H-approach towards a conventional $S(sp^2)$ lone pair. The search itself is distance-based, in this case the primary search constraint is that the hydrogen bond length, d (SH), is less than 2.9 Å, the appropriate sum of van der Waals radii. Other, and secondary constraints can then be applied if required: here $\rho(H)$ was constrained to exceed 90^0 and ϕ was required to be above 60^0.

The software will assemble a table of all geometrical parameters for each occurrence of the fragment located in the CSD; it is then a simple matter (VISTA) to generate visualizations of this data (Figure 5). These provide a basic overview of the characteristics of the interaction: (a) H-bonds to O-H donors are shorter than those to N-H donors, (b) the H-bond angle tends to linearity, (c) H-bonds form to =S acceptors over a wide range of theta values, but (d) there seems to be a clear preference for H-approach at ca. 110^0 to the C=S vector in the 'lone-pair' plane. These data are typical of H-bonds, except for (c) where the wide θ range seems to be due to large numbers of O-H(water) donors, small molecules that can fit multiply into the S coordination sphere. By comparison with analogous data for C=O acceptors, though, N-H...S and O-H...S distances, d(SH), are some 0.25 Å longer than those for the O acceptor, indicative of weaker bonding to S. Also, the ϕ-directionality at S appears to be far more specific than at O which exhibits ϕ-peaks at ca. 130^0 in a continuum that spans $\phi = 180^0$; S shows little inclination to form any H-bonds that are collinear with the C=S vector. What is more interesting is why C=S forms H-bonds at all, given the almost negligible electronegativity difference between C and S. Examination of the R1,R2-C=S substructures available in the CSD show that systems having R1 and/or R2 = nitrogen are dominant. Here, electron donation from the N lone pair(s) (Figure 4b) induces a partial negative charge on S, and produces a lowering of the C=S double bond character. Thus, H-bonding to C=S can be termed resonance-induced, a further development of the concept of resonance-assistance of Gilli et al.[26], that is discussed elsewhere in this Volume. The histogram of C=S bond lengths (Figure 4c) shows quite clearly that only the longer bonds actually take part in H-bonded interactions: the frequency of H-bond formation rises from 5 % at C=S = 1.63 Å to more than 75 % when C=S is longer than 1.70 Å.

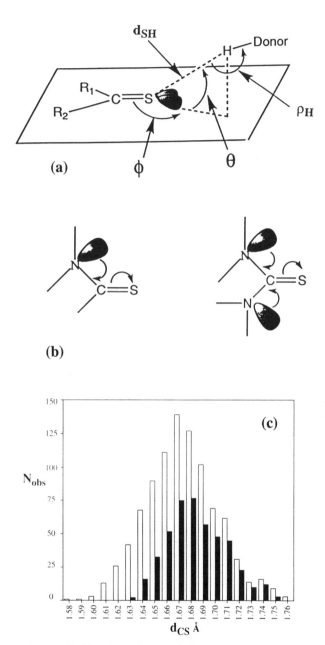

Figure 4. Histograms of geometrical descriptors from the C=S...H-N,O hydrogen bond survey: (a) the S...H distance, (b) the S...H-N,O angle, (c) the directionality parameter θ of Fig.4a, and (d) the directionality parameter φ of Fig. 4a.

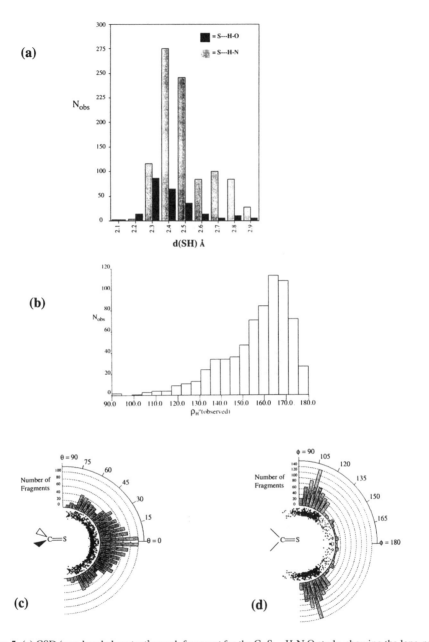

Figure 5. (a) CSD 'non-bonded contact' search fragment for the C=S...H-N,O study, showing the lone-pair directionality parameters (θ and ϕ). (b) Resonance interactions of the C=S bond with the N-lone pair. (c) Histogram of C=S bond lengths that occur in potential H-bonding environments (hatched bars) and those that actually form H-bonds (black bars).

Hydrogen bonds are the 'classic' non-covalent interaction and data from individual crystal structures and from database analyses have been fundamental to the acquisition of knowledge in this area. A recent monograph[27] organises and inter-relates much of this knowledge with a particular accent on the biological importance of the subject.

Weak hydrogen bonds: The C-H...O and X-Hi...π systems

Despite early controversies[28,29] that contradicted a growing body of spectroscopic data, crystallographic evidence for the existence of C-H...X hydrogen bonds is now well established[30] and increasingly well documented[27,31]. The C-H...O system has been most studied, including correlations of $d(C...O)$ versus $v(C-H)$[32] and versus the increasing H acidity as the hybridisation of C changes from sp^3 to sp^1[33]. These studies indicate that acetylenic $C(sp^1)$-H is the most potent C-H donor and the geometrical characteristics of $C(sp^1)$-H...O bonds closely parallel those of much stronger systems.

More recently, there has been a growing interest in the effectiveness of π-systems as H-bond acceptors. The frequencies and geometries of (N or O)-H...π(phenyl or acetylene) interactions have been studied[34] and these studies have extended to the acceptor properties of three-membered rings[35], a possibility first suggested on spectroscopic evidence almost 30 years ago[36]. Steiner[37,38] has examined C-H...π systems, particularly the $C(sp^1)$-H...π(acetylene) self-recognition system, and has been able[38] to engineer crystal structures in which molecular aggregation is mediated by these weak interactions.

No hydrogen bonds at all

Interest in systematic database studies of non-covalent interactions that are not mediated by hydrogen began with the classic and original structure correlation work of Dunitz, Bürgi and co-workers[39], particularly the study of the approach of nitrogen nucleophiles to carbonyl centres. This work generated a plot of the proposed reaction pathway for the reaction and extended our perception of crystal structures from that of static images into the realms of reaction dynamics.

Here we illustrate the application of the modern CSD System in the broad area of 'non H-mediated interactions' through a simple study of dipole-dipole interactions involving the C=O group[40]. The QUEST3D program allows for the location of group...group contacts (Figure 6a) via specification of a limiting contact distance (here 3.6 Å) between any atom in group 1 and any atom of group 2. The torsion angle tor =C=O...C=O, about the non-bonded contact vector, and the angle of elevation (θ) were chosen as additional descriptors of the interaction. Their histograms (Figure 6b,c, 574 observations) show that θ clusters around 90^0 and, remarkably, over 90 % of the tor values are within a few degrees of zero: the two C=O dipoles adopt a strictly antiparallel arrangement whenever they come into close contact in crystal structures. Calculations using the IMPT method[23] indicate an energy of some 20 kJ/mol for the interaction, equivalent to that of a medium strength hydrogen bond.

Other interactions that are of current interest are those involving carbon bound halogens (C-X) and oxygen or nitrogen atoms. The crystallographic data shows that short C-Cl...O=C interactions are linear, i.e. they form preferentially along the C-Cl vector, and recent IMPT calculations[24] show this to be an attractive interaction (ca. 10 kJ/mol) with its directionality due to the anisotropic electron distribution around X. Crystallographic data also shows that C-X halogens interact with nitro groups in two preferred orientations: (a) having X approaching between the two nitro-O atoms, or (b) approaching one nitro-O only. It has proved possible to use interaction (a) to engineer the extended crystal structure of a complex involving 1,4-di-iodobenzene and 1,4-dinitrobenzene[41,42].

The study of non H-mediated interactions is highly important to our understanding of molecular recognition processes. It is an area in which statistical analyses of the available crystallographic data, allied to high-level theoretical calculations, is likely to make crucial contributions in the coming years.

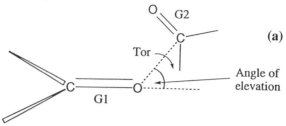

Constraint: atoms of Group 1 must be < 3.6A from atoms of Group2

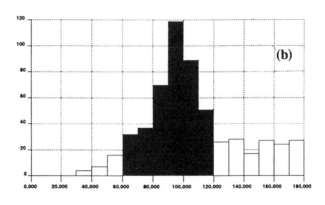

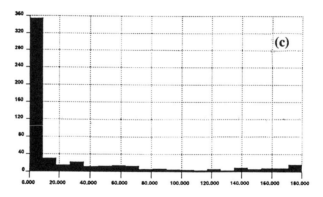

Figure 6. Carbonyl…carbonyl interactions: (a) CSD search fragment and geometrical descriptors, (b) histogram of the angle of elevation, and (c) histogram of the torsion angle C=O…C=O about the non-bonded vector.

The answer "No"

Previous sections have concentrated on the identification and characterization of non-covalent interactions that do exist in crystal structures. Equally important in a molecular modelling environment is knowledge of those interactions that do not form, even though chemical sensibility might indicate otherwise. One example is the thioether fragment, C-S-C, where H-bond formation to the S acceptor might be inferred by chemical analogy with the oxy-ethers C-O-C. A recent study[43] showed that only 4.75 % of more than 1000 thioether-S atoms form contacts to H-donors that are within van der Waals radii sums, and even these contacts are very close to this distance limit. Simple MO calculations show the electroneutrality of S in dialkyl-thioethers, and an increasing delta-positivity of S as the alkyl groups are substituted by unsaturated centres[43,44] due to S-lone pair delocalization in rings such as thiophene, etc. A similar mechanism would appear to operate in furan: out of 118 instances in which furan co-exists with an N-H or O-H donor in CSD entries, it forms H-bonds on only 12 occasions.

THE FUTURE

The crystallographic databases, particularly the CSD and PDB, already play an important role in molecular modelling projects. Allied to search and analysis software of increasing sophistication, these products provide a sound basis for knowledge acquisition from the wealth of experimental structural data that now exists. The joint approach, involving database analyses and modern high level molecular orbital calculations, now seems to offer the chance to answer many of the questions that are relevant in the modelling arena.

However, the databases are just that — reservoirs of raw structural data from which knowledge must be acquired via creative, and sometimes repetitive use of the available software systems. The next development, obviously, is to capture this knowledge itself in an organised and structured form in computerised systems. The development of structural knowledge bases from the underlying databases, and incorporating information from other relevant sources (experimental and computational), will undoubtedly be the flavour of the next decade. Such knowledge bases represent the central memory requirement for intelligent software systems that may then be designed to assist in problem solving in molecular modelling and molecular scene analysis[44].

REFERENCES

1. F.H. Allen, G. Bergerhoff and R. Sievers, eds., "Crystallographic Databases." International Union of Crystallography, Chester, UK (1987).
2. F.H. Allen, J.E. Davies, J.J. Galloy, O. Johnson, O. Kennard, C.F. Macrae, E.M. Mitchell, G.F. Mitchell, J.M. Smith and D.G.Watson, *J. Chem. Inf. Comput. Sci.* 31:187 (1991).
3. F.H. Allen, O. Kennard and D.G. Watson, *in:* "Structure Correlation", H.-B. Bürgi and J.D. Dunitz, eds., Chapter 3. VCH Publishers, Weinheim (1994).
4. R. Taylor and F.H. Allen, *in:* "Structure Correlation", H.-B. Bürgi and J.D. Dunitz, eds., Chapter 3. VCH Publishers, Weinheim (1994).
5. Cambridge Structural Database System User's Manual: Getting Started with the CSD System. CCDC, Cambridge, England (1994).
6. F.H. Allen, O. Kennard, D.G. Watson, L. Brammer, A.G. Orpen and R. Taylor, *J. Chem. Soc., Perkin Trans.* 2:S1 (1987).
7. A.G. Orpen, L. Brammer, F.H. Allen, O. Kennard, D.G. Watson and R. Taylor, *J. Chem. Soc., Dalton Trans.,* page S1 (1989).
8. See many of the individual chapters in "Structure Correlation" H.-B. Bürgi and J.D. Dunitz, eds., VCH Publishers, Weinheim,Germany (1994).
9. H.-B. Bürgi and J.D. Dunitz, *Acta Crystallogr.* B44:445 (1988).
10. F.H. Allen, *Acta Crystallogr.* B36:81 (1980).
11. R. Hoffmann, *Tetrahedron Lett.* 2907 (1970).
12. R. Hoffmann and R.B. Davidson, *J. Amer. Chem. Soc.* 93:5699 (1971)
13. D. Cremer and J.A. Pople, *J. Amer. Chem. Soc.* 97:1354 (1975)
14. R. Taylor, *J. Mol. Graphics* 3:106 (1986).
15. F.H. Allen, M.J. Doyle and T.P.E. Auf der Heyde, *Acta Crystallogr.* B47:412 (1991).
16. F.H. Allen, J.A.K. Howard and N.A. Pitchford, *Acta Crystallogr.* B49:910 (1993).

17. T.P.E. Auf der Heyde and H.-B. Bürgi, *Inorg.Chem.* 28:3960 (1989).

18. F.H. Allen, M.J. Doyle and R. Taylor, *Acta Crystallogr.* B47:51 (1991).

19. F.H. Allen, S.E. Harris and R. Taylor, *J. Comp.-Aided Mol. Design*, Submitted.

20. M.F. Guest, J.H. van Lenthe, J. Kendrick, K. Schoeffel, P. Sherwood and R.J. Harrison, "GAMESS-UK User's Guide and Reference Manual." Computing for Science Ltd., Daresbury Laboratory, Warrington (1993). 21. J.-M. Lehn, *Angew. Chem. (Int. Ed. Engl.)*27:89 (1988).

22. G.R. Desiraju, "Crystal Engineering: The Design of Organic Solids." Elsevier, Amsterdam (1989).

23. I.C. Hayes and A.J. Stone, *Mol. Phys.* 53:83 (1984)

24. J.P.M. Lommerse, A.J. Stone, R. Taylor and F.H. Allen, *J. Amer. Chem. Soc.,* Submitted.

25. F.H. Allen, C.M. Bird and R.S. Rowland, *Acta Crystallogr.,* Submitted.

26. G. Gilli, F. Bellucci, V. Ferretti and V. Bertolasi, *J. Amer. Chem. Soc.* 111:1023 (1989).

27. G.A. Jeffrey and W. Saenger, "Hydrogen Bonding in Biological Structures." Springer-Verlag, Berlin (1991).

28. D.J. Sutor, *Nature* 195:68 (1962)

29. J. Donohue *in:* "Structural Chemistry and Molecular Biology," A. Rich and N. Davidson, eds., W.H. Freeman, San Francisco (1968).

30. R. Taylor and O. Kennard, *J. Amer. Chem. Soc.* 104:5063 (1982).

31. G.R. Desiraju, *Acc. Chem. Res.* 24:290 (1991).

32. G.R. Desiraju and J. Murty, *Chem. Phys. Lett.* 139:360 (1987).

33. V.R. Pedireddi and G.R. Desiraju, *J. Chem. Soc., Chem. Commun.* 988 (1992).

34. M.A. Viswamitra, G. Bandekhar, R. Radhakrishnan and G.R. Desiraju, *J. Amer. Chem. Soc.* 115:4868 (1993).

35. F.H. Allen, J.P.M. Lommerse, J.A.K. Howard, V.J. Hoy and G.R. Desiraju, *Acta Crystallogr.,* Submitted.

36. L. Joris, P. von R. Schleyer and R. Gleiter, *J. Amer. Chem. Soc.* 90:327 (1968).

37. T. Steiner, *J. Chem. Soc., Chem. Commun.* 95 (1995).

38. T. Steiner, E.B. Starikov, A.M. Amado and J.J.C. Teixeira-Dias, *J. Chem. Soc., Perkin Trans.* 2:1321 (1995).

39. The early history of structure correlation is expertly summarised in: J.D. Dunitz, "X-Ray Analysis and the Structure of Organic Molecules." Cornell Univ. Press, Ithaca (1979).

40. R. Taylor, A. Mullaley and G.W. Mullier, *Pestic. Sci.* 29:197 (1990).

41. F.H. Allen, S.B. Goud, V.J. Hoy, J.A.K. Howard and G.R. Desiraju, *J. Chem. Soc., Chem.Commun.* 2729 (1994).

42. G.R. Desiraju, *Angew. Chem. (Int. Ed. Engl.),* In Press for November 1995.

43. F.H. Allen, C.M. Bird and R.S. Rowland, *Acta Crystallogr.,* Submitted.

44. F.H. Allen, R.S. Rowland, S. Fortier and J.I. Glasgow, *Tetrahedron Computer Methodology* 3:757 (1990).

HYDROGEN BONDING MODELS: THEIR RELEVANCE TO MOLECULAR MODELING

Paola Gilli, Valeria Ferretti, and Gastone Gilli

Centro di Strutturistica Diffrattometrica and Dipartimento di Chimica
Universitá di Ferrara
I-44100 Ferrara
Italy

INTRODUCTION

In the last twenty years the words *model* and *modeling* have come into a somewhat acritical use to identify a number of different methods and concepts pertaining to both social and physical sciences. Even restricting our interest to the purely physical aspects, it is not always clear what may have in common things apparently so different as the water-circulation model of the Mediterranean see, the Ising model of phase transitions and critical states, the cell automata as computer models of the developing of life, the molecular modeling of drug-receptor interactions or DNA-protein recognition processes, and the modeling of inter and intramolecular hydrogen bonding interactions (the subject present chapter is actually devoted to).

There is a some degree of confusion about what may be *molecular modeling* as well, and most opinions appear to oscillate between two different but equally naive beliefs: *(i)* molecular modeling is a physico-mathematical simulation of reality which only now is made possible by the availability of modern fast computers and, in this sense, is a novel and promising type of scientific activity; *(ii)* molecular modeling is nothing more than the computer encoding of scientific laws which are already well known but made often ineffective by the need of parametrizing also facts and events of which no certain theoretical understanding has ever been attained. Though none of these extreme positions is likely to be correct, it is a fact that very few critical appraisals of the value and limits of molecular modeling techniques have been so far produced.

For these reasons, the present chapter is divided in two parts. In the second one a model of hydrogen bonding is described which tries to answer the long-lasting question whether such a bond is a simple three-body interaction whose laws are already known and need only to be applied to specific cases, or whether it needs to be modeled according to the infinite variety of its molecular environments. Since, however, the value to be given to any model cannot be independent of what we think of modeling methods in general, the first part of the chapter reports a discussion on the general characteristic models are endowed with and an attempt to frame modeling methods and, in particular, molecular modeling within the accepted scientific methodologies.

PART 1 – MODELS AND MODELING ACTIVITIES: A CRITICAL APPRAISAL

Models, Theories and Natural Laws

Any attempt to define what is modeling in general (and molecular modeling in particular) must inevitably start from an unequivocal definition of what is intended by the word *model*. This may be a not easy task, because there are as many different philosophies as there are philosophers of science. In order to avoid, as far as possible, this sort of complications, we will limit ourselves to pursue a *simply operational definition of model*, that is one which is functional only to the problem treated, and try to base the discussion on scientific common sense considerations rather than epistemological subtilities. In particular, this account makes two important simplifications. It accepts without discussion the idea that natural laws are objective and independent of the observer, though this tenet of positivism has been often questioned (see H. Primas for a recent criticism[1]) and, finally, gives a definition of model as the clue, the original idea of any subsequent physico-mathematical development which is not necessarily canonical, though often used.

Our present aim is then that of defining, in the most precise sense, what is a *model*. This requires the previous definition of two other concepts, i.e. *natural law* and *scientific theory*.

Natural Laws. Empirical observation of nature leads to identify invariant regularities which are called natural laws. Examples could be *(i)* $pV = nRT$, law of ideal gases (or low-pressure real gases); *(ii)* $dU = dq - dw$, first law of thermodynamics; *(iii)* "All men are mortal", one of the most disturbing laws for everybody. Natural laws are widely independent of us and we may discover them but not modify their consequences. They are essentially the constrains that our natural environment puts to our wishes and fantasy and, in this sense, they are the more important the more they actually forbid.[2]

Scientific Theories. Our wish to understand why and how natural laws occur leads to formulate scientific theories, such as the kinetic theory of gases, classical or quantum mechanics, and statistical thermodynamics. For our aims, any complete scientific theory can be thought of as consisting of two essentially independent parts: *the model* and *its physico-mathematical development*. When the latter is missing, the theory is declassed to simple model.

The Model is the original idea of how things may work, a kind of logical nucleus of the theory. In physics, as well as chemistry, mineralogy, biochemistry, molecular biology, and so on along the reductionist chain, models must start from *the intimate constituents of matter*, that is *elementary particles* in nuclear physics and *nuclei and electrons* (or, at least, *atoms and molecules*) in solid state physics, chemistry, molecular biology, and so on. To make some examples, the kinetic theory attempts an explanation of gas laws on the ground of the ideal gas model (non-interacting molecules undergoing perfect elastic collisions among themselves and with the vessel walls); statistical thermodynamics accounts for the canonical energy distribution on the ground of the ergodic principle; and, finally, a model for explaining the natural law "all men are mortal" can be "accumulation of cell duplication errors due to thermal perturbation", though such a model has never been developed in a concluded theory.

The Physico-Mathematical Development of a theory can be defined as the encoding of the model in terms of the laws of a more fundamental theory or discipline (most frequently classical or quantum mechanics, electrodynamics, classical and statistical thermodynamics). The fact that the model can be developed into a coherent theory is to be considered to *validate the model*, because it shows that the model produced to account for (or explain) some new phenomenon is in agreement with what has been previously discovered (of course, within the limits of the approximations almost inevitably done).

From a psychological point of view we naturally tend to *overestimate theories* because of the mathematical effort needed to develop them and to *underestimate models* because they can be often expressed in plain words. Notwithstanding, many scientists and science philosophers might agree that *models are the very contribution of men to scientific theories*, because they are not cast in the experimental data and cannot be inferred from them (this concept has been firstly expressed by David Hume in the 18th Century and never disproved). In other words, the model can only be produced by human imagination and ingenuity.

Modeling and Molecular Modeling. Previous considerations seem to show that there is nothing new in the concept of model and that the activity of making models (*i.e. modeling*) is the traditional way by which scientific investigation tries to explain natural phenomena. In particular, if such modeling is based on the atomistic hypothesis, it seems quite natural to call it *molecular modeling*.

Also molecular modeling is not such a novelty as can be illustrated by the models devised many years ago to account for the heat capacities of gases. A classical-mechanical model can be developed through energy-equipartition considerations to give good predictions for the heat capacities of monoatomic gases but fails for biatomic gases except at higher temperatures; a more efficient (more true?) quantum-mechanical model can be developed within the statistical thermodynamics frame to give quite good heat-capacity estimates for gases at any temperature, at least for fairly rigid molecules; and so on.

In general, we might think that all problems "which have, as a common characteristic, that *complex microscopic behavior* underlies *macroscopic effects*"[3] can be tackled by molecular modeling, and that the dimensions of the tractable problems depend only on the speed of our computers, the mathematical level of our theories and the quality of our software. In other words, we might judge that, at least in principle, the differences between 19th Century and present time modeling are more a fact of quantity than of quality. In practice, however, such quantitative differences are so impressive that we cannot avoid suspecting they will become truly qualitative. To make an example, we presently believe that epilepsy, a *macroscopic* neuronal disease, is due to an imperfect recognition of a neurotransmitter, GABA, by its receptor, an event occurring at *submicroscopic* level; accordingly, modern molecular modeling techniques are actively applied to *design* new drugs, which, because of their better submicroscopic fitting, would be suited to cure the macroscopic illness.

It is at this point, when we try to cross the barriers of complexity (from few molecules to the whole brain), of many length scale (from the molecular to the macroscopic scale), and of time (from molecular vibrations to the time of physiological events) that surprising difficulties of both theoretical and practical nature may start to arise, as it will be briefly discussed in the next section.

Trends in Molecular Modeling

The idea that natural systems can be interpreted and understood in their intimate structure and evolution on the ground of models consisting of interacting atoms and molecules (and, *ultima ratio*, of elementary particles, though this last point is far outside the limits of the present discussion) is clearly extremely ambitious and could not be pursued without an underlying scientific philosophy. As a matter of fact, there are at least three of such philosophies which, for the sake of simplicity, are indicated in the following discussion by the rather arbitrary names of *Extreme Reductionism, Complexity* and *Holistic Schools*. The Holistic School maintains the complete non-separability of nature in independent parts and does not need to be discussed here because, denying separability, rejects any possibility of molecular modeling.

A brief analysis of the other two schools is given below. It makes partial reference to a recent paper by S.S. Schweber on the crysis of physical theory[4] and to the last chapter of the book "Chemistry, Quantum Mechanics and Reductionism" by H. Primas.[1] For what we have called Complexity School, the discussion widely relies on the well known paper "More is Different" published by P.W. Anderson in 1972.[5]

The Extreme Reductionism School can be ideally associated with two famous quotations: "The supreme test of the physicist is to arrive at those universal elementary laws from which the cosmos can be built up by pure deduction" (A. Einstein, 1918); "The underlying physical laws necessary for the mathematical theory of a large part of physics and the whole of chemistry are thus completely known, and the difficulty is only that the exact application of these laws leads to equations much too complicated to be soluble" (P.A.M. Dirac, 1929).[6]

These statements lead directly to the conception that sciences can be arranged in a hierarchical order of decreasing fundamentality, such as

elementary-particles physics $\longrightarrow$ solid state or many-body physics $\longrightarrow$ chemistry $\longrightarrow$ molecular biology $\longrightarrow$ cell biology $\longrightarrow$ physiology $\longrightarrow$ psychology $\longrightarrow$ social sciences $\longrightarrow$ ...,

for which it can be stated that: *(i)* elementary entities of the less fundamental science obey the laws of the more fundamental one (*reductionism*), and *(ii)* the less fundamental science is just applied more fundamental one (*e.g.* chemistry is just applied many-body physics) (*constructionism*). It is clear that, in its extreme manifestations, this line of thought leads to the conclusion that, knowing the *first principles* of elementary particle physics, we should be able to *deduce* the laws of all other sciences; it is just a matter of computability.

In chemistry, the first principles are often shifted downwards to the principles of quantum mechanics, giving origin to what has been called "the dogma of modern numerical quantum chemistry".[1] At present, a remarkable number of scientists rely on this approach and their attitude towards molecular modeling consists in trying to predict the behavior of even more complex molecular systems by computational application of the principles of quantum and statistical mechanics. Problems dealt with by this method include, for instance, structure and dynamics of molecules and macromolecules, structure of solutions and solvent effects, molecular recognition of molecules and biomacromolecules, and reaction pathways in chemical kinetics.

The Complexity School can be summarized by the following quotation: "... the reductionist hypothesis does not by any means imply a 'constructionist' one: The ability to reduce everything to simple fundamental laws does not imply the ability to start from those laws and reconstruct the universe... The behavior of large and complex aggregates of elementary particles... is not to be understood in terms of simple extrapolation of the properties of few particles. Instead, at each level of complexity entirely new properties appear..." (P.W. Anderson, 1972).[5]

In simpler words, the points expressed in this statement can also be presented in the following way. The reductionist hypothesis can also be discussed from a philosophical point of view but it is a fact that it is accepted by the large majority of scientists. The same cannot be said for any form of 'constructivism' because it seems at least improbable that the fantastic complexity of chemistry can be deduced from the laws of many-body physics or that of biology from the more fundamental laws of chemistry. Based on considerations of such nature, a new conception of the interrelations among sciences is emerging, which can be summarized in the following schematic points:

a) there exists a reductionist chain of sciences arranged as an hierarchy of increasing complexity; in physical sciences such an increase of complexity corresponds to an enlargement of the molecular (or particle) aggregates involved, a decrease of their interaction energy, and an increase of the time scale of the relevant events;

b) more complex sciences conform to the more fundamental laws of the less complex ones but, at any level of complexity, new laws may emerge which are peculiar of that specific level;

c) laws emerging at a given level of complexity are decoupled from (and then cannot be predicted by) the laws holding at the lower level; in this case the two levels are irremediably disconnected.

Emergence and *decoupling* are the essential concepts which contribute to delineate this new "picture of the physical world that is hierarchically layered into quasi autonomous domains",[4] and it seems useful to illustrate them by means of two relatively simple examples. The first concerns temperature: being a property of molecular ensembles, it can be said that substances may have a temperature but single molecules do not.[1,7] In other words, temperature is an *emergent property* of molecular aggregates that is not shared by the lower level of organization (the single molecule). The second example concerns the "simple cases [where] the microscopic fluctuations average out when larger scales are considered, and the average quantities satisfy classical continuum equations. Hydrodynamics is a standard example of this, where atomic fluctuations average out and the classical hydrodynamic equations emerge",[3] which seems to implicate that the physics of rheological fluids is to be considered *decoupled* from any molecular model of the fluid itself.

It is clear that the attitude of the two schools towards molecular modeling must be remarkably different. In the extreme reductionist approach it was given for granted, apart from computational problems, that properties of ever more complex molecular aggregates could be inferred from the more fundamental laws holding at a lower organization level, which corresponds to a validation of molecular modeling at an epistemological level. In the complexity approach the situation is much more ambiguous because the role of molecular modeling, intended as a method of prediction and control of the higher levels of molecular organization, is endangered by the possibility of decoupling along the chain of increasing complexity. This seems to lead to the following operational conclusions or suggestion: *(i)* we must be extremely cautious in our attempts at modeling more complex systems starting from the properties of the constituting atoms; some of the most interesting properties emerging at the higher organization level could be lost because of decoupling; *(ii)* laws emerging at any complexity level can only be discovered by traditional empirical methods, that is by inference from the wider and more accurate set of experimental data collected and analysed at that specific level.

Conclusions and Introduction to Part 2

The previous discussion has shown that there are good reasons to be dubious about the real possibilities of a molecular modeling based on simple constructivist ideas. The most reasonable way of dealing with the problem seems that of promoting experimental methods able to give a new insight of the properties of molecular aggregates and decide *'a posteriori'* whether such properties are or are not predictable in terms of those of the constituent atoms.

This suggestion has been applied to the study of H-bond interactions by taking advantage of the fact that we have now at our disposal a number of crystal structure databases (see Table 1) which can be considered a giant archive of all possible inter and intramolecular interactions conceivable, H-bond included. This ensemble of databases can be therefore considered the complex system to deal with in an attempt to single out the emergent natural laws the H-bond is conforming to. The results of such an empirical analysis are described in the second part of the chapter.

Table 1. Crystal Structure Databases

Database	Containing
Cambridge Structural Database (CSD)[8]	≈ 120,000 structures of organic, organometallic and coordination compounds
Inorganic Crystal Structure Database (ICSD)[9]	≈ 32,000 structures of inorganic salts and minerals
Metal Crystallographic Data File (MCDF)[10]	≈ 11,000 structures of metals and intermetallic compounds
Protein Data Bank (PDB)[11]	≈ 650 structures of biological macromolecules

PART 2 – MODELING THE HYDROGEN BOND

Since its discovery in 1921, it has been generally believed that the expression H-bond identifies a unique class of chemical phenomena. After some seventy years it is becoming increasingly clear that such an approach has not been so fruitful because unable to give a chemical interpretation of the extreme variability of properties displayed by a same type of H-bond and well exemplified by the case of the O-H⋯O bond, for which O⋯O distances ranging from 2.36 to 3.69 Å (sum of the van der Waals radii) and experimental bond energies from 31.5 to less than 1 kcal mol^{-1} have been reported. In view of these difficulties we decided to undertake a new analysis of the H-bond problem as independent as possible of previous hypotheses and theories but based, as far as possible, on sheer inference from the extraordinary wealth of experimental data collected in the last thirty years by X-ray and neutron crystallography. The following strategy was then adopted: *(i)* X-H⋯Y interactions were divided according to the nature of the H-bond donor (X) and acceptor (Y) atoms, making a distinction between *homonuclear* X-H⋯X or Y-H⋯Y and *heteronuclear* X-H⋯Y bonds; *(ii)* for each class so identified (e.g. homonuclear O-H⋯O bond or heteronuclear N-H⋯O bond) crystal structure databases were searched for all H-bonds shorter than a prefixed cutoff value and then divided according to the chemical functionalities involved; this procedure was supposed to provide a complete list of chemical groups (and then of chemical factors) able to give rise to strong or very strong H-bonds; *(iii)* this list was analyzed with the aim of singling out a unique chemical model (if any) all strong H-bonds were conforming to; *(iv)* final results were cross-checked by the use of the available spectroscopic (in particular IR ν(X-H) stretching frequencies and ^{1}H NMR chemical shifts) and thermodynamic data.

This work has been so far completed only for the homonuclear O-H⋯O bond, though preliminary results for the heteronuclear N-H⋯O bond are also available. The present exposition summarizes the results obtained in the form of a collection of separate case studies which have contributed to put the bases of the generalized H-bond model illustrated in the last section.

Case Study 1 - H-Bonding in β-Diketone Enols: Evidence for Resonance-Assisted Hydrogen Bonding (RAHB)[12-15]

Preliminary analysis of a number of crystal structures of β-diketone enols (**I**) had shown that these compounds display an interesting behavior: they often form abnormally strong intramolecular (**Ia**) or intermolecular (**Ib**) H-bonds which are associated with an unusual π-delocalization of the O=C–C=C-OH heteroconjugated system. A research project was undertaken[12-15] to establish whether and how H-bond strengthening and π-delocalization were mutually interconnected.

In view of the type of experimental data available, the two effects needed to be quantified in terms of suitable geometrical descriptors: *(i)* the H-bond strength was measured by means of the $O \cdots O$, O-H or $H \cdots O$ distances, hereafter called $d(O \cdots O)$, d(O-H) and $d(H \cdots O)$; *(ii)* the π-delocalization by means of the differences (see **Ia**) $q_1 = d_1 - d_4$ and $q_2 = d_3 - d_2$, being $q_1 = q_2 = 0$ for a totally π-delocalized system. Alternative ways for evaluating the π-delocalization have been used,[12] that is: *(iii)* the antisymmetric vibrational coordinate $Q = d_1 - d_4 + d_3 - d_2 = q_1 + q_2$; *(iv)* the coupling parameter λ, according to which the geometrical state of the heterodiene fragment is a mixture of the keto-enolic **KE (IIa)** and enol-ketonic **EK (IIb)** forms according to $\lambda \cdot \textbf{EK} + (1 - \lambda) \cdot \textbf{KE}$; it is easy to show that $\lambda = (1 - Q/Q_0)/2$, where $Q_0 = 0.320$ Å is the value of Q when pure C–C and C–O single and double bond distances are used, and that $\lambda = 1$ for **KE**, 0 for **EK** and 0.5 for the totally π-delocalized form **IIc**.

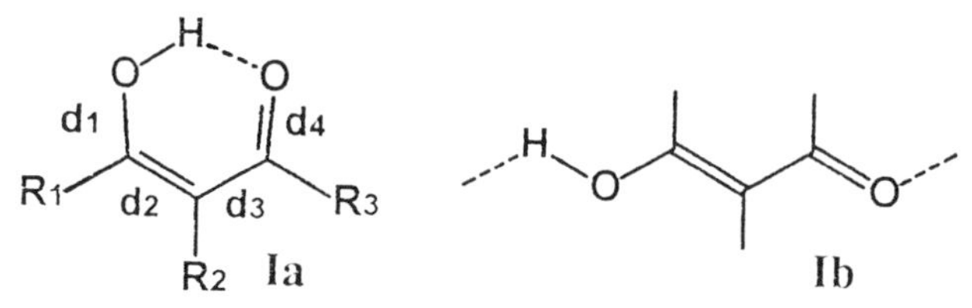

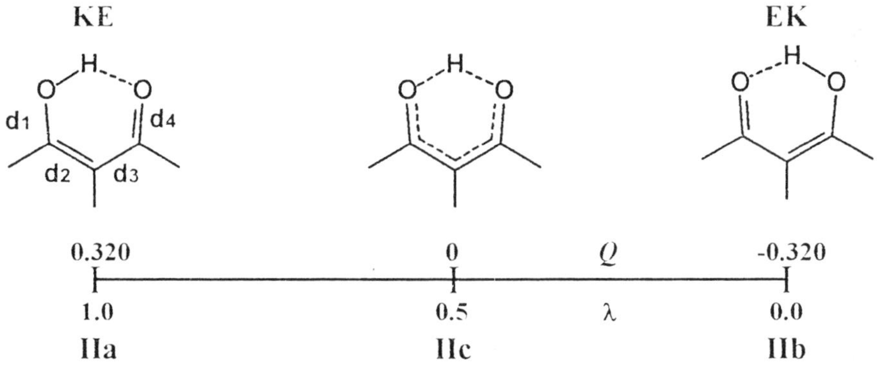

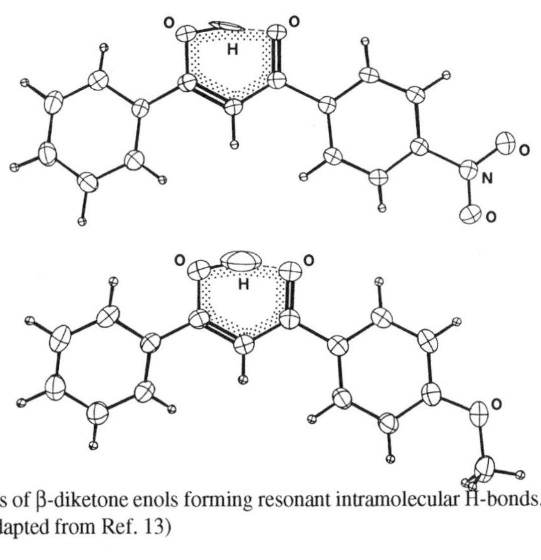

Figure 1. Two examples of β-diketone enols forming resonant intramolecular H-bonds. Resonant rings are marked by shading. (Adapted from Ref. 13)

By searching the Cambridge Structural Database[8] 54 crystal structures of β-diketone enols forming inter and intramolecular H-bonds were retrieved. Examples of such compounds

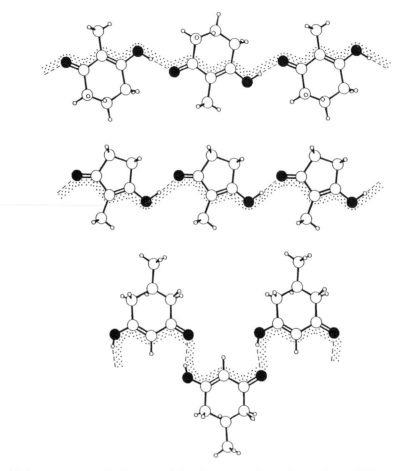

Figure 2. Some examples of β-diketone enols forming resonant intermolecular chains of H-bonds. Resonant chains are marked by shading.

are given in Figs. 1 and 2 for the intra and intermolecular cases, respectively. For each H-bond found the corresponding values of $d(\text{O}\cdots\text{O})$, $d(\text{O-H})$, $d(\text{H}\cdots\text{O})$, q_1, q_2, Q and λ were computed and represented in the multidimensional scatterplot of Fig. 3. Here the vertical axis represents the $\text{O}\cdots\text{O}$ contact distance which decreases going downwards: this direction indicates therefore the increasing strength of the H-bond. The two-dimensional plot on the left represents the concerted variations of $d(\text{H}\cdots\text{O})$ and $d(\text{O-H})$ with $d(\text{O}\cdots\text{O})$ till their progressive equalization for the shortest H-bonds. More interestingly, the three-dimensional $d(\text{O}\cdots\text{O})$ *versus* (q_1, q_2) plot shown on the right makes evident that, in β-diketone enols, very strong H-bonds (*i.e.* those having small $d(\text{O}\cdots\text{O})$ values) are inevitably associated with greatly, or totally, π-delocalized heterodienic systems (*i.e.* having small q_1 and q_2 values). An enlarged view of the $d(\text{O}\cdots\text{O})$ *versus* $Q = q_1 + q_2$ cross-section of the three-dimensional plot is reported in Fig. 4, where the range of $\text{O}\cdots\text{O}$ distances (2.77±0.07 Å) observed in polyalcohols and saccharides (non-resonant systems having $\lambda = 1.0$ or 0.0) has been added for the sake of comparison (grey rectangle in the upper-right corner). The overall plot points to an almost linear dependence of $d(\text{O}\cdots\text{O})$ on λ (or Q), that is of H-bond strengthening on π-delocalization.

It may be concluded that, at least in β-diketone enols, a *synergistic mechanism of H-bond strengthening and π-delocalization enhancement* is operating which, accordingly, can

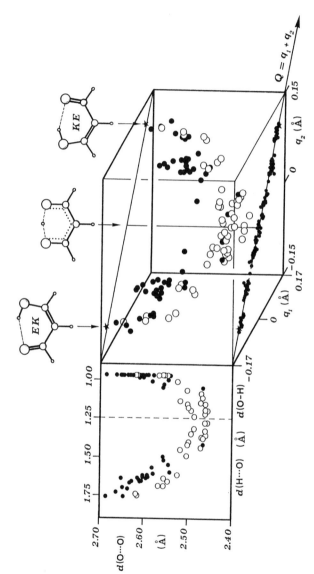

Figure 3. Five-dimensional scatterplot of the geometrical descriptors (see text) for the intra (open circles) or intermolecular (full circles) H-bonds formed by β-diketone enols.

127

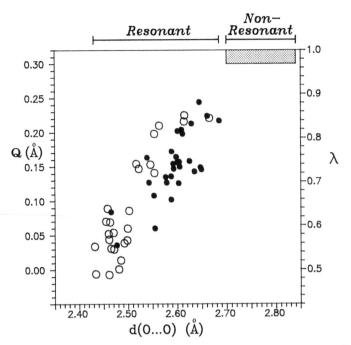

Figure 4. Scatterplot of π-delocalization parameters (Q and λ) *versus* $d(\text{O}\cdots\text{O})$ distances for β-diketone enols forming resonant intra (open circles) or intermolecular (full circles) H-bonds. Non-resonant O–H$\cdots$O bonds are indicated by the shaded rectangle.

be called *RAHB* or *Resonance-Assisted Hydrogen Bond*.[12] This putative new type of bond will be better characterized and extended to other heterodienic systems in the next sections.

Case Study 2 – Effect of Molecular Dissymmetry on the Resonant O-H$\cdots$O Bond[15]

In the previous section RAHB has been identified because of the intercorrelation between H-bond shortening and π-delocalization increase observed in β-diketone enols. No attempt has been made, however, to identify the chemical factors determining the length of each resonant H-bond and, to start with, in the present section the effects of symmetry or dissymmetry of the R_1 and R_3 substituents (see **Ia**) are discussed. The method followed consists in the comparison of structural data retrieved from the Cambridge Structural Database[8] and concerning the enols of β-diketones (**IIIa**), β-ketoesters (**IIIb**) and β-ketoamides (**IIIc**) (R and R'= alkyl or aryl) forming intramolecular H-bonds. 56 β-diketone, 5 β-ketoamide and 37 β-ketoester enols were found for which the usual parameters $d(\text{O}\cdots\text{O})$ and λ were calculated. Amides and esters have been grouped together because statistically indistinguishable.

The results are summarized in Table 2 and in the histograms of Fig. 5. The distributions of $d(\text{O}\cdots\text{O})$ (Fig. 5a) and coupling parameter λ (Fig. 5b) are clearly bimodal, showing that H-bonds in esters and amides are longer and their systems of conjugated double bonds less

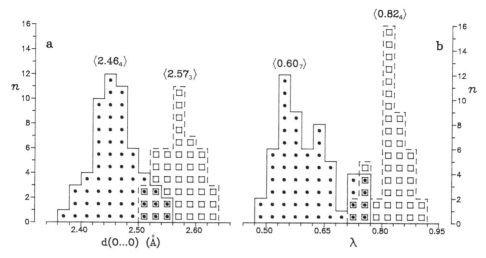

Figure 5. Histograms of O···O contact distances (**a**) and coupling parameters λ (**b**) for a series of intramolecularly H-bonded β-diketone (full points) and β-ketoesters and β-ketoamides (open squares) enols. Average values and e.s.d's are given at the top of each distribution.

delocalized than in ketones. Table 2 reports some average values for the distributions of Fig. 5 together with the estimates of the intramolecular H-bond energies calculated by the semiempirical Lippincott and Schroeder's method.[16] The average value for β-diketone enols is 10 ± 2 kcal mol^{-1}, while that for β-ketoesters and amides assumes the smaller value of 5.4 ± 0.8 kcal mol^{-1}.

It may be concluded that the *dissymmetry of the substituents produces a weakening of the O-H···O bond*. This is a first indication that such bond may have a relevant three-centre-four-electron covalent component, as it will be discussed in more detail in the last section.

Table 2. Ranges and average values of the O···O contact distances (Å), coupling parameters λ and H-bond energies calculated by the Lippincott and Schroeder formula[16] (kcal mol^{-1}) characterizing the intramolecular H-bond in β-diketone, β-ketoester and β-ketoamide enols (n =sample dimension)

$d(O \cdots O)$		λ		E_{HB}[a]	
Range	Average	Range	Average	Range	Average
β-diketone enols (n = 57):					
2.37-2.59	2.46(4)	0.49-0.77	0.60(7)	16.1-5.1	9.6(2.0)
β-ketoesters (n = 37) and β-ketoamide enols (n = 5):					
2.50-2.64	2.57(3)	0.74-0.90	0.82(4)	7.9-3.7	5.4(8)

[a]Calculated for an average O-H···O angle of 149°.

Case Study 3 - Length of the Resonant Chain and Strength of the Resonant O-H···O Bond[15]

All resonant diketone enols can be reduced to the general formula $\cdots O=R_n\text{-}OH\cdots$, where R_n (n odd) is a *resonant spacer* consisting of a chain of alternating single and double bonds. Such resonant spacer is then represented by R_1 for carboxylic acids, which form both dimers (**IVa**) and chains (**IVb**) of resonant nature, R_3 for β-diketone enols (**Va** and **Vb**), and R_5 and R_7 for the δ- and ζ-diketone enols illustrated in **VI** and **VII**, respectively.

A systematic investigation carried out[15] on the crystal structures[8] of compounds belonging to different R_n classes leads to the $d(O \cdots O)$ distribution displayed in the bar-chart of Fig. 6 which appears to indicate that the O-H$\cdots$O bond becomes increasingly shorter while the resonant spacer R_n becomes longer.

Table 3 collects the geometrical parameters of the strongest H-bonds observed for each R_n class together with their energies estimated by the Lippincott and Schroeder method.[16] Such energies are plotted in Fig. 7 as a function of the order n of the resonant spacer R_n.

Two conclusions can be drawn: *(i) resonant O-H$\cdots$O bonds may reach energies as large as some 20 kcal mol^{-1}*, which are about five times greater than the accepted value for the water dimer; *(ii) the H-bond energy increases with the length n of the resonant spacer R_n till the length of $n = 5$ and then remains constant*. Present results can be interpreted in terms of *increasing π-delocalizability of the heteroconjugated system* with the increasing length of the resonant spacer involved.

Table 3. Values of $O \cdots O$ contact distances (Å), O-H$\cdots$O angles (°) and H-bond energies calculated by the Lippincott and Schroeder formula[16] (kcal mol^{-1}) for the shortest O-H$\cdots$O bonds observed in each class of non-charged resonant systems, arranged for increasing length of the resonant spacer R_n. Estimated non-resonant limits given for comparison

	n	$d(O \cdots O)$	O-H$\cdots$O	E_{HB}
ζ-diketone enols	7	2.431	171	18.7
δ-diketone enols	5	2.425	177	20.3
β-diketone enols (intra)	3	≈2.39	≈150	14.8
β-diketone enols (inter)	3	2.465	176	15.5
carboxylic acids (chains and dimers)	1	≈2.62	≈180	6.7[a]
non-resonant	0	≈2.70	≈180	4.7[b]

[a]Experimental thermodynamic values in the range 7-8 kcal mol^{-1} and
[b]3.2-4.0 kcal mol^{-1}.

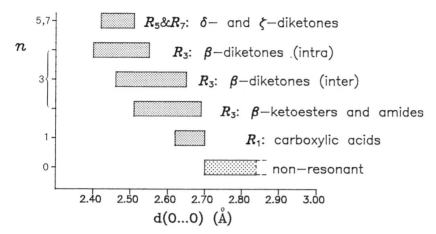

Figure 6. Intervals of O⋯O distances in inter and intramolecular resonant O-H⋯O bonds arranged according to the length of the resonant spacer R_n (non-resonant O-H⋯O bonds interval added for comparison).

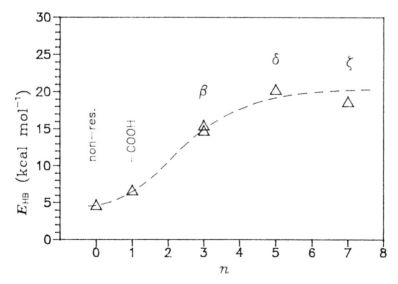

Figure 7. Energy of the strongest H-bond observed in each class of resonant diketone enols as a function of the length n of the resonant spacer R_n (data from Table 3).

Case Study 4 – RAHB Generalization to Other Heteroconjugated Systems

RAHB has been identified in connection with the O-H···O bond but can be easily extended to other H-bonded structures. A generalized system able to form resonant-assisted H-bonds can be written as ···A=R_n-DH···, where A and D are the H-bond donor and acceptor atoms, respectively, and R_n is the resonant spacer previously defined. The most interesting cases, particularly as biochemistry and molecular biology are concerned, arise by substituting oxygen by nitrogen. Some examples are shown in schemes **VIII-X**. Substitution in β-diketone enols (or enolones, **VIIIa**) leads to enaminones (**VIIIb**), enaminoimines (**VIIIc**), and enolimines (**VIIId**), while substitution in the carboxylic acid dimer (**IXa**) can produce the amide dimer (**IXb**) or the amide- amidine complex (**IXc**).

VIIIa	**VIIIb**	**VIIIc**	**VIIId**
ENOLONE	ENAMINONE	ENAMINO-IMINE	ENOLOIMINE

			X
IXa	**IXb**	**IXc**	AMIDE CHAIN
CARBOXYLIC ACID DIMER	AMIDE DIMER	AMIDE-AMIDINE COMPLEX	

Rather interestingly, the amide-amidine complex (**IXc**) plays an important role in the base-pairing of the double-helix structure of DNA and is, in fact, present in both thymine-adenine (**XIa**) and cytosine-guanine (**XIb**) pairing. In this last case cytosine and guanine are also connected by a third N-H···O bond, which is also resonant and closes a much wider cycle of conjugated double bonds. Similar considerations can be made for the H-bonds determining the secondary structure of both fibrous and globular proteins. α-Helices (**XIIa**) accomodate three nearly parallel and homodromic chains of ···O=C-NH··· groups (see **X**) which are obviously resonant, while β-pleated sheets (**XIIb**) display a similar but antidromic pattern. It has been already remarked that "nature itself may have taken advantage of the greater energy of RAHB to keep control of molecular associations whose stability is essential to life."[14]

Case Study 5 – Length of Resonant Chain and Strength of the N-H···O Bond[18]

A generic resonant chain of N-H···O bonds can be represented by the symbol ···O=R_n-NH···, where R_n is the already defined resonant spacer of length n. A systematic analysis of the occurrence of such chains in molecular crystals[8] has been recently carried out,[18] leading to the identification of the typical patterns of resonant H-bonded systems indicated by the shaded and dotted lines of Fig. 8.

A summary of the N···O distances observed in N-H···O bonded chains having different lengths of the R_n spacer is given in the bar-chart of Fig. 9. Its strict similarity with the parallel

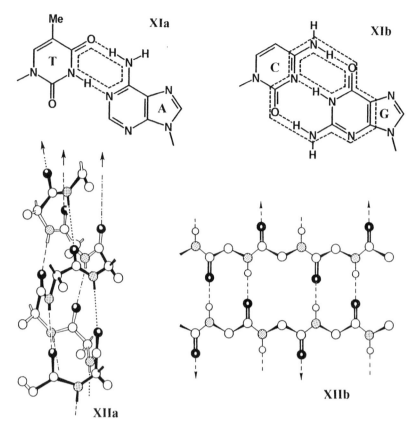

XIa

XIb

XIIa

XIIb

bar-chart obtained for the O-H···O bonds (Fig. 6) seems to indicate that: *(i) the RAHB mechanism is effective for the N-H···O as it was for the O-H···O bonds* as shown by the fact that, in both cases, resonant H-bonds are systematically shorter than non-resonant ones; *(ii) very short resonant systems*, such as amides (R_1 in Fig. 9), *are less susceptible to RAHB strengthening*, so paralleling previous results obtained for carboxylic acids.

Case Study 6 – A General Model for the Strong Homonuclear O-H···O Bond[17]

It has been already remarked in the foreword to Part 2 that the best procedure for seeking a new comprehensive model for the H-bond is probably that of making *tabula rasa* of all previous ideas and try to infer such a model from a set of experimental data as complete, accurate and reliable as possible.

For the specific aim of finding out a general model that all strong O-H···O bonds are conforming to, experimental data analysed were of different nature (structural, spectroscopic, thermodynamic, etc.) but, for the sake of simplicity, only the structural ones are discussed here. These have been obtained from both organic[8] and inorganic[9] crystallographic databases selecting all molecular structures containing O-H···O bonds shorter than a cutoff value of 2.70 Å. Mostly neutron diffraction structures were considered because of their much higher precision in locating protons; a number of accurate X-ray structures were added for those chemical classes not covered by neutron data. Parameters taken into account were O···O, O-H and H···O distances and the coupling parameter λ. The latter was calculated for a generic X_1-O-H···O=X_2 bond as $\lambda = (1 + q/q_0)/2$, where $q = d(X_1 - O) - d(X_2 = O)$ and q_0 is the same quantity but calculated from pure simple and double bond distances. Compounds for which strong and very strong O-H···O bonds occur were found to belong to a limited num-

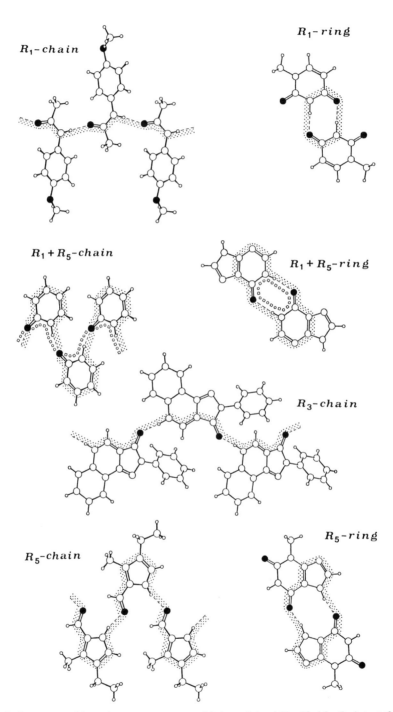

Figure 8. A summary of the main resonant structures (chains and rings) identified for the intermolecular N-H···O bond in crystals. Resonance patterns are marked by shading or by dotted lines. (Adapted from Ref. 14).

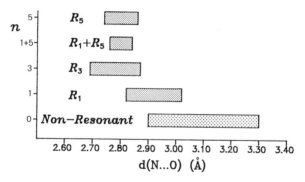

Figure 9. Intervals of N$\cdots$O distances in resonant chains of N-H$\cdots$O bonds arranged according to the length of the resonant spacer R_n (non-resonant H-bonds interval added for comparison). (Adapted from Ref. 14).

ber of chemical classes which are summarized in Table 4 together with the number of structures found for each class (distinguishing neutron, N, and X-ray, X, data) and their ranges of $d(O\cdots O)$ and λ values. The most representative of them are illustrated in Figs. 10 and 11 by their structural formulas.

A systematic analysis of the data collected leads to the following main conclusions:

1. all short H-bonds found (Table 4) can be grouped in three main classes:

   Class A: $-O-H\cdots{}^-O-$ or *negative charge-assisted hydrogen bonding, (-)CAHB*;

   Class B: $=O\cdots H^+\cdots O=$ or *positive charge-assisted hydrogen bonding, (+)CAHB*;

   Class C: $-O-H\cdots O=$ where the two oxygens are interconnected by a system of conjugated double bonds, or *resonance-assisted hydrogen bonding, RAHB*;

2. there is a nearly linear intercorrelation between $d(O\cdots O)$ and λ (Fig. 12), very short O-H$\cdots$O bonds being inevitably associated with $\lambda \approx 0.5$, that is with a situation for which X_1-O and X_2=O distances tend to become identical;

3. most accurate neutron diffraction data show that the strengthening of the H-bond causes a progressive equalization of the O-H and H$\cdots$O distances (Fig. 13) which become identical at the limit of $d(O\cdots O) \approx 2.40$ Å and $d(O-H) = d(H\cdots O) = 1.20$ Å;

4. conclusions 2 and 3 can be joined by saying that all strongest X-O-H$\cdots$O=X bonds display a condition of almost perfect symmetry as far as both X-O and O-H distances are concerned.

A First Conclusion: Covalent Nature of the Strong Hydrogen Bond. It is generally believed that the H-bond is an essentially electrostatic interaction, which is most certainly true for weak and dissymmetrical bonds. The progressive equalization of both O-X and O-H distances associated with the shortening of the H-bond seems to suggest, however, that *very short O-H$\cdots$O bonds are rather to be considered totally delocalized three-centre-four-electron covalent bonds*, that is VB linear combinations $\Psi = a_I\Psi_I + a_{II}\Psi_{II}$ of the resonant forms

$$-O-H\cdots O= \longleftrightarrow -O^-\cdots H-O^+=$$
$$(I) \qquad\qquad (II)$$

which are isoenergetic and then mix with identical a_i coefficients. This is to say that *only chemical situations for which the resonant forms I and II may become energetically, and then*

CLASS A: $O-H--O^-$

CARBOXYLIC ACID-CARBOXYLATES

KH BIS(TRIFLUOROACETATE)

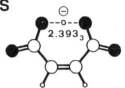

IMIDAZOLIUM MALEATE

METAL OXIMES

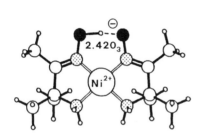

BIS(2-AMINO-2-METHYL-3-
BUTANONE OXIMATO)Ni(II)
CHLORIDE

ALCOHOL-ALCOHOLATES

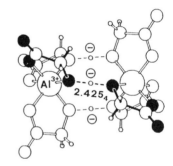

Na_3H_3BIS[TRIS(GLYCOLATO)
ALUMINATE(III)]

INORGANIC ACID SALTS

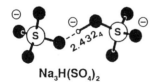

$Na_3H(SO_4)_2$

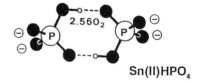

Sn(II)HPO_4

Figure 10. Some of the most representative structures containing strong O-H$\cdots$O bonds according to the classification given in Table 4.

CLASS B: O−−H⁺−−O

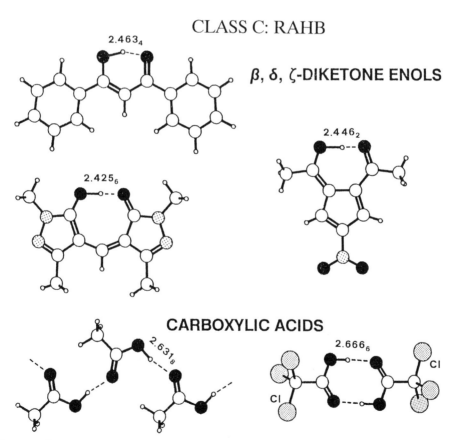

CLASS B: O−−H⁺−−O

2.420₅

TRANS-[H(Me₂SO)₂]
[Rh(III)Cl₄(Me₂SO)₂]

2.41₁

H BIS(PYRIDINE N-OXIDE)
TETRACHLOROAURATE(III)

2.430₃

[V(III)(H₂O)₆][H₅O₂](CF₃SO₃)₄

CLASS C: RAHB

2.463₄

β, δ, ζ-DIKETONE ENOLS

2.425₆

2.446₂

CARBOXYLIC ACIDS

2.631₈

2.666₆

Cl

Cl

Figure 11. Some of the most representative structures containing strong O-H···O bonds according to the classification given in Table 4.

Table 4. Summary of the chemical classes where strong or very strong O-H···O bonds can be observed and their typical ranges of O···O distances (Å) and coupling parameter λ. Class D (non-resonant and non-charged H-bonds) is reported for comparison. HB= i(inter), I(intra); Sample= number of neutron (N) and X-ray (X) structures taken into account

	Class	HB	Sample	d(O···O) range	λ range
A1a.	Carboxylic acid-carboxylates	i	9N, 4X	2.44-2.49	0.50-0.68
A1b.	Carboxylic acid-carboxylates	I	7N	2.39-2.42	0.53-0.60
A2a.	Metal oximes	I	6N, 3X	2.39-2.48	0.56-0.61
A2b.	Metal glyoximes	I	5X	2.44-2.69	0.54-0.78
A3.	Alcohol-alcoholates	i	1N, 1X	2.39-2.43	0.5[a]
A4.	Water-hydroxyl	i	2X	2.41-2.44	0.5[a]
A5.	Inorganic acid salts	i	9N	2.36-2.43	n.c.[b]
B.	O···H+···O	i	3N, 4X	2.36-2.43	0.5[a]
C1a.	β-Diketone enols	I	1N,10X	2.43-2.55	0.51-0.72
C1b.	β-Diketone enols	i	2N,14X	2.46-2.65	0.56-0.76
C2.	β-Ketoester or ketoamide enols	i	1N, 9X	2.55-2.69	0.64-0.88
C3.	δ-Diketone enols	I	2X	2.42-2.44	0.52-0.53
C4.	ζ-Diketone enols	I	4X	2.43-2.51	0.51-0.53
C5a.	Carboxylic acids (chains)	i	1N, 6X	2.62-2.70	0.74-0.84
C5b.	Carboxylic acids (dimers)	i	4N, 6X	2.62-2.67	0.68-0.83
D.	Alcohols and saccharides	any	-	2.77±0.07	1.0

[a] Assumed for symmetry; [b] λ not computable.

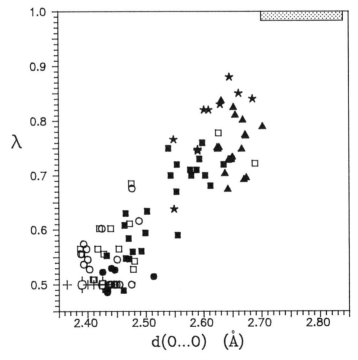

Figure 12. Scatter plot of the coupling parameter λ versus d(O···O) distance for all chemical classes of Table 4. Class symbols: A1= open circles; A2= open squares; A3 and A4= larger open circles; B= crosses; C1= full squares; C2= stars; C3 and C4= full circles; C5= triangles; D= shaded upper right rectangle. (Reproduced by permission from Ref. 17).

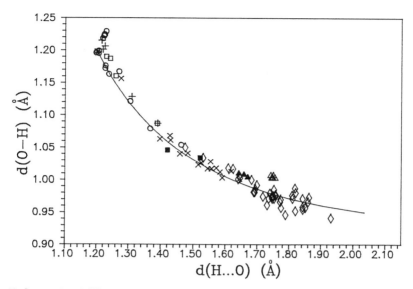

Figure 13. Scatter plot of d(O-H) versus d(H$\cdots$O) distances for selected neutron diffraction data. Class symbols: A1= open circles; A2= open squares; A5= diagonal crosses; B= vertical crosses; C1= full squares; C5= full triangles; Alcohols, saccharides, non-resonant acids and aminoacids= open lozenges; ice Ih= open triangles. (Adapted by permission from Ref. 17).

chemically, equivalent can lead to such very strong H-bonds. It is not difficult to show that such chemical equivalence can be achieved only in the three possible ways displayed in Fig. 14:

A) by adding an electron and giving so rise to $-O\text{-}H\cdots{}^-O-$ bonds;
B) by removing an electron, so producing $=O\cdots H^+\cdots O=$ bonds;
C) by connecting the two oxygens by a chain of conjugated double bonds.

It is evident that the three classes of bonds so defined do correspond to those identified by the empirical analysis of the structural data carried out before, *i.e.*, in the order, (-)CAHB, (+)CAHB and RAHB, confirming beyond any reasonable doubt the simple covalent model of the strong H-bond proposed above.

General Conclusions. Previous considerations can be summarized by saying that, while O$\cdots$O distances shorten from some 2.90-2.80 to 2.40 Å, the H-bond changes from a weak and dissymmetrical O-H$\cdots$O interaction of electrostatic nature to a strong, symmetrical and covalent O-H-O bond and that such a situation of symmetry can be actually achieved only when, in a generic X_1-O-H$\cdots$O=X_2 interaction, the perfect equivalence of the X_1-O and X_2-O bonds is effectively achievable. These findings can be generalized to any H-bond by saying:

1) in any X-H$\cdots$Y bond *the H-bond strength is mainly determined by the degree of symmetry achievable* on the two sides of the hydrogen atom;

2) since the main source of dissymmetry is the difference between X and Y, *homonuclear X-H$\cdots$X or Y-H$\cdots$Y bonds are predicted to be generally stronger than the heteronuclear X-H$\cdots$Y ones;*

3) *the strongest homonuclear X-H$\cdots$X bonds will be associated with a symmetrical substitution on the two sides of the X atoms or, more in general, with a substitution pattern making identical the proton affinities of the H-bond donor and acceptor atoms.*

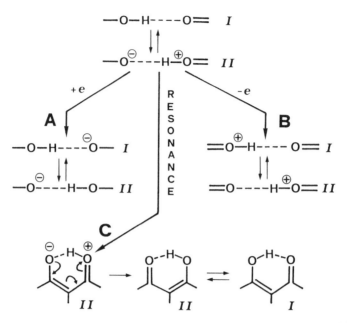

Figure 14. The three ways for making energetically equivalent the two resonant form *I* and *II*. (Reproduced by permission from Ref. 17).

ACKNOWLEDGMENTS

Financial support for this work by Italian Ministry for the University and Scientific and Technological Research (MURST, Rome) is gratefully acknowledged. The authors are indebted to Mr. Stefano Gilli for arranging the LATEX version of this text.

REFERENCES

1. H. Primas. "Chemistry, Quantum Mechanics and Reductionism", Springer-Verlag, Berlin (1983).
2. K.R. Popper. "The Logic of Scientific Discovery", Hutchinson, London (1959).
3. K.G. Wilson, *Rev. Mod. Phys.* 55:583 (1983).
4. S.S. Schweber, *Physics Today* November Issue:34 (1993).
5. P.W. Anderson, *Science* 177:393 (1972).
6. P.A.M. Dirac, *Proc. Roy. Soc. London* A123:713 (1929).
7. H. Primas, *J. Math. Biol.* 4:281 (1977).
8. a) F.H. Allen, S. Bellard, M.D. Brice, B.A. Cartwright, A. Doubleday, H. Higgs, T. Hummelink, B.G. Hummelink-Peters, O. Kennard, W.D.S. Motherwell, J.R. Rodgers and D.G. Watson, *Acta Crystallogr. Sect. B* 35:2331 (1979); b) F.H. Allen, J.E. Davies, J.J. Galloy, O. Johnson, O. Kennard, C.F. Macrae, E.M. Mitchell, G.F. Mitchell, J.M. Smith and D.G. Watson, *J. Chem. Inf. Comput. Sci.* 31:187 (1991).
9. G. Bergerhoff, R. Hundt, R. Sievers and I.D. Brown, *J. Chem. Inf. Comput. Sci.* 23:66 (1983).
10. L.D. Calvert, *Acta Crystallogr. Sect. A* 37:C343 (1981).
11. F.C. Bernstein, T.F. Koetzle, G.J.B. Williams, E.F. Jr. Meyer, M.D. Brice, J.R. Rodgers, O. Kennard, T. Shimanouchi and M. Tasumi, *J. Mol. Biol.* 112:535 (1977).
12. G. Gilli, F. Bellucci, V. Ferretti and V. Bertolasi, *J. Am. Chem. Soc.* 111:1023 (1989).
13. V. Bertolasi, P. Gilli, V. Ferretti and G. Gilli, *J. Am. Chem. Soc.* 113:4917 (1991).
14. G. Gilli, V. Bertolasi, P. Gilli and V. Ferretti, *Acta Crystallogr. Sect. B* 49:564 (1993).
15. P. Gilli, V. Ferretti, V. Bertolasi and G. Gilli, A Novel Approach to Hydrogen Bonding Theory, in: "Advances in Molecular Structure Research", Vol. 2, M. Hargittai and I. Hargittai eds., JAI Press Inc, Greenwich, CT, USA (1995), in the press.

140

16. a) E.R. Lippincott and R. Schroeder, *J. Chem. Phys.* 23:1099 (1955); b) R. Schroeder and E.R. Lippincott, *J. Phys. Chem.* 61:921 (1957).

17. P. Gilli, V. Bertolasi, V. Ferretti and G. Gilli, *J. Am. Chem. Soc.* 116:909 (1994).

18. V. Bertolasi, P. Gilli, V. Ferretti and G. Gilli, *Acta Crystallogr. Sect. B* 51: (1995), in the press.

MOLECULAR ELECTROSTATIC PROPERTIES FROM X-RAY DIFFRACTION DATA

Tibor Koritsánszky

Institute for Crystallography
Free University Berlin
Takustr. 6
14195 Berlin
Germany

INTRODUCTION

Since its discovery, X-ray diffraction has been developed into the most powerful experimental technique of structural research. Characteristic of its recent advances is the increase
in resolution and accuracy, which makes it possible to study supermolecular structures on the
one hand, and to gain information on small molecules at the electronic level, on the other hand.
The phenomenon of X-ray scattering by crystals served as one of the fundamental experimental foundations for the development of quantum mechanics, based on which a new discipline
has emerged and taken an important part of today's chemistry. This is computational chemistry with the main goal of providing relations between energy, structure and chemical behavior of molecules. The link between these physical observables is provided by the wavefunction occurring as a by-product of the numerical procedure of solving stationary Schrödinger
equations. The technical problem of approximating molecular wavefunctions has been connected from the beginning with the theoretical problem of interpreting them. These efforts led
to the development of the molecular orbital theory which has found steady use in modelling
chemical events. The larger the system considered, the less feasible this method becomes and
the stronger the demand is for an approach that can bypass the calculation of the wavefunction. The corresponding theorem is by Hohenberg and Kohn[1] who proved that the ground
state energy is a unique functional of the electron density (ED). It gives a theoretical basis
for relating properties of chemical interest to each other through an observable property, the
charge density. The subjects of the first part of this chapter are the distribution of charge and
its fundamental properties in characterizing a molecular system.

It is evident that quantum chemistry not only provides but also needs structural information. It is less known, however, that the interpretation of a diffraction experiment, as the
main source of structural data, has to make use of assumptions which are directly not deducible
from the experiment and which are necessarily based on quantum theory. A rather general outline on the kinematic theory of diffraction is given in the second paragraph. This is followed
by a short description of the models commonly used for the ED in crystals. Some technical

questions concerning the fitting procedure for estimating the parameters of the model are also discussed.

The chapter is completed by presenting experimental results on small organic molecules, some of which are based on recent measurements interpreted by the most advanced methods.

THE MOLECULAR ELECTRON DENSITY

The electron density $\rho(\mathbf{r})$ is obtained from the many-electron ground-state wavefunction by summing over the spin coordinates and integrating over the spatial coordinates of all the electrons but one $(\mathbf{r}')$:

$$\rho(\mathbf{r}) = \sum_{spin} \langle \Psi | \Psi^* \rangle_{r'} \tag{1}$$

If it is normalized to the total number of electrons it gives the probability for finding electronic charge in an infinitesimal volume element $d\mathbf{r}$. For molecules in the Born-Oppenheimer[2] approximation the electronic (ϕ) and nuclear (χ) wavefunctions can be decomposed and a stationary state can be assigned to each nuclear configuration $\mathbf{R}$. The corresponding ED is a parametric function of the actual spatial coordinates of the nuclei. $\rho(\mathbf{r}, \mathbf{R}_0)$ derived at the equilibrium molecular geometry $(\mathbf{R}_0)$ is referred to as the static electron density.

The ED is of fundamental importance in modern quantum chemistry. On account of the Hellmann-Feynman theorem[3] the electrostatic aspects of the chemical bond can directly be evaluated from it. Hohenberg and Kohn[1] showed that a unique functional relationship exists between the ED and the non-degenerate ground state energy. Based on this theorem numerous density functional methods have been developed and successfully applied even for larger molecular systems.

In Bader's approach[4] the topological analysis of the molecular ED leads to a new interpretation of the chemical bond, a classification of the molecular structure, the determination of all static and reactive properties of the system, and to a quantum mechanical definition of an atom in a molecule.

The Topology of the Electron Density

The ED is considered to be a continuous smooth scalar field defined in three-dimensional space. Its analysis determines the regions of its decrease and increase and locates its extrema where its gradient vanishes. In the theory of "Atoms in Molecules" such stationary points are called *critical points* (CP) which can be characterized by two numbers (ω, s), where ω is the rank being equal to the number of non-zero principal curvatures, the eigenvalues of the Hessian tensor at the CP, and s is the signature being the algebraic sum of the signs of these curvatures. If all curvatures are negative/positive the ED is a local maximum/minimum at the CP in question ((3,-3)/(3,+3) CPs). If two curvatures are negative/positive the ED is a maximum/minimum at the CP in the plane given by their corresponding principal axes and it is a minimum/maximum along the third axis ((3,-1)/(3,+1) CPs).

More details can be revealed by mapping the gradient vector field of the ED. Figure 1 displays such a trajectory map for the plane of the cyclopropane ring. The neighboring point to any given point of such a curve (trajectory), also called gradient path, is obtained by an infinitesimally small displacement in the direction of the gradient vector taken at the given point. As the gradient defines one and only one direction, only one path can be assigned to each point (where the gradient is non-zero) or, in other words, two paths cannot cross each other.

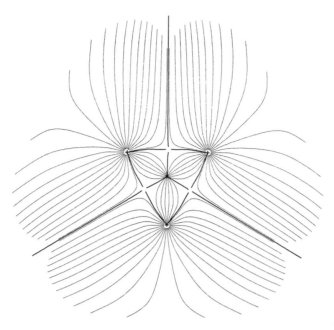

Figure 1. The gradient trajectory map of the electron density of cyclopropane in the plane of the ring

As the gradient points in the direction of greatest increase of the scalar field, a path originates at a point where the function exhibits a minimum at least in one direction. Thus the CPs, where trajectories can only terminate, are the positions of the nuclei where the ED exhibits its local maxima. This property of a (3,-3) CP makes a nucleus a point *attractor* in the gradient field. Analogously, a (3,+3) or *cage* CP can only be an origin of trajectories. Inside the three-membered ring in Figure 1 all paths originate at the (3,+1) or *ring* CP and terminate either at the nuclei or at a (3,-1) or *bond* CP. The later kind of CPs are also the termini of all the trajectories outside the ring which originate at infinity. The graph of the ring is given by those trajectories which originate at the bond CP and terminate at the nuclei, each of these pairs of trajectories form an atomic *interaction line* or a *bond path*. The network of the bond paths linking pairs of neighboring nuclei defines the *molecular graph*.

The special trajectories which terminate at a bond CP span a surface, called the *interatomic surface*, which separates two neighboring atoms. Two of these surfaces for the cyclopropane ring define an open subspace containing a nucleus, a region which is associated to only one attractor and called an *atomic basin*.

It can be demonstrated that the topological analysis of the ED leads not only to the reconstruction of the molecular structure and all of its elements but also to the characterization of atomic interactions, and of chemical bonds. A widely applicable description is provided by local topological properties related to the value of the ED and its principal curvatures at the bond CPs. The former quantity is a sensitive figure of the strength of a bond and an expression for the bond order can be derived from it. The ratio of the negative curvatures of the ED at the bond CP, a quantity giving the extent of departure from cylindrical symmetry, reflects the π character of the bond. A further useful quantity is the length of the bond path which is a direct measure of the bent character of a bond.

The Hessian (the matrix of the second derivatives) of the ED has three scalar invariants of which its trace, known as the *Laplacian*, is also of great importance in characterizing atomic interactions. This function is negative or positive where electronic charge is locally concentrated or depleted, respectively. It is more structured than the ED and provides a faithful map-

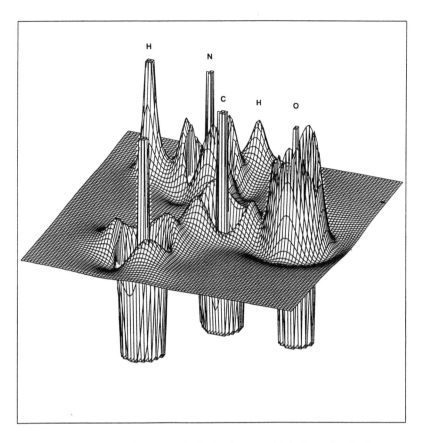

Figure 2. Relief plot of the negative Laplacian distribution for acetamide in the molecular plane

ping of the slight changes in the ED due to bond formations. The topology of the Laplacian is characteristic of a given atom. The region of the outer shell of an atom over which the Laplacian is negative is called the *valence shell charge concentration* (VSCC). The uniform VSCC for an isolated atom gets distorted upon bond formation and its resulting topological structure, given by the (3,-3) CPs in number and location, is in accord with the Lewis and valence shell electron pair repulsion models. To each local maximum in the VSCC a pair of bonded or nonbonded electrons can be assigned. In Figure 2 the negative Laplacian distribution of the acetamide molecule is displayed. Important features are the local maxima in the VSCC of the atoms along the bonds. These *bonded charge concentrations* form contiguous regions between the bonded atom pairs. The two additional maxima in the VSCC of the oxygen atom, the *non-bonded charge concentrations*, correspond to the lone-pair electrons. The concentration of electronic charge in the interatomic surface (relatively large values of the ED and the negative Laplacian at the bond CP) is typical for *shared interactions* (covalent bonds). In the case of *closed-shell interactions* the charge is contracted towards the nuclei leading to a region of positive Laplacian between the nuclei.

The entries in Table 1 can serve as an example for characterizing typical covalent bonds. Here the local topological properties for different carbon-carbon bonds in selected hydrocarbons are summarized.

Table 1. Bond-Topological Parameters (MP2/6-31+G**)[5] in selected hydrocarbons[a]

		C–C		C=C	C≡C	C(r)–C(r)	
Ethane	ρ_b	1.69					
	L_b	−16.0					
	ε	0.0					
Ethylene	ρ_b			2.33			
	L_b			−26.6			
	ε			0.41			
Acetylene	ρ_b				2.69		
	L_b				−27.9		
	ε				0.0		
1,3-Butadiene	ρ_b	1.94		2.31			
	L_b	−20.9		−26.1			
	ε	0.08		0.41			
Propene	ρ_b	1.74		2.32			
	L_b	−16.8		−26.1			
	ε	0.01		0.43			
Cyclopropane	ρ_b					1.66	
	L_b					−12.8	
	ε					0.52	
Vinyl-cyclopropane	ρ_b	1.86		2.42		1.65	1.66
	L_b	−19.1		−28.5		−12.6	−12.8
	ε	0.07		0.43		0.51	0.52
Methyl-vinyl-cyclopropane	ρ_b	1.76	1.87	2.30		1.62	1.62
	L_b	−17.3	−19.4	−25.6		−12.1	−12.1
	ε	0.04	0.07	0.45		0.51	0.51

[a]Units: ρ_b [$e/\text{Å}^3$], L_b [$e/\text{Å}^5$]

The trend shown in the values of both the ED (ρ_b) and its negative Laplacian (L_b) at the bond CP is of chemical significance. As expected, the amount of electronic charge accumulated midway between atoms forming a covalent bond is related to the strength of the bond. The bond ellipticity, defined in terms of the negative principal curvatures ($\varepsilon = \lambda_1/\lambda_2 - 1$), is highest for the bond in the cyclopropane ring and lowest for the single bond in ethane and for the triple bond in acetylene. The charge transfer from an unsaturated bond to the adjacent single bond (conjugation) results in an increase/decrease in the ED, as well as its Laplacian, at the CPs of the adjacent single/double bonds relative to their magnitude in the "isolated" bonds. The π character of this charge delocalization manifests itself in nonzero ellipticity induced in the single bond by the double bonds as illustrated by the bond parameters of 1,3-butadiene, vinyl- and cis-methyl-vinylcyclopropane. The latter two cases indicate that the cyclopropyl group is an effective π-donor like a double bond. Propene is an example for hyperconjugation. Here the H_3C–C bond exhibits partial double bond character transmitted from the adjacent π-donor.

The Laplacian is extensively used to explain reactive properties of molecules. The most intense local charge concentrations/depletions in a molecule are the regions where nucleophilic/electrophilic attacks are most likely to take place.

X-RAY DIFFRACTION AND THE ELECTRON DENSITY

According to the kinematical theory of scattering (first Born approximation),[6] the total (elastic and inelastic) intensity is

$$I_{tot} \sim \sum_m W_m \sum_n |F_{nm}|^2 \tag{2}$$

where

$$F_{nm} = \langle \Psi_n | \sum_j \exp(i\mathbf{Hr}_j) | \Psi_m \rangle \tag{3}$$

is the scattering amplitude (form factor), Ψ_n and Ψ_m are state functions of the crystal representing the initial $|n\rangle$ and final $|m\rangle$ states, W_m is the probability of being in a state $|m\rangle$ (Boltzmann factor), $\mathbf{H}$ is the Bragg vector with integral components h,k,l, relative to the reciprocal axes $\mathbf{a}^*, \mathbf{b}^*, \mathbf{c}^*$. The summation in eq. (3) is taken over all electronic coordinates. To make expression (2) applicable, further simplifications have to be introduced . It is assumed that all excited states are pure vibrational states, i.e. the system stays in an electronic ground state (ϕ_0). In this case the elastic form factor for state $|m\rangle$ in the Born-Oppenheimer approximation is

$$F_{mm} = \langle \chi_m | F_0(\mathbf{H}, \mathbf{R}_0) | \chi_m \rangle \tag{4}$$

where

$$
\begin{aligned}
F_0(\mathbf{H}, \mathbf{R}_0) &= \langle \phi_0 | \sum_j \exp(i\mathbf{Hr}_j) | \phi_0 \rangle \\
&= \int \rho_0(\mathbf{r}, \mathbf{R}_0) \exp(i\mathbf{Hr}) d\mathbf{r}
\end{aligned}
\tag{5}
$$

is the Fourier transform of the ground state ED at the equilibrium nuclear configuration $\mathbf{R}_0$.

The canonical ensemble average over all states (eq. 2) reduces to an average over pure vibrational states. This thermal average can be given in a closed form within the harmonic-convolution approximation. The total ED is assumed to be a superposition of density units (ρ_k), each of which is rigid in the sense of perfectly following the motion of the nucleus (k) it is attached to

$$\rho_0(\mathbf{r}, \mathbf{R}_0) = \sum_k \rho_k(\mathbf{r} - \mathbf{R}_{k0}) \tag{6}$$

It can be seen[7] that at the low temperature limit ($h\nu > kT$, for all vibrational modes) the elastic scattering intensity (I_{el}) is reduced to that of Bragg scattering (I_0)

$$I_{el} \sim |F_0(\mathbf{H})|^2 = |\sum_k F_k(\mathbf{H})|^2 = I_0 \tag{7}$$

where

$$F_k(\mathbf{H}) = f_k(\mathbf{H}) \exp(2\pi i \mathbf{HR}_{k0}) \exp\left(-\frac{1}{2}\mathbf{H}'\mathbf{U}_k\mathbf{H}\right) \tag{8}$$

is the Fourier transform of the kth rigid density unit (pseudoatom). It includes the static scattering factor of the pseudoatom (f_k), a phase shift related to its distance from the origin and the Debye-Waller temperature factor with U_k being the mean-square displacement amplitude tensor of nucleus k. $F_0(\mathbf{H})$ is known as the generalized structure factor and usually introduced as the Fourier transform of the thermally smeared ED. Indeed, a direct space equivalent of equation (8) is a convolution of the static pseudoatomic density with the probability distribution function describing the motion of the kth nucleus in thermal equilibrium:

$$\rho(\mathbf{r}) = \sum_k \int \rho_k(\mathbf{r} - \mathbf{r}_{k0} - \mathbf{u}_k) P(\mathbf{u}_k) d\mathbf{u}_k \tag{9}$$

where P is a normal distribution

$$P(\mathbf{u}) = (2\pi)^{-\frac{3}{2}} (\det \mathbf{U})^{-\frac{1}{2}} \exp\left(-\frac{1}{2}\mathbf{u}'\mathbf{U}^{-1}\mathbf{u}\right) \tag{10}$$

Thus, the diffraction experiment is interpreted as purely elastic scattering from the average crystal structure, since this can be shown to be the main component of the averaged scattering.

The thermally averaged ED can be obtained from a given set of measured structure factors by a Fourier summation:

$$\rho(\mathbf{r}) = \frac{1}{V} \sum_{\mathbf{H}} F_0(\mathbf{H}) \exp(-2\pi i \mathbf{H}\mathbf{r}) \tag{11}$$

This expression shows what is meant by the measurability of the ED: only its thermal average can be approximated by a Fourier summation of a finite number of terms, each of which is affected by experimental errors and by the uncertainty of the phase to be assigned to it.

Modelling the Electron Density

Since the direct evaluation of the ED is subject to these severe limitations, another method has become feasible which bypasses the Fourier calculation. This procedure involves modelling of the ED (a direct-space parametrization of ρ) and optimizing its parameters by adjusting the model-predicted structure factors to those measured. For the optimization and error estimation the method of least squares is applied.

The Atomic Electron Density

For a single Slater determinant atomic wavefunction composed of orthogonal spin-orbitals (Φ_i), the electron density is given by

$$\rho = \sum_i n_i |\Phi_i|^2 \tag{12}$$

where n_i's are the orbital occupation numbers (1 or 2). An orbital is composed of a radial and an angular part. The former is expanded in terms of basis functions (O_{lj}), the latter is represented by spherical harmonics (y_{lm}):

$$\Phi_{lm} = \left[\sum_j C_{lj} O_{lj}\right] y_{lm} \tag{13}$$

149

The corresponding orbital density is

$$\rho_{lm} = \left[\sum_{j,k} D_{jk} O_{lj} O_{lk}\right] y_{lm} y_{lm} = R_l y_{lm} y_{lm} \tag{14}$$

The spherical harmonics form a complete basis set, thus their product can be expanded in terms of spherical harmonics:

$$y_{lm} y_{lm} = \sum_{L,M} C_{LlMm} y_{LM} \tag{15}$$

where C_{LlMm}'s are the Clebsch-Gordon coefficients[8].

Making use of this relationship, eq. (12) can be rearranged with respect to l and m leading to the total atomic ED:

$$\rho = \sum_l R_l \sum_m P_{lm} y_{lm} \tag{16}$$

It follows that the orbital product representation of the atomic density is equivalent to an expansion over spherical harmonics (multipole expansion). The corresponding form factor is[9]

$$f(\mathbf{H}) = \sum_l \langle J(H)_l \rangle \sum_m P_{lm} y_{lm}(\mathbf{H}/H) \tag{17}$$

where $\langle J(H)_l \rangle$ is the lth-order Fourier-Bessel transform of R_l :

$$\langle J(H)_l \rangle = 4\pi i^l \int j_l(2\pi H r) R_l(r) r^2 dr \tag{18}$$

with j_l being the lth order spherical Bessel function.

Conventional Formalism

The zero-order function ($l = 0$) in eq. (17) is the conventional scattering factor which corresponds to the isolated, spherical atomic density. In Figure 3 such a function is displayed for the carbon atom derived from its ground-state Hartree-Fock wavefunction.[10] The total scattering factor is the occupation weighted sum of the Fourier-Bessel transforms of different orbital products:

$$f(\mathbf{H}) = 2\langle J(H)_0 \rangle_{1s} + 2\langle J(H)_0 \rangle_{2s} + 4\langle J(H)_0 \rangle_{2p} \tag{19}$$

The conventional model disregards the static deformations and thus the chemical bonding. Consequently, a least-squares refinement based on it leads to a bias in the variables. Errors in positional parameters, often called "asphericity shifts", usually manifest themselves in significantly shorter bond distances relative to the neutron values. The accuracy of the thermal parameters is even more doubtful, as the anisotropic displacements can "absorb" density deformations.

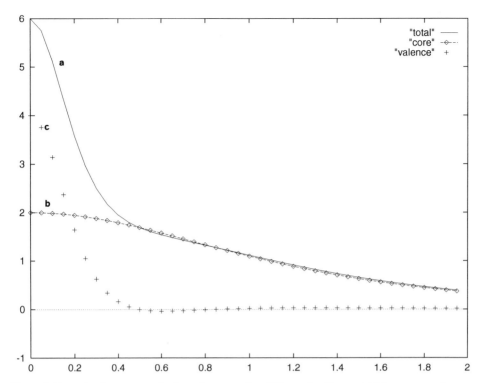

Figure 3. Scattering factor curves for the carbon atom from HF wavefunction: a: total, b: core, c: valence

One of the simplest methods for overcoming the inadequacy of the isolated atom model is the so-called "high-order refinement"[11] . In Figure 3 both the core and the valence contributions to the total scattering factor are depicted. This example illustrates clearly that in the atomic regions where the electron density is less affected by the bonding, the isolated atom model can be expected to be a fair approximation. The sharp density peaks localized at the nuclei (core density) have appreciable contributions to reflections at high Bragg angles where the scattering by the more diffuse valence or bond density is negligible. For this reason a refinement emphasizing the high-order data is expected to yield atomic parameters less biased by the inadequacy of the spherical-atom model. The success of this method lies in the appropriate selection of the cutoff value of H_{min} which is usually underestimated when obtained on the basis of the free valence scattering factors. An optimal angle separation is usually troublesome.

The accuracy of the parameters can significantly be increased by accounting for the valence asphericities in the atomic scattering factors and making use of all measured data.

The Aspherical-atom Formalism

To account for the density deformations due to chemical bonding several methods have been developed and applied.[12,13] One of the most successful refinement techniques is based on the nucleus-centered finite multipole expansion of the ED . The formalism developed by Hansen & Coppens[14] is outlined below. The aspherical atomic electron density $\rho(\mathbf{r})$ is divided into three components:

$$\rho(\mathbf{r}) = \rho_c(r) + P_v \rho_v(\kappa r) + \rho_d(\kappa' \mathbf{r}) \tag{20}$$

where ρ_c and ρ_v are the spherical core and valence densities, respectively, and

$$\rho_d(\kappa' \mathbf{r}) = \sum_l R_l(\kappa' r) \sum_m P_{lm} y_{lm}(\mathbf{r}/r) \tag{21}$$

is the term accounting for valence deformations and being formally equivalent to that of eq. (16). The y_{lm} are density normalized real spherical harmonics:

$$|y_{lm}| = 1 \text{ for } l = 0 \text{ } mbox and \text{ } |y_{lm}| = 2 \text{ for } l > 0 \tag{22}$$

The isolated-atom valence density and the radial functions R_l are modified by the screening constants (κ and κ', respectively) to account for the radial expansion or contraction of the valence shell.

The corresponding scattering factor is:

$$f(\mathbf{H}) = f_c(H) + P_v f_v(H/\kappa) + \sum_l \langle J(H/\kappa')_l \rangle \sum_m P_{lm} y_{lm}(\mathbf{H}/H) \tag{23}$$

The core and spherical valence density can be calculated from the Hartree-Fock atomic wavefunctions[10] expanded in terms of Slater-type basis functions:

$$O_l = [(2^{n(l)})!]^{-\frac{1}{2}} (2z_l)^{n(l)+\frac{1}{2}} r^{n(l)} e^{-z_l r} \tag{24}$$

where the z_l are energy optimized orbital exponents. The radial functions of the deformation density are also taken as simple Slater functions:

$$R_l(r) = \frac{(\alpha_l^{n(l)+3})}{(n(l)+2)!} r^{n(l)} \exp(-\alpha_l r) \tag{25}$$

with $n(l) \geq l$ to obey Poisson's equation[13] and with values for α_l as deduced from single-zeta wavefunctions. Besides the conventional parameters, the P_v, P_{lm}, κ and κ' are the variables of the least-squares procedure. The angular dependence of the valence deformation density is described in local frames centered at each atomic site. The number of parameters to be refined can be considerably reduced by imposing local symmetry in agreement with the connectivity and hybridization. The symmetry restrictions for real spherical harmonics are given in.[15] Chemically equivalent atoms can be constrained to have the same deformation density.

TOPOLOGICAL ANALYSIS OF EXPERIMENTAL ELECTRON DENSITIES

In this section some applications of chemical interest are given for experimental static EDs based on the rigid pseudoatom model, the parameters of which were extracted from low-temperature, high-resolution X-ray diffraction data. The topological properties of the experimental ED are compared to those obtained from wavefunctions at different levels of theory.[5] Interesting aspects of such studies are the transferability of these properties, the effects of first neighbors and of substituents, the reliability of the experimental Laplacian function in showing the effect of the crystal field and in revealing "closed shell" interactions, and last but not least, the temperature dependence of the accuracy. In the course of these studies special care was taken to maintain as much as possible similar experimental conditions and refinement strategies. Table 2 shows some characteristic data of the molecules for which results are presented. The data collections for bullvalene, serine and the crownether complexes were carried out with a SIEMENS four-circle single-crystal diffractometer equipped with a nitrogen gas stream device. In all other cases the measurements were performed with a HUBER four-circle single-crystal diffractometer using a double-stage, closed-cycle He refrigerator for the cooling system.

The resolution of the data (its extent in the reciprocal space) is typical for MoK_α radiation. Symmetry equivalent reflections were always measured and the internal merging indices for their averaging were typically lower than 2%.

Table 2. Experimental and refinement conditions

	T(K)	NREF/NVAR	$\sin\theta/\lambda_{max}$	R_w	Neutron	C/A[a]
Bullvalene	100	33	1.05	0.023	Yes	C
Acetamide	100	21	1.03	0.027	Yes	A
	23	21	1.03	0.019	Yes	
Serine	100	29	1.07	0.015	Yes	C
Aspartic acid	20	15	1.08	0.020	No	C
18-Crown-6· 2 cyanamide	100	13	1.08	0.015	Yes	C
18-Crown-6· K^+ N_3^-	120	16	1.06	0.018	Yes	C

[a]C/A means centric/acentric

The refinements and topological analyses of the data were carried out with XD[16], a computer program package developed recently. In all cases the function of

$$\sum_{\mathbf{H}} w_{\mathbf{H}}(|F_{obs}(\mathbf{H})| - k|F_{cal}(\mathbf{H})|)^2$$

was minimized using the statistical weight of

$$w_{\mathbf{H}} = \frac{1}{\sigma^2(|F(\mathbf{H})|)}$$

Only those structure factors were included in the calculation which met the criterion of $|F(\mathbf{H})| > 3\sigma(|F(\mathbf{H})|)$. The indices of the final fit provided in Table 2 are given in terms of the weighted R-factors

$$R_w = \left[\frac{\sum_{\mathbf{H}} w_{\mathbf{H}} ||F_{obs}(\mathbf{H})| - k|F_{cal}(\mathbf{H})||^2}{\sum_{\mathbf{H}} w_{\mathbf{H}} |F_{obs}(\mathbf{H})|^2} \right]^{\frac{1}{2}}$$

The hydrogen atoms were treated in the following way. Their positional and thermal parameters were fixed at the values obtained by neutron diffraction, when such data were available (see Table 2). An overall scaling of the neutron anisotropic displacement parameters (ADP) was applied to account for the temperature difference between the two data collections, when it was necessary. In the lack of neutron data the ADPs of the hydrogen atoms can be estimated by fitting the rigid-body or segmented rigid-body[17] model to the motion of the non-hydrogen atoms. The thermal parameters, obtained in such a way, are then fixed during the refinement of charge density and all other conventional parameters. If significant changes occur in the ADPs of the non-hydrogen atoms, the procedure can be repeated until convergence is reached. The density asphericity of a hydrogen atom was represented by a bond-directed dipole for which the population was refined together with the charge and the positional parameters. For those involved in strong hydrogen bonds, an additional quadrupole (Y_{20}) was introduced. The contracted function given by Stewart[18] was used as the monopole scattering factor.

The valence deformation of the non-hydrogen atoms were described up to the hexadecapolar level in the expansion. To decrease the number of variables to be refined local site symmetries were adopted and constraints on the valence deformations for atoms of the same type were applied. These restrictions are based on preconceptions on the hybridization or the chemical equivalence of the atoms to be considered. The constraints were either introduced in a stepwise manner or systematically released in subsequent cycles of refinement. The change in the least-squares residuals, the significance of the obtained populations and the examination of the correlation coefficients served as the basis for accepting or rejecting a restriction. In critical cases the residual maps, evaluated for the two models in question, were compared and the results were judged by the significance of density deformations left after the different refinements. In this way it was possible to maintain relatively high ratios of observed reflections to refined variables.

The Carbon–Carbon Bonds in Bullvalene

The bond topological properties discussed earlier bear significance only in relation to idealized bonds, and their analysis should include the comparison of results obtained for different systems. Such a comparison is straightforward if it is based on wavefunctions. In the case of experimental ED the data collection for a reference system would require different conditions. In addition, properties characteristic of the isolated molecules are not expected to be preserved in the crystal. The effect of the crystal field is less important for neutral molecules lacking nonbonded valence electrons such as hydrocarbons. An ideal situation occurs when bonds to be compared are formed within the same molecule. Bullvalene (tricyclo-[3.3.2.0$^{2.8}$]deca-2,5,8-triene)[19] is such an example having four types of C–C bonds (Figure 4): single, double, bent bonds and an additional single bond with a strong conjugation from the neighboring π bonds.

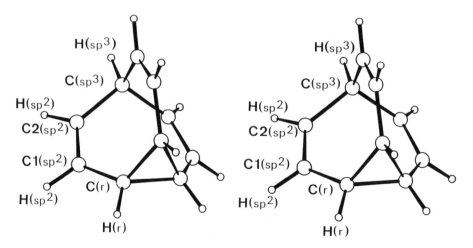

Figure 4. Stereo plot of the bullvalene molecule.

In the refinement model the C_{3v} molecular symmetry was adopted. Accordingly, four types of chemically different carbon atoms (C(r), C1(sp^2), C2(sp^2) and C(sp^3)) were considered, each of them with the appropriate constraint for the corresponding local symmetry (m, m, m, C_{3v}).

The ab-initio calculations were performed at the HF and MP2 level of theory, utilizing the standard basis sets of 6-31+G** and 6-311++G($3df, 3dp$), in the former and that of 6-31+G**, in the latter case. The neutron diffraction data averaged to C_{3v} symmetry were taken as starting parameters for the optimization. The bond distances obtained (Table 3) show the need either for the MP2 level of theory or for the use of diffuse and multiple polarization functions in the HF optimization. The good agreement between the stationary (theoretical) and the thermally averaged geometry (experimental) is important for the direct comparison of local topological properties obtained by different methods.

Table 3. Ab-initio Optimized and Experimental C–C Bond Distances for Bullvalene

	HF	MP2	X-ray	Neutron
6-31+G**	6-311++G($3df,3pd$)	6-31+G**	Multipole	
C1(sp^2)=C2(sp^2)				
1.3255	1.3369	1.3529	1.345(2)	1.342(1)
C(r)–C1(sp^2)				
1.4831	1.4845	1.4677	1.473(2)	1.473(1)
C2(sp^2)–C(sp^3)				
1.5192	1.5244	1.5102	1.516(2)	1.516(1)
C(r)–C(r)				
1.5178	1.5366	1.5326	1.535(2)	1.533(1)

Table 4 gives a comparison of the results in terms of different bond properties.

Table 4. Topological Parameters for the C–C Bonds in Bullvalene[a]

	HF		MP2	X-ray
	6-31+G**	6-311++G($3df,3pd$)	6-31+G**	Multipole
C1(sp^2) = C2(sp^2)				
ρ_b	2.44	2.30	2.31	2.36(2)
L_b	−28.2	−25.9	−25.2	−24.8 (1)
ε	0.44	0.41	0.43	0.30
C(r)–C1(sp^2)				
ρ_b	1.86	1.81	1.91	1.95(2)
L_b	−18.5	−17.2	−19.6	−19.8 (1)
ε	0.07	0.07	0.08	0.10
C2(sp^2)–C(sp^3)				
ρ_b	1.76	1.69	1.79	1.79(1)
L_b	−16.9	−15.3	−17.5	−16.2 (1)
ε	0.02	0.02	0.02	0.04
C(r)–C(r)				
ρ_b	1.62	1.51	1.57	1.59(1)
L_b	−11.7	−9.5	−11.0	−8.3 (1)
ε	0.50	0.54	0.47	0.90

[a]Units: ρ_b [$e/Å^3$], L_b [$e/Å^5$], $\varepsilon = \lambda_1/\lambda_2 - 1$

155

The theoretical parameters show only a moderate basis set dependence, and also only a slight change occurs when the electron correlation is included at the MP2 level. The quantities deduced from the experimental ED are averages calculated according to the assumed symmetry. The general agreement between theory and experiment is excellent for ρ_b reflecting the well known feature that the experimental method has its highest accuracy in the bonding areas. The trend shown in the strength of the different bonds is as expected on the basis of simple orbital theory. The double bond is the strongest with the highest charge concentration, while the bonds forming the cyclopropane ring are the weakest with the least charge accumulation in the bonding area. The latter bond is more elongated perpendicular to the interatomic vector than the former. There are significant differences in the topological properties of the two formal single bonds; the bond next to the cyclopropane ring ($C(r)$–$C1(sp^2)$) appears to be stronger and it exhibits more π character than that formed by the $C2(sp^2)$ and $C(sp^3)$ atoms. This is in accordance with theoretical and experimental observations on the conjugation effect that is caused by the unsaturated character of the cyclopropane ring[20].

The bond path lengths are practically equivalent to the geometrical bond lengths for all bonds but that of the $C(r)$–$C(r)$ in the cyclopropane ring indicating its bent character. Even in this case there is only a slight difference between the two quantities. The displacement of the experimental deformation density (spherical atomic densities are subtracted from the molecular ED) peak from the bond center (Figure 5) is found to be 0.12 Å which gives an exaggerated estimation for the bent character of the bond if compared to the corresponding distance of the bond CP in the total ED (0.02 Å).

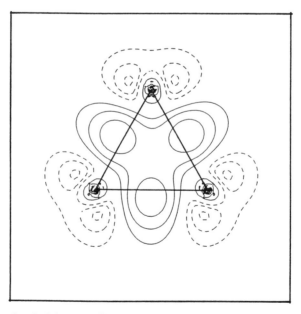

Figure 5. Experimental static deformation ED map in the plane of the cyclopropane ring. Contour intervals are 0.1 e/Å^3, positive solid, negative dashed.

Figure 6 displays relief plots of the negative Laplacian in the plane of the cyclopropane ring and in the plane containing the atoms of the ethylenic wing. These plots demonstrate how the experimental Laplacian function compares in detail with that derived from theory. Characteristic of a covalent interaction is the continuous negative area between a bonded atom pair, which is clearly reproduced by the experimental function for all bonds. The locations of bond charge concentrations (the maxima in the negative Laplacian) are in good agreement, especially for the C–C bonds.

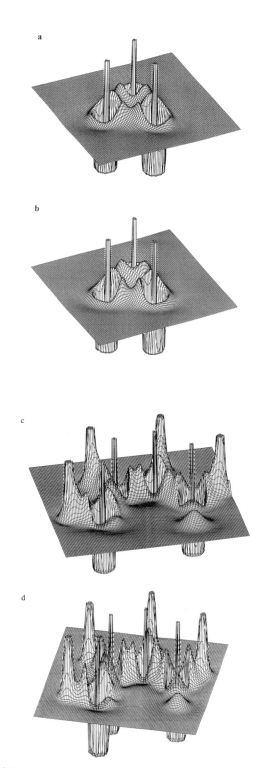

Figure 6. Relief maps of the negative Laplacian functions in the plane of the cyclopropane ring (a,b) and of the ethylenic wing (c,d). (a,c) experimental, (b,d) theoretical.

The Effect of Temperature on the Experimental Electron Density: A Study on Acetamide

As argued above, lowering the temperature improves the kinematical conditions and reduces the effect of thermal smearing. The results are increased, resolvable and relatively accurate intensities at high scattering angles. The increased number of observed reflections leads to "statistically balanced" data and in the weighted refinement the high-angle reflections become as important as those measured at low angle.

In the acetamide molecule for the nitrogen and oxygen atoms $mm2$ while for the carbon atom in the methyl group $3m$ site symmetries were adopted[21]. Two types of hydrogen atoms were assumed: one for the NH_2 and one for the CH_3 groups. The discrepancies in the geometrical parameters obtained by X-ray and neutron diffraction are not significant but the ADPs based on the former experiment are much higher than those observed by neutron diffraction. To account for this difference the neutron ADPs were scaled by fitting an overall thermal parameter against the X-ray data.

In Figure 7 the two experimental model deformation ED maps are compared. The largest differences between the experimental maps appear in the vicinity of the nuclei. These discrepancies are mainly due to the differences in the monopole populations. Significantly different valence charges are obtained depending on which data set is used. These differences can be attributed to the changes in the monopole populations of the hydrogen atoms. This indicates that the ADP's of the hydrogen atoms were underestimated.

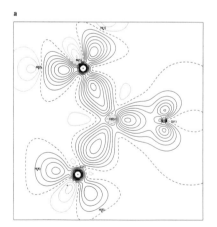 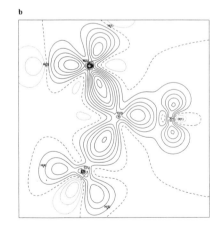

Figure 7. Static deformation ED maps for acetamide: a. experimental (100K), b. experimental (23K). Contour intervals are 0.1 e/Å^3.

One way to reveal the gain in significance of the ED by lowering the temperature is to compare the error distributions obtained from the two data sets. In Figure 8 such a comparison is made in terms of the relief plots of the corresponding error maps. They turn out to be topologically equivalent, with sharp peaks at the nuclei and with less pronounced bonding areas.

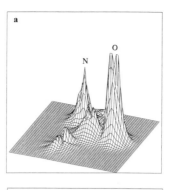

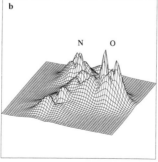

Figure 8. Relief plot of the error function of the experimental EDs of acetamide based on: a. 100K and b. 23K data.

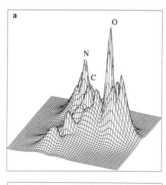

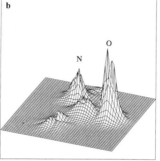

Figure 9. Relief plot of the error function of the Laplacian distribution of acetamide based on: a. 100K and b. 23K data.

The ED extracted from the 23K data is clearly shown to be more accurate than that based on the 100K measurement. The comparison of the Laplacian densities in Figure 9 shows even more pronounced differences.

Table 5 compares local topological properties for the different bonds in acetamide. The theoretical values were obtained at HF level with different standard basis functions and at the experimental molecular geometry.

Table 5. Theroretical and experimental bond topological properties for acetamide[a]

| | HF | | | X-RAY / MULTIPOLE | |
	6-31+G**	6-311+G**	6-311+G(3df,3dp)	23K	100K
C–C					
ρ_b	1.83	1.83	1.81	1.82(1)	1.90(2)
L_b	−18.9	−19.3	−17.9	−14.8 (1)	−18.9 (2)
C=O					
ρ_b	2.90	2.65	2.68	2.67(5)	2.79(8)
L_b	6.2	−8.6	−18.4	−26.2 (5)	−33.2 (6)
C–N					
ρ_b	2.22	2.29	2.33	2.52(5)	2.42(8)
L_b	−19.6	−18.3	−28.8	−29.8 (4)	−24.6 (6)

[a]Units: ρ_b $[e/\text{Å}^3]$, $L_b[e/\text{Å}^5]$

An interesting observation is that the ED and the Laplacian at the bond CPs for the carbon-heteroatom bonds show considerable basis set dependence. However, there is a clear tendency for these properties to approach their experimental values as the wavefunction is approaching the HF limit. This is especially pronounced for the C=O bond, in which case the lowest basis set does not even reveal the expected covalent character of the bond. The differences in the experimental Laplacian values suggest model inadequacies in the refinement, which are most likely due to insufficient deconvolution of the thermal smearing. This problem is often encountered, especially for acentric structures.

Preliminary Results for two Amino Acids

As part of our ongoing project to derive experimental electrostatic properties for amino acids, data have been collected for D,L-serine and aspartic acid at 100K and 20K, respectively. In spite of the difference in the temperature maintained during the measurements the results obtained in terms of global and local topology of the EDs are comparable. Table 6 lists properties at the bond CP for selected bonds for both molecules obtained from ab-initio, as well as experimental static densities.

Concerning the theoretical values the same conclusion can be drawn as for acetamide; high angular basis functions are needed to approach the experimental result. The values based on the 20K data for aspartic acid are uniformly higher than those obtained from the 100K experiment for serine. It is doubtful, at the present stage of the analysis, whether these discrepancies are of chemical nature (not supported by the ab-initio data) or due to the weakness of the thermal motion model. The result of the study on aspartic acid certainly suffers from the lack of accurate neutron diffraction data. In the comparison one should not neglect the effects of the hydrogen bonded environment and the crystal field, which are considerably different for the two structures.

Table 6. Topological properties for selected bonds in serine and aspartic acid[a]

	Serine			Aspartic acid	
	6-31G**	6-311++G(3df,3dp)	X-ray	HF/6-31G**	X-ray
C=O					
ρ_b	2.58	2.71	2.66(9)	2.62	2.96(7)
L_b	−7.63	−32.2	−34.3 (4)	−2.5	−48.2 (4)
C=O					
ρ_b	2.59	2.81	2.71(9)	2.64	2.96(6)
L_b	−3.84	−37.7	−37.3 (3)	−4.15	−45.6 (4)
OOC–C					
ρ_b	1.77	1.76	1.95(2)	1.78	2.12(5)
L_b	−18.8	−16.5	−21.4 (1)	−18.9	−25.7 (1)
C–C					
ρ_b	1.814	1.77	1.77(2)	1.74	2.12(5)
L_b	−18.24	−16.9	−16.1 (1)	−16.7	−22.7 (2)

[a]Units: $\rho_b\ [e/\text{Å}^3]$, $L_b[e/\text{Å}^5]$

ELECTROSTATIC POTENTIAL

One of the most important properties directly deducible from the ED is the electrostatic potential (EP):

$$V(\mathbf{r}) = \sum_j \frac{Z_j}{|\mathbf{r} - \mathbf{R}_j|} - \int \frac{\rho(\mathbf{r})}{|\mathbf{r} - \mathbf{r}'|} d\mathbf{r}' \qquad (26)$$

where $\mathbf{R}_j$ and Z_j are the position and charge of the jth nucleus, respectively. Its negative gradient at a given point gives the force acting on a positive point charge placed at that point. Regions in space with negative/positive potential are attractive/repulsive for a positive point charge. This function has a widespread use in molecular modelling, recognition and design. Interacting molecules can be described by representing one of them with its potential and the other with its charge density. A detailed mapping of the EP can help in interpreting and predicting chemical reactivity, molecular solvation and crystal packing. Given a multipole representation of the ED the EP can be approximated by an expansion over multipole moments of individual pseudoatoms[22]. Beyond the Van der Waals envelope this approximation has proven to be appropriate and several applications have been reported[23,24]. The results presented here are based on the method of Su and Coppens[25], which does not rely on such an approximation and which allows to evaluate the EP at any position in space.

Figure 10 shows relief plots of the experimental EP for 18-crown-6 molecules extracted from their crystal environments. One of them forms an ionic complex with potassium[26] and the other one encloses two cyanamide molecules as neutral guests.[27] The calculations are based on experimental EDs extracted from 100K X-ray data under very similar refinement conditions. The maps are generated for the least-squares plane through the ring atoms and include the contributions of all pseudoatomic densities, up to the hexadecapolar level of the multipole expansion.

Both maps visualize the form of the cavity (negative range of the potential) where the cation is to be captured but for the neutral compound this basin is less extended and flatter than that found for the ionic molecule (without the contribution from the potassium cation).

The comparison clearly reveals the polarization induced by the cation on the field of the ring in the macrocation. It is important to note that this difference cannot be seen in the charge density which exhibits very similar features for the two complexes.

a

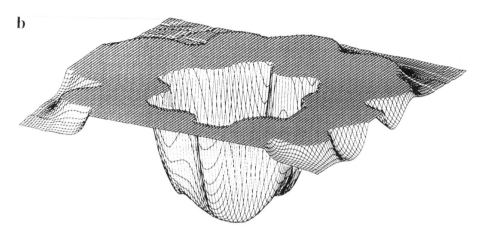

b

Figure 10. Relief maps of the EP in the least-squares plane of the 18-crown-6 rings extracted from their crystal environment: a. neutral complex, b. ionic complex without the contribution of the cation.

Figure 11 compares experimental EP contour maps in the plane of the $^-OOC–CN^+$ fragments for serine and aspartic acid derived from their static multipole densities discussed above. The dominant features of both maps are the negative/positive regions around the COO^- / NH_3^+ groups. The comparison of these contour diagrams shows a high degree of topological equivalence.

a

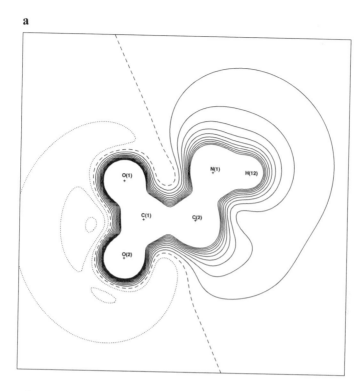

b

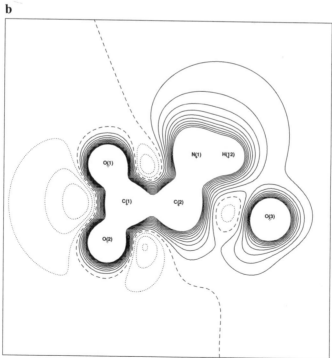

Figure 11. Contour maps of the experimental electrostatic potential for a. serine and b. aspartic acid. Contour intervals are 0.1 e/Å, negative dotted, positive solid and zero dashed.

163

MOLECULAR ELECTRIC MOMENTS

The classical electrostatics has proven to be a feasible approach to the estimation of interaction energy at large separations of the interacting molecules. A widely used approximation to calculate the molecular electrostatic energy is based on a multipole expansion in which the electric moments are involved. The components of the mth rank electric multipole moment can be derived from the ED:

$$\mu_{ijk}(m) = \int \rho(\mathbf{r}) x^i y^j z^k d\mathbf{r} \tag{27}$$

with $i + j + k = m$.

X-ray diffraction data are important sources of experimental electric moments. Spackman[28] gives an excellent review of such studies.

The dipole moment (first order multipole moment) is one of the most important molecular properties and can easily be obtained by direct integration of the experimental ED described in terms of pseudoatoms. We note that the diffraction experiment provides all three components of the first moment. The entries in Table 7 serve to illustrate, through a couple of examples of small organic molecules, the capacity of the method in comparison to other techniques as well as to theoretical results. With a few exceptions, the values given by X-ray diffraction are larger than those obtained in liquid or gas phase by other techniques or by theoretical calculations. The enhancement of molecular dipole moments in crystals is a quite general phenomenon which is an inductive effect and expected to be more pronounced for hydrogen bonded systems.

Table 7. Estimated values for the dipole moments (10^{-30} Cm) of selected organic molecules

Molecule	X-ray	Other experiment	HF/6-31G**
Water	$5.3 - 8.1$	6.19	7.29
Formamide	16.1(17)	12.8(1)	14.18
Parabanic acid	11.7(60)		8.92
L-alanine	43.0(23)	41.0	41.4
Uracil	14.7(43)	13.9(1)	16.22
p-Nitropyridine-N-oxide	1.3(33)	2.3(1)	1.00
Acetamide	20.7(15)		15.16
Serine	43.7(31)		46.03
Aspartic acid	33.0(28)		38.43

The effect of the crystal field is expected to be reflected also in the values of second moments (Table 8). Molecules are not only polarized upon intermolecular interactions but also contracted in certain directions. The reliability of the diffraction estimates of second moments seems to be very sensitive to the treatment of the thermal motion. This is mainly because of the similarity between the quadrupole deformation density and the probability distribution function of harmonic vibration. Anisotropic displacement parameters based on neutron diffraction should be incorporated into the multipole refinement to reduce this bias.

Table 8. Estimated values for the diagonal components of second moments ($10^{-40} Cm^2$) of selected organic molecules

Molecule	Method	m_{11}	m_{22}	m_{33}
Water	X-ray	−15.4	−21.4	−14.6
	HF/6-31G**	−18.93	−20.67	−17.13
Urea	X-ray	−64.6	−86.1	−36.7
	HF/6-31G**	−68.9	−84.7	−40.5
Imidazole	X-ray	−100.8	−95.8	−67.5
	HF/6-31G**	−101.45	−85.06	−89.96
Cytosine	X-ray	−129.7	−60.1	−149.8
	HF/6-31G**	−173.94	−100.45	−160.75
Benzene	X-ray	−116.2	−105.7	−116.0
	HF/6-31G**	−117.72	−106.70	−117.87
p-Dicyanotetrafluoro-benzene	X-ray	−329.7	−287.6	−236.2
	HF/6-31G**	−359.5	−264.95	−239.62

SUMMARY

Diffraction experiments can provide reliable estimates of molecular electronic properties provided accurate intensity data are available and their interpretation is based on careful modelling of the static density deformations as well as the thermal motion of the molecule. The applicability of the kinematic theory is compromised by the scattering object, the wavelength of the radiation used and the method of data collection. Molecules containing atoms of high atomic number ($Z > 18$) are not well suited for charge density study when a standard X-ray source is used. Bonding effects are likely to be invisible for atoms with a small ratio of valence to core electrons. The intensity data should be corrected for systematic errors due to dynamic scattering. To reveal these effects equivalent reflections should be measured. To minimize their influence the measurement should be carried out at low temperature ($T <$ 100K).

The scattering theory applied relates the total dynamic ED to the diffracted intensity distribution. The averaged ED obtained may well account for the limited data, but a property being a function of only a subset of the refined variables, can be unreliable. Model inadequacies result in bias in the least-squares variables. A typical bias of such a nature is that introduced into the static density deformations by the inadequate decomposition of the thermal smearing. In this respect the importance of the proper treatment of the hydrogen atoms has to be emphasized. The flexibility of the model and the limited number of observations make it necessary to limit the optimization to a subset of parameters or to their combinations. The variables are selected on the basis of simple chemical arguments, therefor on the basis of preconceptions.

The outcome must be tested to judge on its physical significance. A plausible approach to reduce model ambiguities is to introduce constraints into the refinement.

The standard deviation of the total static ED obtained in a careful study can be smaller than 0.05 e/$Å^3$ in the bond regions formed by first row atoms. It is considerably higher in the vicinity of the nuclei. The estimated error of the Laplacian distribution is around one order of magnitude higher in the bond CPs. This accuracy is sufficient for characterizing chemical bonds and atomic interactions.

Both the magnitude and the direction of the dipole moment are reliably obtained from X-ray diffraction data. The typical errors in the dipole moment are $2 - 4 \cdot 10^{-30}$ Cm in magnitude, and $\pm 10°$ in direction. For molecules participating in weak intermolecular interactions in the crystal the second moments can be determined with a typical error of $3 - 10 \cdot 10^{-40} Cm^2$ (5–10% for the diagonal components).

ACKNOWLEDGEMENT

The results presented here are based on the X-ray data collected by Dr. J. Buschmann and Dr. D. Zobel. Their contribution is gratefully acknowledged. I also thank to R. Flaig for converting the text into LaTeX format.

REFERENCES

1. P. Hohenberg and W. Kohn, *Phys. Rev.* B136:864 (1964)
2. M. Born and R. Oppenheimer, *Ann. Phys.* 84:457 (1927)
3. H. Hellmann, "Einführung in die Quantenchemie", pp. 285 ff. Franz Deuticke, Leipzig und Vienna (1937)
4. R.F.W. Bader, "Atoms in Molecules", Oxford Science Publications, Clarendon Press, London. (1990)
5. M.J. Frisch, G.W. Trucks, M. Head-Gordon, P.M.W. Gill, M.W. Wong, J.B. Foresman, B.G. Johnson, H.B. Schlegel, M.A. Robb, E.S. Replogle, R. Gomperts, J.L. Andres, K. Raghavachari, J.S. Binkley, C. Gonzalez, R.L. Martin, D.J. Fox, D.J. Defrees, J. Baker, J.J.P. Stewart, J.A. Pople, (1992) Gaussian 92, Revision C, Gaussian, Inc., Pittsburgh PA.
6. M. Born, *Zeitschr. für Physik* 38:803 (1926)
7. R.F. Stewart and D. Feil, *Acta Cryst.* A36:503 (1980)
8. P. Coppens and P.J. Becker, International Tables for X-ray Crystallography, Kluwer Academic Publishers, Vol. C. 627 (1992)
9. R.F. Stewart, *J. Chem. Phys.* 51:4569 (1969)
10. E. Clementi and D.L. Raimondi, *J. Chem. Phys.* 38:2686 (1963)
11. G.A. Jeffrey and D.W.J. Cruickshank, *Quart. Rev. Chem. Soc.* 7:335 (1953)
12. F.L. Hirshfeld, *Israel J. Chem.* 16:226 (1977)
13. R.F. Stewart, *Israel J. Chem.* 16:124 (1977)
14. N.K. Hansen and P. Coppens, *Acta Cryst.* A34:909 (1978)
15. K. Kurki-Suonio, *Israel J. Chem.* 16:115 (1977)
16. T. Koritsánszky, S. Howard, R.P. Mallinson, Z. Su, T. Richter, N.K. Hansen: (1995) XD, A Computer Program Package for Multipole Refinement and Analysis of Charge Densities from X-ray Diffraction Data.
17. V. Schomaker and K.N. Trueblood, *Acta Cryst.* B24:63 (1968)
18. R.F. Stewart, E.R. Davidson and W.T. Simpson, *J. Chem. Phys.* 42:3175 (1965)
19. T. Koritsánszky, J. Buschmann and P. Luger, *J. Phys. Chem.* , submitted (1995)
20. D. Nilveldt and A. Vos, *Acta Cryst.* B44:296 (1988)
21. D. Zobel, P. Luger, W. Dreißig and T. Koritsánszky, *Acta Cryst.* B48:837 (1992)
22. K.E. Laidig, *J. Phys. Chem.* 97:12760 (1993)
23. G. Moss and D. Feil, *Acta Cryst.* A37:414(1981)
24. R. Destro, R. Bianchi and G. Morosi, *J. Phys. Chem.* 93:4447 (1989)
25. Z. Su and P. Coppens, *Acta Cryst.* A48:188 (1992)
26. T. Koritsánszky, J. Buschmann, P. Luger, A. Knöchel and M. Patz, *J. Am. Chem. Soc.* 116:6748 (1994)
27. T. Koritsánszky, J. Buschmann, L. Denner, P. Luger, A. Knöchel, M. Haarich and M. Patz, *J. Am. Chem. Soc.* 113:8388 (1991)
28. M.A. Spackman, *Chem. Rev.* 92:1769 (1992)

MODELING OF STRUCTURAL AND SPECTROSCOPIC PROPERTIES OF TRANSITION METAL COMPOUNDS

Peter Comba
Anorganisch-Chemisches Institut
Universität Heidelberg
Im Neuenheimer Feld 270
69120 Heidelberg
Germany

INTRODUCTION

Chemistry is the art of isolating known and preparing new materials and compounds, of their purification, their elemental analysis, and of analyzing and interpreting their properties. The ability to thoroughly interpret and understand properties of materials such as stabilities, reactivities, solubilities and colors, the knowledge of why exactly compounds behave as they do, would enable scientists to create at their own will new materials with desirable properties. This clearly is not possible. The missing link is an unambiguous interpretation of macroscopic properties on a molecular and submolecular level. The way to bridge this gap is to invent and refine laws that are based on patterns in empirically observed properties within a group of compounds, and thus enable us to relate species within this group, whether known or unknown, to one another. Such a model and the emerging instrument, the theory, may be more or less efficient, rigorous or accurate, but the model never is right or wrong. And a molecular model is always limited in its applicability by and to the set of observations from whose regularities it was derived. Using a molecular model has nothing to do with the truth. Molecular modeling is just an instrument to explain observations and predict new ones. It is not important that a particular model is sophisticated and rigorous in terms of modern theories, it rather is the question of whether and how accurate the observable properties may be reproduced in the well defined limits of application that determines the quality of a model.

The generally accepted basis for molecular modeling is the molecular structure that by itself is a model rather than a fact. A molecular structure is the three-dimensional arrangement of the atoms of a molecule in space. On a quantum mechanical basis a molecular structure may be understood using the Born-Oppenheimer approximation. This allows the definition of the nuclear positions as structural parameters and using them for the calculation of the electronic Schrödinger equation. In contrast, molecular mechanics modeling assumes the arrangement of electrons to be fixed and the positions of the nuclei are the calculated variables.

Structure determination and structure prediction are the starting point for understanding the properties of a known or designing a new material since all molecular properties, stabilities, reactivities and spectroscopic observables, may be related to the molecular structure (see Scheme 1). Enhanced reactivities resulting from steric crowding, the effect of ring strain on

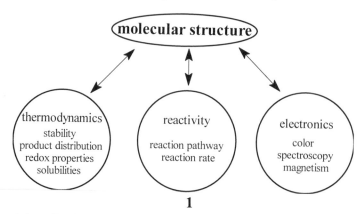

stabilities and that of bulky substituents on redox potentials and spectroscopic properties are a few commonly discussed examples. Thus, molecular modeling involves the computation of molecular structures and that of thermodynamic, kinetic and electronic molecular properties, based on these structures.

Quite often environmental effects, i.e. the influence of the environment on the molecular structure and properties, and that of the molecule on the environment, are neglected in molecular modeling. Neglecting environmental effects, such as solvation, ion-pairing and crystal lattice effects, not only means that molecular rather than bulk properties are modeled, it also means that specific interactions to and from the environment are neglected as minor perturbations. I will discuss this point in more detail below and restrict the discussion mainly to isolated molecules.

In the first part of this article, I will further discuss the *real* basis for molecular mechanics modeling and present some theoretical background, in particular for modeling transition metal compounds. A number of methods for appreciating electronic effects in these systems will be analyzed in terms of their general applicability, performance and limitations. For a more detailed discussion on theoretical aspects of molecular mechanics modeling I refer to a number of recent reviews[1-3]. I will also discuss the general theoretical basis for a number of methods that allow the computation of molecular properties, such as ligand field and EPR spectra, and redox properties, from molecular structures (see Scheme 1). Again, for detailed theoretical discussions I refer to the original literature[1e,1g,3]. In the second part, I will present some selected applications of molecular modeling in coordination chemistry. The specific examples presented are all related to the determination of structures of coordination compounds in solution and to the modeling of ligand field and redox properties.

THEORY

The Molecular Mechanics Model

The basic concept of molecular mechanics modeling is that the positions of the atoms of a molecule, of an ion or of an array of molecules or ions are determined by forces between pairs or groups of atoms (Figure 1). An empirical set of potential energy functions U_i and corresponding parameters (the force field) are used to calculate the energies E_i resulting from these forces. The total strain energy of a molecule, E_{total}, defined as the sum over all individual potential energy functions, and over the whole molecule, depends on the compound

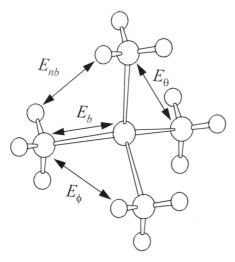

Figure 1. The molecular mechanics concept

(stoichiometry, connectivity and geometry) and on the force field (potential energy functions and their parameterization). Minimization of the strain energy E_{total} by rearrangement of the atoms leads to an optimized structure and a value for the minimized strain energy.

Any set of potential energy functions that is able to represent a molecule in a way similar to that shown in Figure 1 may be used in a molecular mechanics model, and a number of conceptually different approaches have been used[3]. However, usually the total strain energy, E_{total}, of a molecule is represented by the set of functions given in Eq. 1. These include bonding (E_b), valence (E_θ) and torsional angle (E_ϕ) interactions, as well as nonbonded interactions (E_{nb}), electrostatic interactions (E_ε) and various additional terms such as out-of-plane bending (E_δ), hydrogen bonding (E_{hb}), and others.

$$E_{total} = \sum_{molecule} (E_b + E_\theta + E_\phi + E_{nb} + E_\varepsilon + E_{hb} + E_\delta + \cdots) \tag{1}$$

$$E_b = \frac{1}{2}k_b(r_{ij} - r_0)^2 \tag{2}$$

$$E_\theta = \frac{1}{2}k_\theta(\theta_{ijk} - \theta_0)^2 \tag{3}$$

$$E_\phi = \frac{1}{2}k_\phi(1 + \cos[m\{\phi_{ijkl} + \phi_{offset}\}]) \tag{4}$$

$$E_{nb} = Ae^{-Bd_{ij}} - Cd_{ij}^{-6} \tag{5}$$

$$E_\delta = \frac{1}{2}k_\delta\delta^2 \tag{6}$$

$$E_\varepsilon = \frac{q_i q_j}{\varepsilon d_{ij}} \qquad (7)$$

$$E_{hb} = \frac{F}{d_{ij}^{12}} - \frac{G}{d_{ij}^{10}} \qquad (8)$$

The generally used potential energy functions are those of Eq. 2 – Eq. 8 (r_{ij}, θ_{ijk}, ϕ_{ijkl}, d_{ij} and δ are the specific bond distances, valence angles, torsional angles, nonbonded distances and out-of-plane angles, respectively; r_0 and θ_0 are the corresponding ideal parameter values; q_i are charges, and ε is the dielectric constant; A,B,C,F,G are adjustable parameters for the van der Waals and hydrogen bonding terms). Some programs use different functional forms, such as Morse potentials for the bonding terms, the Lennard-Jones instead of the Buckingham potential for van der Waals interactions, and in some cases cross-terms are added to account for the fact that various potentials are, as one would expect, not fully decoupled.

The parameterization of the potential energy functions is one of the most demanding problems in molecular mechanics modeling. It might be argued that some of the parameters are directly available from experiment, such as stretching force constants from normal coordinate analyses of vibrational spectra. However, there are fundamental differences between molecule specific spectroscopic and molecule independent molecular mechanics force fields[3]. Also, the whole set of force constants is highly cross-correlated. Thus, even where similar functional forms have been used, spectroscopic force constants or parameters from another molecular mechanics force field may only be used as initial values for fitting to experimental data. Figure 2 shows the real basis for molecular mechanics modeling, and this relates to the basic and general idea of a molecular model, as discussed above: regularities in a set of empirically observed data are used to setup and parameterize potential energy functions which may then be used to predict similar properties of similar compounds. Molecular mechanics is an interpolative method, and its limits clearly are defined by the data used for setting up the force field.

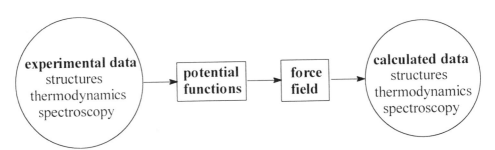

Figure 2. The real basis for empirical force field calculations

As with any other interpolation, the accuracy of molecular modeling may be enhanced by an increasing degree of parameterization. The disadvantages of doing so are clearly that, with increasing complexity, molecular mechanics is losing much of its appeal and, on a more fundamental basis, the accurate determination of every additional parameter has to be based on some independent observables. Examples for an increasing degree of parameterization are cross-terms, the substitution of harmonic potentials by a Morse function and the perturbation due to electronic effects exerted by transition metal ions that will be discussed below.

It is immediately clear that, for any particular problem, there is not the one and only correct force field. Parameterization schemes used with transition metal compounds range from the universal force field, using generic force constants which allow the refining of structures across the periodic table[4], over rather general conventional force fields[5] to parameter sets that have been created to solve a specific problem[6]. So far, generic force fields have, not unexpectedly, not matched the performance of conventional parameter sets. With force fields that have been developed for specific problems, on the other hand, one needs to be very careful that the limits of interpolation are not exceeded. A problem that usually is not appreciated is that a force field is not only limited by the type of compound but also by the type of data to which it has been fitted. That is, a parameter set fitted to a number of structures may be a good model to predict new structures of similar molecules, but it may not be able to accurately compute isomer distributions or vibrational frequencies, even within the limited set of compounds. This is because structural parameters are related to potential energy minima while thermodynamic parameters and vibrational frequencies are dependent on energy differences and on the curvature of the potential energy surfaces, respectively. This is a particular problem for specialist force fields, where the parameters have been adjusted to a limited range of very similar structures. More general force fields are often also mainly based on structural data, but some of the highly cross-correlated parameters may originate from thermodynamic and/or spectroscopic data, and some of these more general force fields have therefore been used successfully in the area of conformational analysis.

The challenge of transition metal ions

The versatility of transition metal compounds is not only based on the number of elements and oxidation states, but also on the large variety of possible coordination numbers, geometries and electronic states that each of them can adopt. This variability may be attributed to partly filled d-orbitals, and it is the main challenge in the modeling of transition metal compounds. The prediction of coordination numbers and electronic ground states has not yet been attempted by conventional molecular mechanics methods, and more advanced techniques may also not lead to an easy and general solution. The remaining considerable problem of modeling coordination centers with well defined coordination numbers and electronic structures is that of flexible coordination geometries.

Valence angles around transition metal centers are generally soft, and the coordination geometry is to a considerable extent dependent on the metal-ligand distances, the ligand geometries and ligand atom-ligand atom repulsion. This is the basis of the very simple and crude ligand-ligand repulsion model that computes coordination geometries by minimizing the ligand atom-ligand atom repulsion energy, using rigid ligands and fixed metal-ligand distances[7]. A similar approach has been adopted in molecular mechanics models, where the usual harmonic angle bending potential of Eq. 3 has been replaced by 1,3-nonbonded interactions around metal centers[5,8]. Among others, such an approach has the advantages that less metal dependent parameters are needed and that the "unique labeling problem" (e.g. differentiation between *cis*- and *trans*-angles in octahedral complexes) does not exist[9]. However, the main advantage of this approach is that the observed plasticity of transition metal centers is not restricted by forces defined in the starting model, i.e. tetrahedral and square planar (tetracoordination), trigonal bipyramidal and square pyramidal (pentacoordination), and octahedral and trigonal prismatic (hexacoordination) geometries may be modeled with the same respective force fields (see discussion below for an important modification of this oversimplified view). Obviously, such an approach does neglect electronic influences excerted by the metal center, and the good quality of results obtained so far might be astonishing. However, it has been argued[3,8b] that part of the reason for the unexpected success of this type of parameterization is the high degree of cross-correlation in any force field, i.e. the fact that the seemingly neglected electronic factors are accounted for in other parts of the parameterization scheme.

"Electronically doped" conventional force fields have been one recent approach to correct remaining inaccuracies:

- The harmonic sine function of Eq. 9, 10, with a generic force constant that depends on the d-orbital occupancy F and the metal-ligand breathing modes k_{ML} and $k_{ML'}$ (spectrochemical series) has been added as a ligand field based perturbation to a force field using 1,3-nonbonded interactions around the metal center[5a,10].

$$E_\theta^M = \frac{1}{8} k_{LML'} \sin^2(2\theta) \tag{9}$$

$$k_{LML'} = cF \frac{k_{ML} + k_{ML'}}{2} \tag{10}$$

One of the advantages of molecular mechanics modeling of transition metal compounds with this electronically doped force field is, that only one fitted parameter is required for hexacoordinate complexes of the entire first transition metal row. The resulting modified force field leads to more accurate coordination geometries, and this is particularly important, when spectroscopic properties, based on these structures, are computed. An interesting observation is that the force constants $k_{LML'}$ are roughly an order of magnitude smaller than those used in a conventional molecular mechanics approach. This is not unexpected since here the harmonic angle function is only a small perturbation to a force field that computes coordination angles with 1,3-nonbonded interactions. Therefore, this is a good example for demonstrating the fact, discussed above, that force constants are not directly relatable to physical properties.

- For the trigonal twist of octahedral compounds (Bailar twist) and the tetrahedral twist of square planar species, approaches were developed that account for the electronic preferences of octahedral over trigonal prismatic and of square planar over tetrahedral geometries (Figure 3)[5a,11]. Again, these functions are adopted as perturbations to a lig-

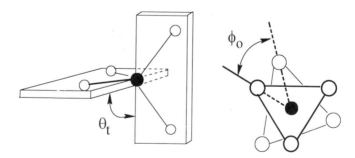

Figure 3. The twist functions

and atom-ligand atom repulsion based force field. An important point here is that the metal-ligand stretching modes are also ligand field dependent. That is, the metal centered force field parameters of a tetrahedral complex are different from those of a square planar complex, and the parameters of an octahedral species are different from those of a trigonal prismatic complex. Therefore, all of these parameters have to be adjusted as a function of the corresponding twist angles in each cycle of the structural refinement[5a,11].

- The computation of Jahn-Teller distorted metal complexes is a difficult problem to solve with conventional molecular mechanics. A number of approaches have been used, but most of them are unsatisfactory since they require some knowledge of the structure before starting the refinement procedure. This is not the case for a recently developed method for the accurate computation of copper(II) hexaamine complexes[12]. The method is based on a harmonic first-order model to compute the Jahn-Teller stabilization energy. The electronic term is added to the strain energy that is computed conventionally (see Figure 4). The method is general and not restricted to copper(II) and amine ligands, and similar techniques must be applicable to coordination numbers other than six.

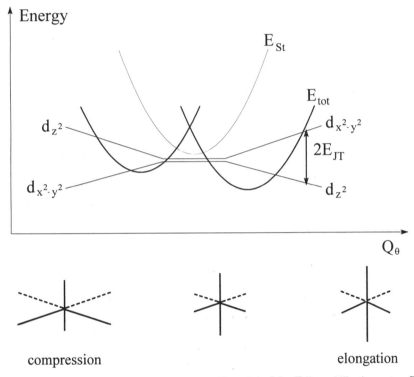

Figure 4. Total energy as a function of the strain energy E_{ST} and the Jahn–Teller stabilization energy E_{JT}

- Two promising new methods to model the geometry of transition metal complexes are the VALBOND[1i,13] and the CLFSE/MM[14] approaches. Both use an electronic term to compute valence angles. The VALBOND model needs two atom based parameters for a valence bond type angle function and, for organic molecules and metal hydrides, it has produced results that are as accurate as data obtained with conventional models[13]. In the CLFSE/MM model angle bending around the metal center is computed with a combination of 1,3-nonbonded interactions and an angular overlap model based electronic term. Thus, it is not dissimilar to the approach involving 1,3-nonbonded interactions and a harmonic sine or a twist function (see above). The advantage of the CLFSE/MM method is that it is more general since it is not restricted to a specific coordination geometry. Moreover, Jahn-Teller distortions are also treated with the same approach. In order to keep the calculations computationally as efficient as possible, a one electron coupling scheme had to be used. Therefore, the d-orbital configuration and the spin state have to be well defined. Unfortunately, at this stage, this precludes a straight forward treatment of spin-equilibrium compounds. Also, some of the approximations for the ligand field parameters are leading to values of the electronic energies that do not allow a direct determination of spectroscopic properties (see the MM–AOM model below).

Modeling of Molecular Properties

The determination of a molecular structure by molecular modeling is then of importance, when the determination by diffraction methods is not possible or not relevant, i.e. when no single crystals are available or when the solution structure rather than that in the solid state are of interest[1g]. For two reasons molecular mechanics alone may not be sufficient for an accurate structure determination: (i) environmental effects (solvation and ion-pairing in the case of solution structures) are often neglected and (ii) the unambiguous determination of the global minimum, i.e. of the lowest energy structure, is only possible for rather simple molecules, where the conformational space may be scanned deterministically. An elegant way to solve both problems is to determine experimentally a molecular property that may be related to the molecular structure, and to develop and apply an algorithm that computes this property based on the calculated structure, thus allowing the determination of the structure related to the global energy minimum in the medium of interest, e.g. in solution.

The problem of neglecting environmental effects in molecular mechanics is not primarily that isolated molecules are refined. In fact, it is a misconception to understand the generally used molecular mechanics approach as "gas phase calculations". This is because molecular mechanics parameterization schemes are usually based on experimental data obtained from molecules that are interacting with an environment. Often, these are crystal structures, where the molecular structure is perturbed by counter ions, hydrogen bonding and other interactions with the environment. In the process of parameter fitting these specific effects are averaged, and the result of a molecular mechanics refinement is that the computed structure is usually more symmetrical than that experimentally observed in the solid state. It however is a reasonable model for the molecule in an anisotropic medium, i.e. in solution.

Thus, the determination of a (solution) structure by molecular modeling is based on the same techniques and algorithms as the design of new materials with desirable properties: computation of the structure followed by that of a molecular property (see Scheme 1). Any molecular property, for which an algorithm for its computation based on a molecular structure is found, may be used in such a scheme. The types of data that have been used in the area of coordination chemistry are shown in Scheme 2.

The technique that possibly has seen the widest application so far is to derive structural properties from NMR data. Structural information derived from NOE data and Karplus relations has been widely used for the modeling of biomolecules[15,16]. For various reasons, coordination chemists have not extensively applied this technique so far. Some of the few examples that have appeared in the literature have been reviewed elsewhere[1g,3], and theoretical assumptions as well as applications in organic and biochemistry are also well covered in the literature[15,16].

Vibrational frequences may be directly extracted from molecular mechanics calculations if a full matrix second derivative method, such as the Newton-Raphson procedure, is used for the refinement[3]. However, there are only few reports that include the comparison of experimentally observed and calculated IR or Raman data, particularily in the area of coordination chemistry, and these have been reviewed elsewhere[3].

Thermodynamic data may be directly related to strain energies if these are corrected by entropic effects – which in principle is possible though seldom appreciated in practice[3] – and

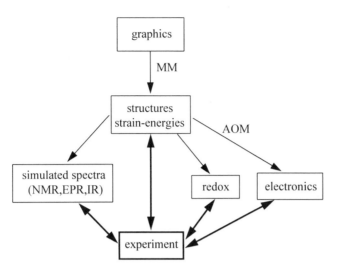

environmental effects are accounted for. Applications include conformational analyses, the determination of stereospecificities and that of metal ion selectivities[1,3]. In rare cases, the comparison of experimentally determined and calculated isomer distributions has been used for the determination of structures in solution[1g,3].

In this article I will concentrate on some recently developed techniques which have been applied successfully to coordination compounds. These involve the computation of redox potentials, that of EPR spectra of weakly coupled dinuclear copper(II) compounds and the combination of molecular mechanics with AOM calculations to compute UV-vis-NIR and EPR spectroscopic parameters.

MM-Redox. The redox potential is a function of the energy difference between the ground states of the oxidized and reduced forms and may therefore be related to the relative strain energies (Figure 5). This approach neglects effects other than steric strain to $\Delta G°$

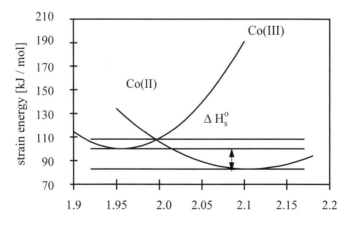

Figure 5. Molecular mechanics and redox potentials

($E° = f(\Delta H°_{strain})$), and one needs to be careful, even in the case of pure interpolation. In Figure 6 the experimentally determined potentials of a series of hexaaminecobalt(III/II) couples are plotted as a function of the strain energy differences between the oxidized and reduced forms[17]. There is a linear dependence of the redox potential on the free enthalpy (Eq. 11),

$$\Delta G^\circ = -nFE^\circ \qquad (11)$$

where F is the Faraday constant ($F = 96.5$ kJmol^{-1}) and $n = 1$. Thus, the slope of the

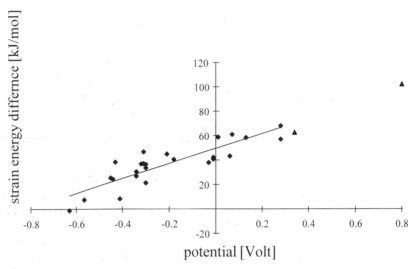

Figure 6. The modeling of redox potentials (the two couples [Co(dmpn)$_3$]$^{3+/2+}$ and [Co(dmpncage)]$^{3+/2+}$, with extreme structures falling outside the interpolation limit, appear as triangles)

line in Figure 6 should be 96.5 kJmol^{-1}V^{-1}, if strain relaxation would be the only contribution to the variation of redox potentials. The observed slope of 61 kJmol^{-1}V^{-1} indicates that, as expected, this is not the case. Other possible contributions to the thermodynamics of electron transfer are electronic effects, specific hydrogen bonding, ion-pairing, solvation and hydrophobicity[3,17]. Similar studies with other systems (e.g. copper(II/I)[18]) are in progress, and the computation of redox potentials with different methods, though not involving molecular mechanics refinement of molecular structures, has also been reported[19].

MM-EPR. EPR spectra of weakly coupled dicopper(II) systems (dipole-dipole interactions) can be simulated using the spin hamiltonian parameters (anisotropic g- and A-values of each metal center), line shape parameters and the four geometric parameters (r, τ, η, and ξ) shown in Figure 7, unambiguously defining the relative orientation of the two g-tensors[20]. For copper(II) complexes this generally is in good agreement with the orientation of the chromophores, but it does not give any information on the overall shape of the molecule. Also, the number of parameters involved in the spectra simulations is usually too large with respect to the number of spectroscopic features to be fitted, to expect a unique solution. However, the combination of the simulation of dipole-dipole coupled spectra with molecular mechanics calculations has been used successfully to determine solution structures of dicopper(II) complexes (Figure 8)[21].

MM-AOM. The term energies of transition metal ions are dependent on the ligand atoms and their geometric arrangement around the metal center. The ligand field theory and in particular the angular overlap model (AOM) are powerful tools to interpret the electronic properties (UV-vis-NIR and EPR spectroscopy and magnetism) of coordination compounds. The application of AOM calculations to predict spectroscopic and magnetic properties involves a constant and transferable parameter set. This is not a priori a given property of the AOM model,

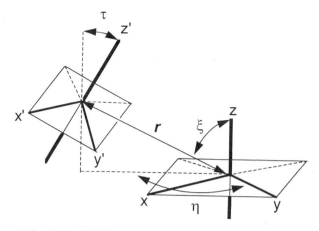

Figure 7. Simulation of EPR spectra of weakly coupled dicopper(II) compounds

and the question of transferable ligand field parameters has been discussed extensively[22]. We have successfully used constant and transferable e-values, adjusted only by the theoretically and experimentally justified variation of $1/r^n$ (r is the metal-ligand distance; $n = 5$ or 6) and an empirical correction for misdirected valencies[10,22]. Clearly, this does not mean that ligand field parameters are transferable, it just means that the putative error due to using transferable parameter sets is neglectable. A similar approximation was used in the CLFSE/MM approach discussed above[14]. The combination of AOM calculations with molecular mechanics, i.e. the prediction of spectroscopic and magnetic data by AOM calculations, based on structures computed by molecular mechanics (Figure 9) is a powerful tool for designing new materials and for the determination of solution structures. The electronic parameters used in MM-AOM calculations are presented in Table 1.

Table 1. Electronic parameters for MM-AOM calculations

parameter [cm^{-1}]	Cr(III)	ls Fe(III)	Co(III)	Ni(III)	Cu(II)[d]
F_2[a]	1211	786	1046	1243	-
F_4[a]	101	57	85	79	-
e_σ(NH$_3$)[b,c]	7200	-	7245	3582	-
e_σ(RNH$_2$)[b,c]	7400	7500	7433	3857	6400
e_σ(R$_2$NH)[b,c]	8000	7500	7715	4133	6700
e_σ(R$_3$N)[b,c]	8700	-	8186	4592	-
e_σ(pyridine)[b,c]	-	7900	-	-	-
e_π(pyridine)[b,c]	-	−500	-	-	-

[a] Condon-Shortley interelectronic repulsion parameters
[b] For low spin iron(III) a single e_σ for the various amines was used
[c] Normalized for Cr-N = 2.080, Fe-N = 1.985, Fe-N$_{py}$ = 1.968, Co-N = 1.980, Ni-N = 2.130, Cu-N = 2.027, Cu-O = 2.440 Å; for the calculations the values are adjusted with $1/r^6$
[d] $k = 0.7$, $\xi = -580$, $K = 0.43$, $P = 0.036$ cm^{-1}, $\alpha^2 = 0.74$

An interesting observation is that for an accurate prediction of spectroscopic parameters a differentiation between different degrees of alkyl substitution on amines is required. This is not unexpected, but in practice the change in nucleophilicity is usually not detected experimentally, because the increasing nucleophilicity due to alkyl substitution is generally counterbalanced by increasing repulsion. The separation of steric from electronic effects in the compu-

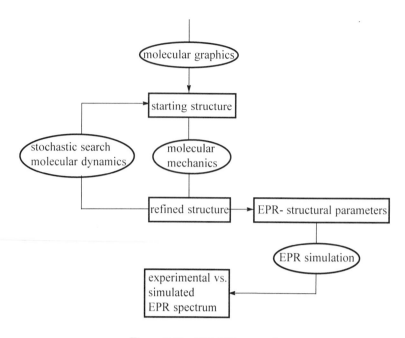

Figure 8. The MM–EPR approach

Figure 9. The MM–AOM approach

tation of electronic properties by MM-AOM calculations was the first efficient way to demonstrate this effect. Recently, a similar conclusion was drawn from MO calculations and experimental investigations of N-alkylated macrocyclic ligands and their copper(II) complexes[23]. It might be argued that the conclusions drawn from Table 1 are an overinterpretation, i.e. they might be believed to be an artefact of the parameterization schemes used, which are not directly related to real molecular properties (see above). As a matter of fact, based on my interpretation of the data in Table 1, it could be argued that the force field parameters for metal-ligand stretching modes should follow the same pattern, i.e. different parameters ought to be used for varying degrees of ligand alkylation. I believe, the interpretation given here is appropriate because the observed variation of the nucleophilicity as a function of the degree of alkylation is not the result of the empirically derived force field or electronic parameter set, but that of separating electronic from steric effects. The molecular mechanics and the AOM calculations are both able to independently predict accurate structural and accurate spectroscopic data, respectively. I therefore assume that a differentiation of the degree of alkylation in the force field should lead to improved accuracies. However, our experience so far has shown that an increase of the degree of parameterization is not necessary, and thus not appropriate in this case.

APPLICATIONS[*]

MM-Redox

One of our aims is to correlate redox potentials of transition metal compounds with their structures. A possible result emerging from these studies is a tool for the design of new reagents for preparatively interesting redox processes. A successful correlation between redox potentials and coordination geometries may also produce useful data for the determination of structures in solution. Primarily, however, it may help us to understand why, in the limited area of our studies, compounds react in redox processes the way they do. In a first stage we have restricted ourselves to cobalt(III/II) hexaamine couples in aqueous solution[17], and we are now extending our investigations to copper(II/I) tetraamine couples[18]. I will report here only on the cobalt hexaamine systems, a seemingly unspectacular area. However, it will allow me to demonstrate the accuracy of the method and its limits. One reason for the choice of cobalt(III/II) couples is the large range of available, accurately determined redox potentials[24], and the well documented correlation with structural and other properties[24a,b]. Reduction potentials of hexaamine cobalt(III/II) couples span a range of roughly 1.5 V (see Figure 6 and Table 2), and recent studies have shown that, with well planned experiments, this might be con-

Table 2. Observed and calculated (MM-Redox) redox potentials of some cobalt(III/II) hexaamines.

compound	total strain energy [kJ mol^{-1}]		ΔH_s^0 [kJ mol^{-1}]	$E_{1/2}^{obs}$	$E_{1/2}^{cal}$
	Co^{III}	Co^{II}			
[Co(en)$_3$]$^{3+/2+}$	45.12	4.48	40.64	−0.18	0.10
[Co(tmen)$_3$]$^{3+/2+}$	154.33	86.22	61.11	+0.28	+0.28
[Co(trans-diammac)]$^{3+/2+}$	98.56	82.00	3.01	−0.63	−0.63
[Co(sar)]$^{3+/2+}$	127.73	79.73	25.52	−0.45	−0.32
[Co(sep)]$^{3+/2+}$	115.58	60.95	33.63	−0.30	−0.20
[Co(tacn)$_2$]$^{3+/2+}$	109.57	83.23	8.45	−0.41	−0.55
[Co(NH$_3$)$_3$]$^{3+/2+}$	29.38	−8.78	38.16	+0.06	−0.14
[Co(dmpn)$_3$]$^{3+/2+}$	121.76	59.01	62.75	+0.21	+0.34

siderably extended[25]. One way to achieve this, i.e. to obtain couples with extremely high and very low potentials, is to prepare ligands enforcing very long or short metal-ligand bonds, respectively. This emerges from Figure 5 and the fact that enforced long metal-ligand bonds will stabilize the reduced (large ionic radius), while short bonds will stabilize the oxidized (small ionic radius) form of the couple. Thus, *trans-diammac*, leading to the lowest cobalt(III/II) potential in the series, is a hexaamine ligand that generally enforces rather short bonds and consequently leads to strong ligand fields[22b,26]. On the other end of the scale, we find sterically crowded ligands such as *tmen*,[27] leading to long bonds and weak ligand fields.

It is immediately clear that the design of new ligands with extreme properties means stretching the model to or over its limit: Using a force field that has been fitted to Co^{III}-N$_{amine}$ bonds from about 1.94 Å to 1.99 Å, to predict structures and properties of compounds with Co^{III}-N$_{amine}$ distances considerably longer than 2.0 Å, is no longer interpolation. This does not necessarily mean that one should not use the original model in these cases. It just means that one should be aware of the possibly limited accuracy of the predictions. Therefore, here it is especially important to support the predictions with experimental data and, if necessary, to modify the model accordingly.

---

[*]The results presented in this section are based on studies from my lab, where molecular mechanics calculations are performed with MOMEC[5] and the corresponding force field[5,10].

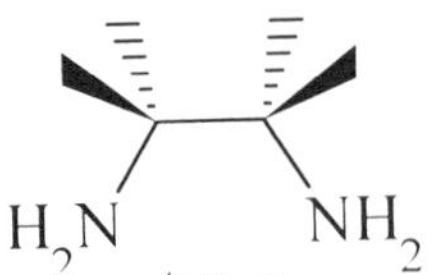

tmen

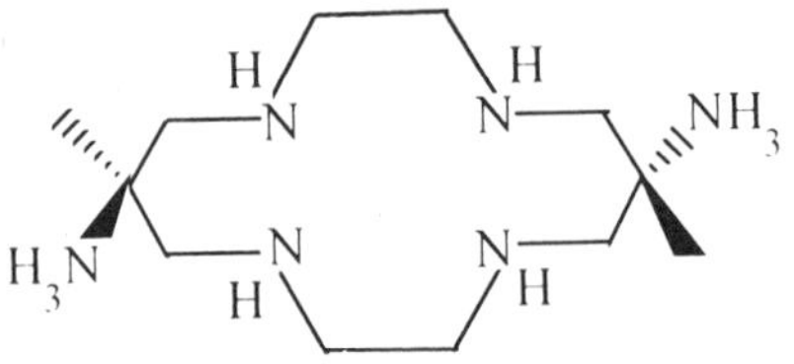

trans-diammac

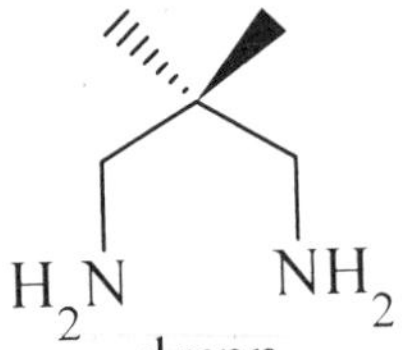

dmpn

The $[Co(dmpn)_3]^{3+/2+}$ couple was calculated to have a redox potential of ca. +0.3 V, based on Co^{III}-N_{amine} bonds of ca. 1.98 Å, and leading to a purple hexaaminecobalt(III) compound with λ_{max} (MM-AOM, see below) of ca. 500 nm[28].[†] Subsequent experimental studies have defined the redox potential to be +0.34 V, and the compound was purple with λ_{max} of 520 nm, suggesting a Co^{III}-N_{amine} distance of 1.99 Å. It is not unexpected that extremely long bonds are underestimated when modeling them with a harmonic potential. The only other case where the harmonic approximation for hexaaminecobalt(III) complexes was demonstrated to be insufficient, is that of a cage complex with an extended cavity[25]. In that case, two isomers could be isolated, one of them having an extremely high redox potential and a spectacular

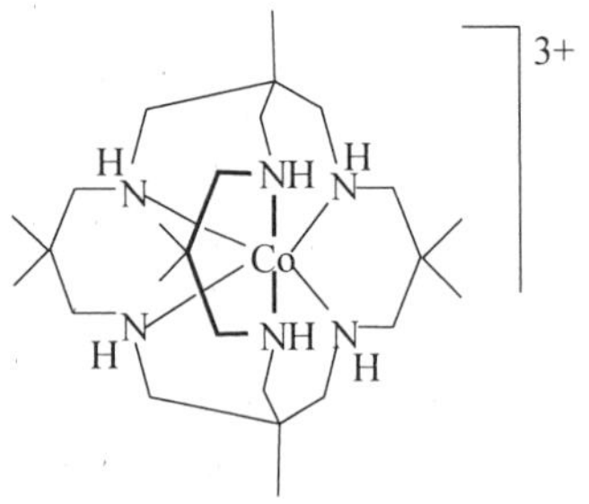

dmpncage

blue color, suggesting very long Co^{III}-N_{amine} bonds[25]. A combination of molecular mechanics calculations and modeling of the redox potentials (MM-Redox) and ligand field spectra (MM-AOM) allowed the determination of the crystallographically not analyzed structure of the blue compound[30].

[†] A full conformational analysis leads to a range of cobalt(III) and cobalt(II) isomers related by equilibria, and consequently to a range of redox potentials. This problem has been evaluated in detail in the literature[29] and is not discussed here.

MM-EPR

The two structurally related bis(macrocyclic)octaamine dicopper(II) complexes [Cu$_2$(*dinitrobm*)]$^{4+}$ and [Cu$_2$(*diazabm*)]$^{4+}$ are prepared in a simple one-step template

$$[Cu_2(dinitrobm)]^{4+}$$

$$[Cu_2(diazabm)]^{4+}$$

reaction[21b,31]. Interestingly, the two compounds give strikingly different EPR spectra (Figure 10). While the spectrum of the diaza compound is typical for a weakly dipolar coupled dicopper(II) species, that of the dinitro compound does not show any coupling[21b]. A conformational analysis indicates that there are two low energy structures each, and the MM-EPR analysis (see Figure 8) indicates that for the dinitro dimer the stretched form predominates in solution, while for the diaza dimer the folded structure is more abundant (Figure 11; the simulated spectra are also shown in Figure 10). The usual molecular mechanics analysis does not immediately lead to this result: for both compounds the folded form seems to be more stable. This clearly demonstrates the necessity to combine molecular mechanics with the modeling of a molecular property. But what is the reason for the apparent failure of molecular mechanics alone? As is often the case, the computations did not include any electrostatic interaction energy. The actual charge distribution does not seem to be a critical structure determining factor in this case[21b]. However, in these dinuclear compounds partial charge neutralization by axial coordination of the counter ions and/or by ion-pairing is of considerable importance. Model calculations indicate that the charge per copper chromophore is in the range of 1.4 to 1.8 (Figure 12).

The dicopper(I) complex of the 24-membered macrocyclic tetra-Schiff-base ligand *metaim* is highly oxygen sensitive, showing tyrosinase type reactivity[32], while the dicopper(I) complex of the isomeric 26-membered macrocycle *paraim* is air stable[21c]. A key

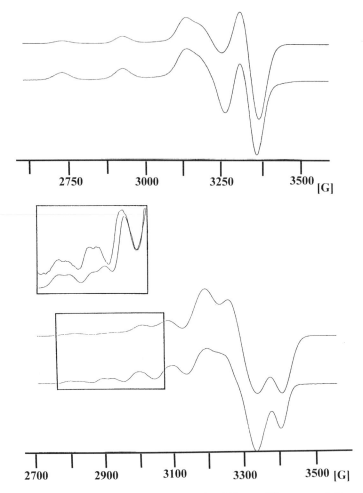

Figure 10. EPR spectra (observed and simulated) of two related dicopper(II) compounds (top: $[Cu_2(dinitrobm)]^{4+}$, bottom: $[Cu_2(diazabm)]^{4+}$)

issue in understanding this drastic difference in reactivity is the structural difference between the two dicopper(I) complexes. While the X-ray structure of $[Cu_2(paraim)]^{2+}$ has been determined, no useful experimental data for any structural modeling was obtained for $[Cu_2(metaim)]^{2+}$. Therefore, the two dicopper(II) complexes of the related hexaamine ligands *metaam* and *paraam* were structurally characterized by MM-EPR. The whole set of data, i.e. the X-ray structure and the two MM-EPR structures support the structural characterization of the tyrosinase model compound $[Cu_2(metaim)]^{2+}$ (Figure 13)[21c]. Important factors are that a constant force field was used, and in all four cases an identical procedure for the conformational analysis lead to the relevant minimum energy structure. The distinct difference in reactivity is interpreted as being the result of the significant structural difference: the copper-copper distance of the air stable compound $[Cu_2(paraim)]^{2+}$ is 7.04 Å while that of the tyrosinase model compound $[Cu_2(metaim)]^{2+}$ is 3.4 Å, comparable to the copper-copper distance reported for hemocyanine and tyrosinase[33].

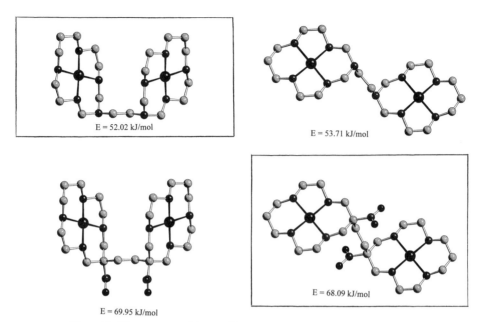

E = 52.02 kJ/mol

E = 53.71 kJ/mol

E = 69.95 kJ/mol

E = 68.09 kJ/mol

Figure 11. The structures of the two bismacrocyclic dicopper(II) complexes

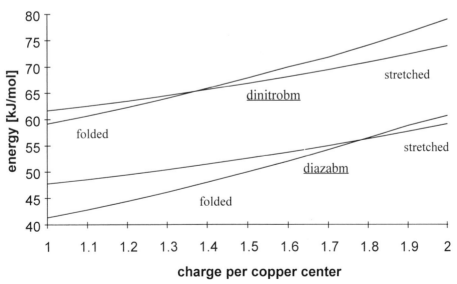

Figure 12. Charge compensation by ion–pairing

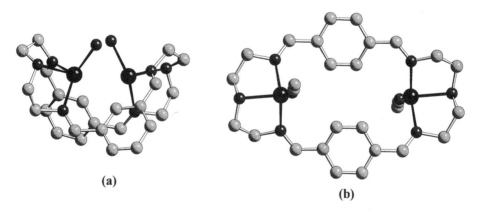

paraim

metaim

paraam

metaam

(a)

(b)

Figure 13. The calculated and the X–ray structure of two structurally related dicopper(I) compounds; (a) calculated structure of the tyrosinase model compound (metaim); (b) X–ray structure of the air stable dicopper(I) compound (paraim)

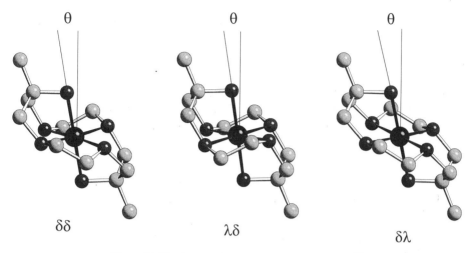

Figure 14. The three conformers of [Fe(*trans*–diammac]$^{3+}$

MM-AOM

We have developed and used transferable AOM parameters for the prediction of d-d spectra of chromium(III), cobalt(III) and nickel(II) hexaamines, for the prediction of the EPR spectra of low spin iron(III) hexaamines and for computing electronic and EPR spectra of copper(II) tetraamines, based on structures obtained from molecular mechanics calculations (see Table 1 above)[10,22]. The examples involving hexaaminecobalt(III) compounds with elongated Co^{III}-N_{amine} bonds, given in the MM-Redox section, demonstrate that the accuracy of the computed structure is a critical factor in the MM-AOM approach. In fact, the recently developed perturbation of our original force field with an electronic term was also inspired by unacceptable inaccuracies of some of our earlier MM-AOM computations[10,22]. In this article I will restrict the discussion to a few low spin iron(III) compounds[22a].

Three cations of $[Fe(trans-diammac]^{3+}$ have been investigated by crystal structure analysis[34]. In all three cases, the five-membered chelate rings have been disordered, making an assignment of the structures to one of the three conformers (see Figure 14) impossible. A conformational analysis indicated that two of the conformers have a similar strain energy while the third was less stable by more than 20 kJ/mol (Table 3)[22a]. Thus, the observed disor-

Table 3. Observed and calculated (MM-AOM) EPR spectra of [Fe(*trans*-diammac)]$^{3+}$

isomer	θ	g_1	g_2	g_3	strain energy [kJ/mol]
structure A	11.0	1.42	2.61	2.81	
structure B	9.5	1.51	2.61	2.78	
structure C	11.3	1.62	2.48	2.84	
solution		1.63	2.46	2.84	
calculated (MM-AOM)					
$\delta\delta$	9.2	1.45	2.54	2.95	90.7
$\lambda\lambda$	3.5	0.95	1.96	3.29	91.5
$\delta\lambda$	15.7	1.89	2.05	2.46	112.2

der might have involved $\delta\delta, \lambda\lambda$ and $\lambda\delta$ conformations. A study involving single crystal EPR experiments and AOM calculations indicated that the g-values are strongly dependent on θ[35].

From the MM-AOM analysis it was concluded that the experimentally determined g-values are due to the $\delta\delta$-conformer (see Table 3)[22a]. Therefore, the observed structural disorder is due to $\delta\delta$ and $\lambda\lambda$ conformations. A minor contribution by the $\lambda\delta$ conformer cannot be excluded.

ACKNOWLEDGMENT

Our own work is supported by the German Science Foundation (DFG), the Fonds of the Chemical Industry (FCI) and the VW-Stiftung. I am grateful for that, and for the help from Marlies von Schönebeck-Schilli and Brigitte Saul in preparing this manuscript.

REFERENCES

1a. Brubaker, G.R., and Johnson, D.W. *Coord. Chem. Rev.* (1984) 53:1.

1b. Boeyens, J.C.A., *Struct. Bonding* (Berlin) 63:65 (1985)

1c. Hancock, R.D., *Prog. Inorg. Chem.* 37:187. (1989)

1d. Hambley, T.W., *Comm. Inorg. Chem.* 14:1. (1992)

1e. Comba, P., *Coord. Chem. Rev.* 123:1. (1993)

1f. Hay, B.P., *Coord. Chem. Rev.* 126:177. (1993)

1g. Comba, P., *Comm. Inorg. Chem.* 16:133. (1994)

1h. Deeth, R.J. *Struct. Bonding* (Berlin) 82:1 (1995)

1i. Landis, C.R., Root, D., and Cleveland, T., *in* "Reviews in Computational Chemistry" Lipkowitz, K.B., and Boyd, D.B., eds., VCH, Weinheim. 6:73 (1995)

1j. Zimmer, M., *Chem. Rev.* in press

2. Burkert, U., and Allinger, N.L., "Molecular Mechanics" ACS Monograph, Washington D.C. (1982)

3. Comba, P., and Hambley, T.W., "Molecular Modeling of Inorganic Compounds," VCH, Weinheim. (1995)

4a. Rappé, A.K., Casewit, C.J., Colwell, K.S., Goddard III, W.A., and Skiff, W.M., *J. Am. Chem. Soc.* 114:10024 (1992)

4b. Casewit, C.J., Colwell, K.S., and Rappé, A.K., *J. Am. Chem. Soc.* 114:10035 (1992)

4c. Casewit, C.J., Colwell, K.S., and Rappé, A.K., *J. Am. Chem. Soc.* 114:10046 (1992)

4d. Rappé, A.K., Colwell, K.S., and Casewit, C.J., *Inorg. Chem.* 32:3438 (1993)

5a. Comba, P., Hambley, T.W., and Okon, N., "MOMEC, a molecular modeling package for inorganic compounds", Altenhoff & Schmitz, Dortmund (1995)

5b. Bernhardt, P.V., and Comba, P., *Inorg. Chem.* 31:2638 (1992)

6a. Charles, R., Coanly-Cummingham, M., Warren, R., and Zimmer, M., *J. Mol. Struc.* 265:385 (1992)

6b. Tueting, J.L., Spence, K.L., and Zimmer, M., *J. Chem. Soc. Dalton Trans.* 551 (1994)

6c. Munro, O.Q., Bradley, J.C., Hancock, R.D., Marques, H.M., Marsicano, F., and Wade, P.W., *J. Am. Chem. Soc.* 114:7218 (1992)

6d. Marques, H.M., Munro, O.Q., Grimmer, N.E., Levendis, D.C., Marsicano, F., Pattrick, G., Markoulides, T., *J. Chem. Soc., Faraday Trans.* 1741 (1995)

6e. Marques, H.M., Brown, K.L., *Inorg. Chem.* 34:3733 (1995)

7a. Kepert, D.L., *Prog. Inorg. Chem.* 23:1 (1977)

7b. Kepert, D.L., "Inorganic Chemistry Concepts" Springer, Berlin Vol. 6 (1980)

8a. Hambley, T.W., Hawkins, C.J., Palmer, J.A., and Snow, M.R. *Aust. J. Chem.* 34:45 (1981)

8b. Comba, P., *Inorg. Chem.* 28:426 (1989)

9. Comba, P., and Zimmer, M., *J. Chem. Educ.* in press

10. Comba, P., Hambley, T.W., and Ströhle, M., *Helv. Chim. Acta.* in press

11. Comba, P., Lienke, A., and Okon, N., work in progress

12. Comba, P., and Zimmer, M., *Inorg. Chem.* 33:5368 (1994)

13. Root, D.M., Landis, C.R., and Cleveland, T., *J. Am. Chem. Soc.* 115:4201 (1993)

14a. Burton, V.J., Deeth, R.J., *J. Chem. Soc., Chem. Commun.* 573 (1995)

14b. Burton, V.J., Deeth, R.J., Kemp, C.M., Gilbert, P.J., *J. Am. Chem. Soc.* 117:8407 (1995)

15. Wüthrich, K., "NMR of proteins and nucleic acids" Wiley, New York (1986)

16. Wüthrich, K., *Acc. Chem. Res.* 22:36 (1989)

17. Comba, P., and Sickmüller, A., in preparation

18. Comba, P., and Jakob, H., work in progress

19a. Lever, A.B.P., *Inorg. Chem.* 29:1271 (1990)

19b. Lever, A.B.P., *Inorg. Chem.* 30:1980 (1991)

19c. Masui, H., Lever, A.B.P., *Inorg. Chem.* 32:2199 (1993)

19d. Dodsworth, E.S., Vlcek, A.A., Lever, A.B.P., *Inorg. Chem.* 33:1045 (1994)

20. Pilbrow, J.R., and Smith, T.D., *Coord. Chem. Rev.* 13:173 (1974)

21a. Bernhardt, P.V., Comba, P., Hambley, T.W., Massoud, S.S., and Stebler, S., *Inorg. Chem.* 31:2644 (1992)

21b. Comba, P., Hilfenhaus, P., *J. Chem. Soc., Dalton Trans.* 3269 (1995)

21c. Comba, P., Hambley, T.W., Hilfenhaus, P., Richens, D.T., *J. Chem. Soc., Dalton Trans.* in press

22a. Comba, P., *Inorg. Chem.* 33:4577 (1994)

22b. Bernhardt, P.V., and Comba, P., *Inorg. Chem.* 32:2798 (1993)

22c. Comba, P., Hambley, T.W., Hitchman, M.A., Stratemeier, H., *Inorg. Chem.* 34:3903 (1995)

23. Golub, G., Cohen, H., Paoletti, P., Bencini, A., Messori, L., Bertini, I., Meyerstein, D., *J. Am. Chem. Soc.* 117:8353 (1995)

24a. Hambley, T.W., *Inorg. Chem.* 27:2496 (1988)

24b. Ventur, D., Wieghardt, K., Nuber, B., Weiss, J., *Z. Anorg. Allg. Chemie* 551:33 (1987)

24c. Bond, A.M., Lawrance, G.A., Lay, P.A., Sargeson, A.M., *Inorg. Chem.* 22:2010 (1983)

24d. Hendry, P., Ludi, A., *Adv. Inorg. Chem. (Sykes, A.G., Ed.)*35:117 (1990)

25. Sargeson, A.M., unpublished observations

26a. Bernhardt, P.V., Comba, P., *Helv. Chim. Acta.* 74:1834 (1991), 75:645 (1992)

26b. Bernhardt, P.V., Comba, P., Hambley, T.W., *Inorg. Chem.* 32:2804 (1993)

27a. Hendry, P., Ludi, A., *J. Chem. Soc., Chem Commun.* 891 (1987)

27b. Hendry, P., Ludi, A., *Helv. Chim. Acta* 71:1966 (1988)

28. Comba, P., Nuber, B., and Sickmüller, A., in preparation

29a. Bond, A.M., and Oldham, K.B., *J.Phys. Chem.* 87:2492 (1983)

29b. Bond, A.M., and Oldham, K.B., *J.Phys. Chem.* 89:3739 (1985)

29c. Bond, A.M., Hambley, T.W., and Snow, M.R., *Inorg. Chem.* 24:1920 (1985)

29d. Bond, A.M., Hambley, T.W., Mann, D.R., and Snow, M.R., *Inorg. Chem.* 26:2257 (1987)

30. Comba, P., and Sickmüller, A., unpublished observations

31. Rosozha, S.V., Lampeka, Y.D., and Maloshtan, I.M., *J. Chem. Soc. Dalton Trans.* 631 (1993)

32. Menif, R., Martell, A.E., Squattrito, P.J., Clearfield, A., *Inorg. Chem.* 29:4723 (1993)

33. Magnus, K.A., Ton-That, H., Carpenter, J.E., *Chem. Rev.* 94:727 (1994)

34a. Bernhardt, P.V., Comba, P., Hambley, T.W., and Lawrance., G.A., *Inorg. Chem.* 30:942 (1991)

34b. Bernhardt, P.V., Hambley, T.W., and Lawrance., G.A., *J. Chem. Soc. Chem. Commun.* 553 (1989)

35. Stratemeier, H., Hitchman, M.A., Comba, P., Bernhardt, P.V., and Riley, M., *Inorg. Chem.* 30:4088 (1991)

CONFORMATIONAL ANALYSIS OF LONG CHAIN SECO-ACIDS USED IN WOODWARD'S TOTAL SYNTHESIS OF ERYTHROMYCIN A - CONFORMATIONAL SPACE SEARCH AS THE BASIS OF MOLECULAR MODELING

Eiji Ōsawa,* Eugen Deretey, and Hitoshi Gotō

Department of Knowledge-Based Information Engineering
Toyohashi University of Technology
Tempaku-cho, Toyohashi
Aichi 441, Japan

Abstract:

The crucial step in Woodward's total synthesis of erythromycin A, published in 1981, was the macro-cyclization of appropriately substituted α-hydroxy-ω-carboxylic acid, called seco acid. After a number of unsuccessful attempts, they discovered that only one seco acid cyclized into lactone with an acceptable yield. Since then, it has been assumed that the seco acid that cyclized takes a folded conformation favorable for cyclization but this assumption has never been proved. This lecture reports the progress in our computational work aimed at identifying the structural elements that generated the folded conformation in this particular seco acid but not in the other, similarly substituted seco acids that Woodward and his coworkers studied. While the conformational space search of all of the Woodward's seco acids have not been completed, we have so far found a cluster of folded conformers having close distance between the reacting hydroxy and carboxylic groups and have been able to determine a transition state of the cyclization reaction that arises from one of the folded conformers by using semi-empirical molecular orbital calculations.

INTRODUCTION

Principles of molecular modeling are generally derived from the studies, either experimental, theoretical or computational, of small molecules. Molecules chemists are interested are sometimes much larger than those model molecules. For example, consider a problem of predicting the higher-order structure of a peptide molecule. Factors that determine the final conformation of peptides are so numerous and they interact so heavily among each other that it is extremely difficult to predict on *a priori* basis the outcome of the balance of these factors. Consequently, knowledges on the small molecules alone are often insufficient to be able to predict favorable conformations of a peptide or other large and flexible molecules.

The computer modeling techniques like molecular mechanics and molecular dynamics calculations are certainly capable of solving such a complex problems.[1] However, it is generally difficult to understand from the result of such calculations how the final structure was obtained. This situation resembles the difficulties in understanding details of interactions among

Fundamental Principles of Molecular Modeling
Edited by Werner Gans *et al.*, Plenum Press, New York, 1996

the primitive molecular orbitals that eventually led to the final molecular orbitals set from the results of SCF calculations alone. We always wish to grasp the global situation and to understand the mechanism of interactions. We are also interested to know if there are new conformational principles in large molecules that we have never encountered before in small molecules. We are attracted to analyzing the results of theoretical calculations in order to draw some general rules behind the maze of numerals.

A particularly interesting case in point is the macrocyclization reaction in the famed synthesis of erythromycin A (**1**) by Woodward and his coworkers,[2] in which they prepared seventeen seco-acids by painstaking multistep procedures, but only one (**2**) of them cyclized to the desired macrolide in an acceptable yield while the rest failed mysteriously. The working hypothesis employed by these workers for this key step is that the reactive species must have a folded conformation

2   **1**

in which the reacting end groups (OH and ester) are disposed close to each other already in the ground state conformation and that such a folded conformation may be realized by appropriate patterns of protecting groups.[2] However, as mentioned above, the latter hypothesis turned out to be extremely difficult to achieve. Although eventually they found the correct protecting pattern by a sheer luck, the goal had to be fought over by the heroic Edisonian approach of trial and error.

The working hypotheses they took were, in a sense so dangerous that no one dared to take the same strategy until today. It would have been a terrible disaster if they failed in such a big project after consuming gigantic amounts of cost and manpower. Clearly, this is the kind of work which only Woodward could have done. One can readily imagine that the most difficult part of the work would have been to persuade the proud and critical coworkers to get involved in such a risky project, who gathered from all around the world as a sort of representative from their own native countries in order to add some experience of working in the prestigious laboratory in Harvard to their already brilliant records. Only a man like Woodward with so high confidence in himself and great past achievements could have driven as many as 49 of them to such a demanding job.[3] To be true, we owe a great deal to these brave people for their courage and devotion. We really are lucky to share the seventeen cases of success and failure, from which one should be able to extract the causes of such a long linear chain molecule to take nicely folded conformations.

Macrocyclization reactions are inherently difficult to achieve. Once a friend of mine, a noted natural products chemist, replied to my question that these reactions must be easy because we see so many of them in the publication. His reply was that they were certainly difficult to achieve and there should be more failures than successes, the only thing being that they never publish failures. The reaction must work against the thermodynamic law (large loss of entropy and increase in strain). While the high dilution method is certainly a viable solution, it is basically for the laboratory experiments and its industrial application would be highly disadvantageous, especially in these days when large amounts of solvents pose such serious environmental problems.

For reasons mentioned above in some length, the cases of macrolactonization encountered in Woodward's erythromycin A synthesis are worthwhile studying. The folding of seco acid appears to occur by a subtle change in remote substitution. Whereas it is difficult to predict if there would be any generality in the conformational feature of such a special molecule like erythronolide A seco acids **2**, it is still challenging and interesting to identify the structural conditions for producing the folded conformation of **2**, which has been believed to populate only in this but not in the other seco-acids. We analyze the Woodward case by computational conformational analysis technique. It is hoped that this work, when completed, will benefit the molecular modeling of macrocyclic molecules or the chemistry of macrocyclization in general.[4]

METHODOLOGY

So far no one has ever attempted the conformational analysis of **2**, clearly due to the too many number of possible conformers that have to be covered for this large and flexible molecule. As we show below, we took into accounts of 25 bonds or torsional angles as the variables in the conformational space search of **2**. If all of the rotamers were to be systematically generated by locating three energy minima (trans, gauche and gauche') for each rotatable C–C and C–O bond, the total number of conformers to be geometry-optimized would be 3^{25} or approximately 10^{11}. If it takes an average of 10 seconds to optimize each geometry by molecular mechanics, the job will need about 20,000 years to complete!

We began preliminary calculations of seco acids in early 1980s and tested several available conformational space search programs. It soon became evident that the random search programs that were popular at those times were incapable to cope with these molecules. Later in the course of testing our own conformational search program, we learned that conformational spaces of organic molecules are much more crowded in the high-energy region than in the low-energy region. This means that the conformation search by the random method has to spend much of the computing time in geometry-optimizing those unstable and difficult conformations which we do not need. We actually need only the low-energy conformers and can disregard high energy ones.[5]

Before realizing the pitfall of random search, we had begun to exploit an entirely new algorithm of conformational space search by imitating elemental processes of thermal molecular vibration in 1988. This work developed into a new category of conformational space search which we call an *efficient* search method and was completed in 1993.[6] In the efficient method, *only the low-energy regions of conformational space are exhaustively exploited.* The way our program does conformer searching resembles filling an empty reservoir with a small stream of water, which first goes down to the lowest point (global energy minimum) and then fill the space slowly but never missing any voids while going upwards. For this reason we named the method 'reservoir filling' algorithm.

In addition to the reservoir algorithm, we have implemented several other new strategies to increase the efficiency of the conformation search. These are (1) adaptation of fast molecular mechanics calculation using the block diagonal Newton-Raphson geometry-optimization, (2) quick removal of duplicate conformations by a pre-check option, in which key dihedral angles of the conformation being optimized are compared with those of the all unique conformers already found, (3) 'stepping stone' strategy to find global energy minimum, in which the lowest-energy conformer among the unique conformers that have been found is used to the starting conformation in the next perturbation cycle, and (4) 'variable search limit' strategy, in which the height of reservoir is increased gradually as the search progresses, so that the search can be terminated when no new low-energy structure appeared any longer in the course of the two successive increases in the search limit.[6]

It is important to emphasize here that the previously inaccessible conformational problems of very large molecule like **1** and **2** became amenable for the first time with the appearance of our program CONFLEX[7] embodying all these algorithms.

GLOBAL SEARCH OF FOLDED CONFORMATION

We have initially replaced the seco acid **2** and its cyclized product with somewhat simpler model structures **3** and **4**, respectively, and subjected them to the CONFLEX search in combination with MM2(91)[8] as the geometry-optimizer to find thousands of conformers within 10 kcal/mol from the global energy minimum.[9]

3

4

These conformers were first classified into clusters based on the skeletal conformation. A pair of conformation A and B are judged to belong to the same cluster if *conformational distance*, d_{AB}, is less than 10^0. d_{AB} is a measure of conformational similarity, and is defined here as the root-mean-square differences of the seventeen endocyclic dihedral angles along the atoms and bonds being perturbed:

$$d_{AB} = \sqrt{\frac{\sum_{i=1}^{17}(\omega_i^A - \omega_i^B)^2}{17}} \qquad (1)$$

where ω_i^A and ω_i^B are the i-th dihedral angle of conformers A and B, respectively. The summation covers the backbone chain and ring bonds being compared as shown in **5**.

Among several hundreds of conformational clusters of **3** thus created, we did find several clusters which indeed had close average distances between the centers of ring closure, C1 and O13, namely folded conformations. However, many of them are simply unsuitable for the cyclization, either having bad orientation or too large separation between the end groups, or

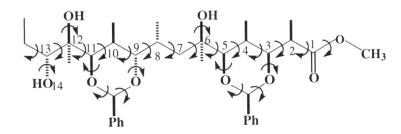

5

having no resemblance to the product conformation.[10] One particular cluster, whose stability rank among the clusters of **3** is well within the ten most stable, has a remarkably short average C1-to-O13 distance of 3.4 Å. The most stable member of this cluster, **3a**, is shown below in ORTEP stereo drawing for crossed-eyes view (all stereo drawings in this paper are for the crossed view). Furthermore, this cluster has a *cluster distance* (similarly defined as equation (1) except that the dihedral angles ω_is are the population-weighted average over the whole component conformers in the cluster) of only 4.7° from the most stable and dominant (96.8 %) cluster **4a** of the ring closed product, of which the most stable member is shown. This means that the folded seco acid conformers in the cluster **3a** are almost superimposable with those of lactone cluster 4a and that the ring closure of **3a** does not need much nuclear movements. It is hence likely that the cluster **3a** is 'it' which chemists have been looking for more than 15 years.[9]

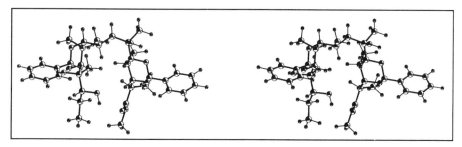

3a

4a

There are, however, three problems regarding **3** that must be rationalized. The first question is the fact that the equilibrium concentration of the folded cluster **3a** seems to be very low, very probably less than 2 %.[11] Regarding this point, we found at least two stable clusters of **3** having the combined equilibrium population of about 40 % are close to **3a**: intercluster distances between **3a** and these dominant clusters are 27^0 to 28^0.[12] This means that, while **3a** may be quickly cyclized into **4a** and hence consumed quickly, it will be promptly replenished from these populating clusters.

Two other concerns regarding the suitability of **3a** to be the reactive cluster is that if the C1-O13 distance of 3.4 Åis really short enough to induce the intramolecular esterification reaction and that **3a** is 10.4 kcal/mol higher in the MM2-strain than **2a**. While this distance is certainly shorter than the threshold value of ca. 4.2 Å, the largest allowed intermolecular separation of two parallel double bonds for the photochemical $[\pi 2 +_\pi 2]$ cycloaddition reactions in the crystal of cinnamic acid to take place,[13] we are not certain of the 'reactive distance' for a different reaction, namely nucleophilic substitution taking place in solution. The reaction is likely to be an endothermic reaction. If the loss of entropy is taken into account, how high would be the activation energy of this reaction? The two last questions can be best answered by actually locating the reaction transition state that exists between **3a** and **4a**, which is described below.

COMPUTATIONAL TECHNIQUE OF TRANSITION STATE

PM3 modification of the MNDO Hamiltonian was used within the framework of Dewar's MOPAC semi-empirical MO package.[14] PM3 has been specifically designed to handle sulfur atoms, henceforth this parameters set are suited for our present purpose of calculating sulfur derivatives. The transition state was searched first by gradually decreasing the distance between the reactive centers, namely the hydroxylic oxygen atom and ester carbonyl carbon atom, and subjecting each point to the GEO-OK geometry-optimization. The candidate structures thus obtained were then subjected to the NLLSQ option to locate the first order saddle point having one imaginary frequency. The TS option implemented in MOPAC package failed in the model Corey system, but this option proved useful when applied to the almost but never completely finished NLLSQ result of the transition state calculation of the seco acid system. Intrinsic reaction coordinate (IRC) calculations were then performed starting from the first order saddle point to see if it leads to both the starting and the product energy-minima.

The same MM2(91) molecular mechanics force field[8] as used in the global conformation search was employed whenever additional MM calculations became necessary.

6   7

194

MODEL CALCULATIONS ON COREY'S 2-PYRIDYL THIOLATE LEAVING GROUP

Our purpose here is to find the transition state of the reaction of **2** to **1** by means of semi-empirical calculations, by utilizing the folded conformation known from the MM calculations of model **3**. Woodward used Corey's leaving group, 2-pyridyl thiolate,[15] to effect the macrocyclic esterification (see **2**). Because our preliminary conformation search was carried out with a model **3** which contained methoxycarbonyl group as the leaving group instead of 2-pyridyl thiolate, we must first find out the transition state structure of the Corey's system. For this purpose, we used a simple model system, 2-acetylthiopyridine (**6**) and methanol.

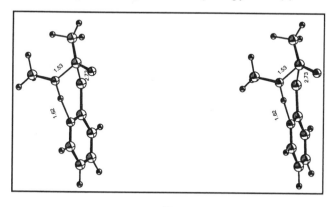

7

We have tested various initial orientations of **6** and methanol, but only one of them gave a first-order saddle point which gave back the starting system as well as the ester by the IRC calculation. The transition state (**7**, see also the inset stereo picture) thus obtained had a sofa-like six-membered ring involving S, CO, O, H, N and a pyridyl carbon atoms with the CO group somewhat protruding out of the plane of other five atoms. Thus, the pyridine ring plane is almost coplanar with the incipient ester C-O single bond. **7** has one imaginary frequency of $191.5i \text{ cm}^{-1}$ whose main mode is the carbonyl carbon-methanol oxygen stretch. Pertinent interatomic distances in **7** are as follows: S-CO 2.73 Å, O-CO 1.53 Å, O-H 1.05 Å, N-H 1.62 Å. As these values indicate, in the six membered ring of transition state, S-CO bond is almost broken while O-CO bond is half formed and the hydroxyl proton is about in midway from alcoholic oxygen to the pyridine nitrogen. This picture is consistent with the mechanism of double activation of hydroxy and carboxyl functions simultaneously by internal proton transfer as envisaged by Corey.[16] The final gradient norm is 0.96 kcal/mol/, and the calculated activation enthalpy 50 kcal/mol.

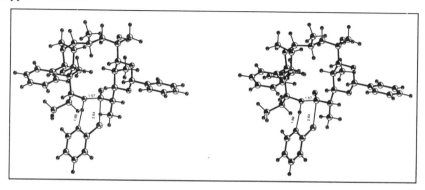

8

TRANSITION STATE OF THE MACROCYCLIZATION REACTION OF 2 TO 4 BY PM3 CALCULATION

We located the transition state (**8**) of the cyclization from the folded cluster **3a** by (1) fitting the model transition state structure 7 into one of the conformations in **3a**, which has a similar rotation of ester group as in **7**, (2) optimizing the whole molecule with PM3 except for the reacting portion which was fixed, and then (3) performing the NLLSQ transition state search option. In this way, we were able to locate the transition state **8**, which was further confirmed to give the final product **4** by IRC calculations. Re-optimized with MM2, this product structure proved to be identical with one of the conformers in the dominant cluster of lactone **4a**.

8 has one imaginary frequency of $238.1i$ cm^{-1} whose main mode is the C1-O14 stretch. Pertinent interatomic distances are as follows: S-CO 2.64 Å, O-CO 1.57 Å, O-H 1.02 Å, N-H 1.69 Å. The final gradient norm is 0.60 kcal/mol/Å, and the both calculated activation enthalpy 65 kcal/mol. All these values are reasonably close to those obtained for the model 2-acetylthiopyridine/methanol system. This means that the large and flexible 'ligand' of **8** can adjust itself by utilizing its large freedom of nuclear movements to bring the structure of reaction site near the ideal transition state geometry of the open model system **7**. It seems that, once a good conformation is taken by the ligand, then the macrocyclization may have a reasonably low energy barrier.

Thus, we believe that the folded conformation **3a** that we have found in the global conformational space search of **3** is indeed a highly likely candidate for the reactive conformation for macrocyclization, on the ground that it produced the desired lactone by way of a transition state which has been characterized by rigorous mathematical treatment. While all these have been done by computation, it should be noted that there is no experimental equivalent to these techniques and at the moment theoretical approach is the only viable tool.

CONCLUDING REMARKS

Many more things remain to be done, all waiting for the CPU time of our workstation system. First we must find out why the rest of the sixteen seco acids failed. We wish to prove that either they do not produce ideally folded conformations like **3a** in significant equilibrium concentration or even if there were folded conformations there is no efficient way to supply the consumed conformation. Here again we expect in the latter case that only a limited number of conformational interconversion pathways are available in these highly substituted acyclic systems. Thereupon, we wish to extract informations on the cause of folded conformation from the careful analysis of the computed results.

Acknowledgment. Generous financial support given by the Organizing Committee of INDABA is gratefully acknowledged.

REFERENCES AND NOTES

1. H.U. Suter, D.M. Maric, J. Weber, and C. Thomson, *Chimia* 49:125 (1995).
2. R.B. Woodward et al., *J. Am. Chem. Soc.* 103:3210, 3213, 3215 (1981).
3. See A. Nickon, and E.F. Silversmith, "Organic Chemistry — The Name Game, Modern Coined Terms and Their Origins", Pergamon Books, New York (1987) p. 67.
4. E.J. Thomas, *Acc. Chem. Res.* 24:229 (1991); N.A. Porter, B. Giese, and D. P. Curran, *Acc. Chem. Res.* 24:296 (1991); Q. Meng, and M. Hesse, *Topics Current Chem.* 161:107 (1991); P. Guerriero, P.A. Vigato, D.E. Fenton, and P.C. Hellier, *Acta Chem. Scand.* 46:1025 (1992).
5. H. Gotō, E. Ōsawa and M. Yamato, *Tetrahedron* 49:387 (1993).

6. (a) H. Gotō, and E. Ōsawa, *J. Mol. Struct. (THEOCHEM)* 285:157 (1993). (b) S. Tsuzuki, L. Schäŀfer, H. Goōt, E.D. Jemmis, H. Hosoya, K. Siam, K. Tanabe, and E. Ōsawa, *J. Am. Chem. Soc.* 113:4665 (1991); (c) H. Gotō, and E. Ōsawa, *J. Chem. Soc., Perkin Trans.* 2:187 (1993); (d) H. Gotō, Y. Kawashima, M. Kashimura, S. Morimoto, and E. Ōsawa, *J. Chem. Soc., Perkin Trans.* 2:1647 (1993).

7. CONFLEX Version 3, JCPE, P40, P55; QCPE No. 592. The address of JCPE (the Japan Chemistry Program Exchange) has changed as of Oct. 1, 1995 to Nishi-nenishi 1-7-12, Tsuchiura-shi, Ibaraki-ken 300, Japan. Fax x-81-298-30-4162.

8. N.L. Allinger, R.A. Kok, and M.R. Imam, *J. Comput. Chem.* 10:591 (1988).

9. T. Hata, H. Got, E. Ōsawa, T. Hamada, and O. Yonemitsu, *Electron. J. Theor. Chem.* in press.

10. There are only two significantly populating clusters for **4**. Ref. 9.

11. At the time of writing this Proceeding, we have not completely covered all of the significantly low-energy conformational space of **3**.

12. A distance of 30 means that the pair of conformers convert into each other by the rotation of only one bond.

13. G. Desiraju, "Crystal Engineering," Elsevier, Amsterdam (1989), p. 4.

14. (a) J.J.P. Stewart, *J. Comput. Chem.* 10:221 (1989); (b) M.B. Coolidge, J.E. Marlin, and J.J.P. Stewart, *J. Comput. Chem.* 12:948 (1991).

15. E.J. Corey and K.C. Nicolaou, *J. Am. Chem. Soc.* 96:5614 (1974).

16. E.J. Corey, D.J. Brunelle, and P.J. Stork, *Tetrahedron Lett.* 38:3408 (1976).

PACKING MOLECULES AND IONS INTO CRYSTALS

Leslie Glasser

Centre for Molecular Design
Department of Chemistry
University of Witwatersrand
P.O. Wits 2050
Johannesburg, South Africa

Abstract

It is a matter of great interest and significance to be able to predict the packing of molecules, especially of flexible molecules, and of ions into crystals. Packing may affect physical properties such as rate of solution, hardness, crystal shape, and so forth; these properties may be of crucial consequence in chemical reactivity and pharmacological activity.

The first attempts at packing predictions (due to Kitaigorodskii) relied on general considerations regarding the overall shapes and volumes of the molecules to be packed. Presently, work is proceeding by using force fields to describe the details of species interactions, and minimizing the resultant energy. Sophisticated minimization procedures are needed to overcome the "multiple minimum" problem. Procedures for these packing calculations will be described, and successes and failures thereof noted.

INTRODUCTION

Crystallization is one of the fundamental unit operations of chemistry; it is central in the purification of materials, in the study of chemical structure by X-ray and other methods, and in the design, preparation, and processing of matter in the solid state.

It was early appreciated that the observed regularities of crystals, with flat faces, fixed interfacial angles, cleavages, and so forth, imply an underlying periodicity of structure, in the form of packing of identical units in regular arrangements. Haüy's structural mode[1] was a little too *ad hoc*, however, consisting simply of packing of identical blocks with appropriately chosen symmetry.

EQUAL SPHERES

The packing problem of less conveniently shaped particles, that is, of equal spheres, was tackled first by Kepler (1611) whose purpose was an exploration of the symmetry of ice crystals, later followed by Gauss (1831) and many others. These early workers described dense systems with spheres in closest-packed arrangements, leading to cubic and hexagonal closest-packing. Gauss proved that such packing was the densest possible for a periodic array (1831), but it was only in 1991 that Hsiang was able to prove that these are the densest of all packings, even allowing for non-periodic (quasi-crystalline) arrangements[2].

Gauss's method of proof (which will be relevant to us later) was to use the intuitively (but not mathematically) obvious result that a layer of spheres in hexagonal packing (each sphere having six immediate neighbours) is the densest possible in a plane. Then, laying the bumps of a second layer (B) in the hollows of the first (A), we get a densest-layered system. Finally, we can produce a regular continuation of the ...AB.... sequence, yielding hexagonal closest-packing (hcp), each sphere with twelve nearest neighbours in contact, or of an equivalently dense layering, ...ABC....., corresponding to cubic closest-packing (ccp). Of course, any combination of these – regular or irregular, such as ...ABACBA... – also yields a closest-packing of identical density, since each sphere always has twelve nearest neighbours in contact.

The validity in atomic terms of these packing patterns of equal spheres became appreciated when the structures of metals and alloys were determined, and the ideas were also extended to describing the structures of simple ionic materials, such as NaCl or ZnS, by allowing for the insertion of smaller spheres into the gaps between adjacent closest-packed layers. Even complex systems, such as $NaNO_3$, can be related in their packing to the simple ionic systems – $NaNO_3$ has a similar pattern of packing to NaCl[3].

Large molecules may occur with roughly spherical shapes (buckyballs, some globular proteins), and may adopt a packing structure related to one or another of the closest packings.

IRREGULAR OBJECTS: GENERAL PRINCIPLES

Most organic molecules have irregular shapes. In spite of this, their crystals have similar densities (~ 0.8–0.9 g cm^{-3}), and even polymorphs hardly differ in their densities (for example, anthraquinone[4] has volume packing coefficients: 0.773 and 0.778).

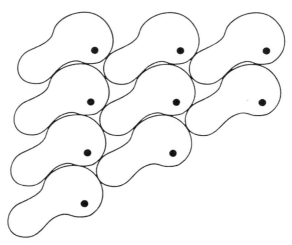

Figure 1. Closest-packing of arbitrary shapes in a layer; each object has six nearest neighbours (after A.I. Kitaigorodsky, "Molecular Crystals and Molecules", Academic Press, New York (1973)).

This strongly suggests that even these irregular molecules pack in a close-packing so as to best occupy space, without specific interactions which would control alignments. Indeed, as early as 1929/30, B.P. Orelkin was suggesting that the "bumps and hollows" principle applied, with bumps in one molecule fitting into complementary hollows in a neighbour[4]. The Rus-

200

sian crystallographer, A.I. Kitaigoradskii, developed this idea of close-packing to its limits by considering which space groups (or packing arrangements) were compatible with irregularly shaped objects, with or without symmetry.

Kitaigorodskii[4] started with the same premise as did Kepler and Gauss, that the number of nearest neighbours in a plane, even for an irregular object, must be six for closest packing. This requires a low-symmetry oblique or right-angled cell (Fig. 1 to 3), so limiting the choices of lattice (to nine of the seventeen plane groups). Developing this theme, he concluded that, for six-fold coordination in a closest-packed layer to be possible, the objects can have no symmetry in the lattice, or must occupy inversion centres.

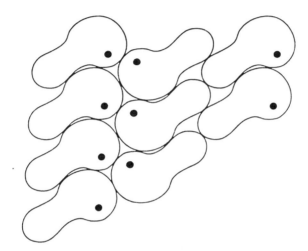

Figure 2. Closest-packing of arbitrary shapes in a layer, utilizing centres of symmetry (after A.I. Kitaigorodsky, "Molecular Crystals and Molecules", Academic Press, New York (1973)).

The Dutch artist, Maurits Escher[5], has most deeply explored the packing of irregular objects in the plane. He has had the advantage of being able to adjust his objects to the space available (Fig. 4) – so that his planes are always closest packed (at 100 %) – but still subject to the limitations noted by Kitaigorodskii.

IRREGULAR OBJECTS: THREE DIMENSIONS

Kitaigorodskii carried his analysis through from plane groups to determine which of the 230 space groups can accommodate irregular objects, and established that only 44 space groups are permissible, only thirteen are probable, and only eight are closest-packed (Table 1). This analysis was supported by an extensive discussion of many crystal structures, tending to support the conclusions.

PACKING CALCULATIONS

In more recent years, emphasis has, of course, been directed to computational assessment of the packing of molecules into crystals. The force fields which are used are of the same form as used in molecular mechanics for isolated molecules (indeed, such force fields are often derived using crystal structures for reference). However, the calculations required are much more extensive; not only must one satisfy the internal forces to establish the stable conformation of the molecule itself, but that conformation is subject to the "packing forces"

Table 1. Closest-Packed Space Groups[a], and their Frequencies[b]

Molecular Symmetry in Crystal	Space Group[c]		Frequency	
	Number	Symbol	Number	%
1	2	$P\bar{1}$	-	
	4	$P2_1$	2 488	8.3
	14	$P2_1/c$	13 877	46.2
	29	$Pca2_1$	275	0.9
	33	$Pna2_1$	600	2.0
	19	$P2_12_12_1$	4 466	14.9
$\bar{1}$	2	$P\bar{1}$	-	–
	14	$P2_1/c$	13 877	46.2
	15	$C2/c$	2 354	7.8
	61	$Pbca$	1 645	5.5
m	62[d]	$Pnma$	502	1.7
TOTAL			30 024	100

[a]from A.I. Kitaigorodsky, "Molecular Crystals and Molecules", Academic Press, New York (1973), Table 2, p. 34

[b]Data reported by A.J.C. Wilson, paragraph 9.7, p. 792, of A.J.C. Wilson, ed., International Tables for Crystallography, vol. C, Kluwer Academic Publishers, Dordrecht (1995).

[c]T. Hahn, ed., International Tables for Crystallography, vol A, 3rd revised ed., Kluwer Academic Publishers, Dordrecht (1992).

[d]The space group $Pnma$ is the only one remaining with a frequency greater than 350; it is listed by Kitaigorodskii[a] as being "of maximum density".

of neighbours in the conformation. So, each molecule in the packed array must be a copy of the prototype molecule, and each copy must symmetrically follow any changes in the conformation, orientation and position of the prototype.

Such calculations are performed by setting up a unit cell with a basis (or asymmetric unit), which is replicated within the cell by the symmetry operations of the space group, and the cell is repeated by translation in three dimensions. Each atom in the asymmetric unit now interacts with each of its intra-molecular atomic neighbours, and also with the atoms in the molecular copies. In principle, the energy summation extends to infinity; fortunately, the strength of the interaction falls off quite rapidly with distance, and one need only sum to a radius of 10 Å, or so, to have the sum converge quite satisfactorily. Nevertheless, there may be ten thousand or so terms to be summed for a crystal.

Charge interactions, following Coulomb's law, are of much longer range, and special procedures need to be invoked to obtain convergence with a reasonable number of terms (the reciprocal lattice is used, in an Ewald summation procedure) – it may be recalled that the Madelung sum is related, and is known to be very slowly convergent indeed.

Once this lattice energy is determined for a given conformation in a structure, the structural variables (such as lattice constants, positional and torsional variables) are adjusted by an optimizer towards an energy minimum. Optimizers adjust the variables so as to reduce the energy and this, almost invariably, means that a local energy minimum will be reached rather than the desired global energy minimum. Procedures such as simulated annealling are now available for the optimizer to surmount small energy humps, but there can never be a guarantee that the global minimum is reached; indeed this "multiple minimum" problem has a similar status in chemical computation[6] as the notorious "phase problem" has had in X-ray crystallography.

The standard application of force fields in crystal packing calculations has been, initially, simply to reproduce observed structures. Latterly, workers have extended packing calculations: to predict conformations of flexible molecules perhaps in novel environments; to assist in solving X-ray crystallographic problems; or to investigate the behaviour of novel materials. As a very simple example, we have studied the structure and predicted the lattice energy of $NaCl \cdot 2H_2O$[7].

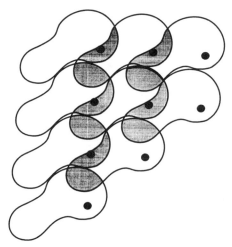

Figure 3. Demonstration that tetragonal symmetry is unsuitable for packing arbitrary shapes in a layer (after A.I. Kitaigorodsky, "Molecular Crystals and Molecules", Academic Press, New York (1973)).

USING KITAIGORODSKII'S PRESCRIPTIONS

A number of attempts have been made to see if molecules can be fitted into structures other than those adopted by Nature; this is not as odd as it might seem at first sight, because of the existence of polymorphs and the possibility of some being as yet unobserved. Hagler and Leiserowitz[8] studied the packing of adipamide, which has an anomalous hydrogen-bonding structure. They showed that, for this particular amide, the formation of the usual cyclic dimers was less favourable than the observed hydrogen-bond chain, by virtue of a closer packing of layers using the chain structure. They also attempted packing in structures with various imposed initial monoclinic and orthorhombic symmetries, and note that "orthorhombic space groups often impose severe restraints ... (leading) to 'bad' contacts, which destabilize these structures R". This conclusion accords well with Kitaigorodskii's analysis to which (strangely enough, since his work appears as one of their references) they do not refer, nor did they take account of it in selecting their initial space groups.

A little while ago, Scheraga and I undertook to implement Kitaigorodskii's prescriptions by attempting to find stable packings of the flexible pentapeptide, leu-enkephalin[9]. This molecule is known in four different crystal polymorphs, both in extended and bent (hairpin) conformations, each molecule in the asymmetric unit having a different conformation (ten different conformations in all). The conformation of the isolated molecule has also been the object of many studies, yielding a (global?) lowest-energy bent structure, which is quite different from that observed in the crystals. Clearly, packing forces are at play here in altering the conformation of the flexible leu-enkephalin. The question was: could our force-field (ECEPP/2) reproduce the known structures, and could we find some, as yet unknown but stable, crystal structures? Reproduction of the known structures proved quite straight-forward although the calculations were very time-consuming, at 5hr/run of an IBM vectorizing supercomputer. In

Figure 4. M.C. Escher's packing of irregular objects in a plane; each object has six nearest neighbours, and the space is fully occupied (with permission, from C.H. MacGillivray, "Symmetry Aspects of M.C. Escher's Periodic Drawings", A. Oosthoek, Utrecht (1965)).

fact, the system of 332 atoms and 176 independent variables was the largest molecular optimization to have been reported at that time[10].

In order to find other stable structures, we used conformations obtained in other minimizations of isolated molecules, and placed them (with or without accompanying water molecules) in cells of space-groups $P2_1$ or $P2_12_12_1$ as suggested by Kitaigorodskii's analysis. Many trials yielded a smooth curve of packing energy versus volume per molecule (Fig. 5), with the known crystals having the smallest volume (highest densities) – in other words, we did not find better packings than Nature had chosen!

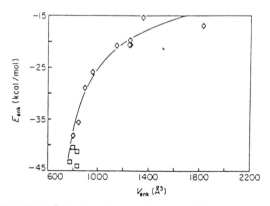

Figure 5. Packing energy versus volume per molecule of Leu-enkephalin, for a number of known and hypothetical crystal structures (with permission, from, L. Glasser and H.A. Scheraga, J. Mol. Biol., 199, 513 (1988)).

The slope of this curve of energy vs volume yields the interaction energy as prescribed in ECEPP/2 for H...H atoms. This suggests that the determining factor in the packing is H...H interaction, the flexible molecule simply altering its conformation to adjust to local packing requirements – a beautiful confirmation of Kitaigorodskii's intuition, and in accord with other observations that the mode of stabilization of a crystal is non-specific – except that H-bonds are formed where possible which further stabilize the corresponding structures. (Such bonds may not be determining[8]: *cf.* Kálmán's observation to the same effect in this Symposium; and the observation that disulphides in proteins serve a similar function[11].

Perlstein[12] has developed what he terms the Kitaigorodskii aufbau principle (KAP) method which uses a simulated annealling technique to first generate one-dimensional structures (with translation, screw, glide or inversion offsets). These structures tend to be local energy minima, visually very different from the global minimum, due to non-complementarity of surfaces (rather different from Allen's observation in this Symposium that crystal structures are generally not far from the global minimum). Then, in the *aufbau* spirit, Perlstein packs the one-dimensional chains into layers, where the molecular orientations prove to have been already well-defined in the one-dimensional case, while conformation is altered to the final condition already at the stage of packing into two dimensions, to form a layer. Coupled with limited experimental observation, one may then choose the local minima which are expected to be experimentally observable.

The extension to a full three-dimensional prediction is discussed, but not yet published.

AB INITIO PACKING METHODS

The crude methods of trial-and-error have now been superseded by more sophisticated, so-called *ab initio* methods – though, as yet, these are too crude to deal with a very flexible molecule of the nature of an enkephalin.

The new methods do not rely on quantum mechanics; rather, *ab initio* implies that only molecular information is used, and little or no prior crystal information is invoked in the modelling. One method[13] commences with small clusters (2 – 4 molecules) constructed with the most common of the symmetry elements: centre of symmetry; screw axis; and glide plane. These "nuclei" are built by a grid search of possible positions of the symmetry operators. Next, again by a grid search, a chain of structures is constructed by translational search, then a layer, and finally a three-dimensional structure. Such procedures yield a number of trial structures within the most common space groups. It is claimed that "moderate success was achieved in reproducing the observed crystal structures of test compounds", given the ambiguities involved in comparisons. Basically, structures are obtained with appropriate densities and with potential energies which differ among themselves by less than 10 % – unfortunately, these results are insufficiently precise to distinguish among polymorphs.

Gibson and Scheraga[14] have reported on studies of benzene structures, where they commenced with Pbca symmetry for the structure, using sixty different starting points, but optimized the structures without constraints. The known crystal structures were obtained with two force-fields, but a third force-field yielded some reversals in energy.

A more completely *ab initio* approach, with no assumptions with respect to likely symmetries is, that of Karfunkel and Gdanitz[15]. Their procedure operates in a number of stages, as follows, where groups of degrees of freedom are optimized sequentially, rather than attempting a full global optimization from the start – which, in any case, is likely to become trapped in a local minimum.

Stage 1: A single molecule or pair with frozen internal degrees of freedom (fixed bond lengths, angles, and torsion angles) is placed in a unit cell of no symmetry, P1. A Monte Carlo simulated annealling is performed, by allowing only molecular rotations to be optimized for a series of cell constants. Then, with fixed angular variables, the spatial constants are optimized.

Stage 2: The large number of crude crystal structures is now grouped into classes of similar structures by a "cluster" analysis.

Stage 3: Representative structures from each cluster are subjected to a full crystal optimization, optimizing all degrees of freedom simultaneously. The lowest energy structures, within a limited energy window, are regarded as the solutions to the problem.

In this way, Karfunkel and Gdanitz were able to generate the ethylene structure with essentially no error. A number of other materials produced crystal structures which agreed satisfactorily with known structures, while a last produced very different results – with the crystal structure being regarded as likely to be incorrect!

INORGANIC STRUCTURES

The problems in packing inorganic solids are no smaller than for organic; although the ions may have simpler shapes, their multiplicity in a structure and the long range of the forces renders the problem no more tractable. In one case, Freeman and Catlow[16] studied the structures of TiO_2 which is polymorphic, with non-cubic unit cells and having several formula units per cell. Because of the complexity of the problem, the starting point chosen was the known lattice constants and cell content, but with the latter placed in random positions. A process of successive refinements of fifty initial configurations (including lattice constant refinement) yielded the rutile structure in 41, with the others being of too low density.

Recently, Catlow and colleagues[17] have attempted a more ambitious project, the solution of the structure of the previously unknown lithium ruthenate, Li_3RuO_4. The crystallography did not yield a definitive structure. Hence, the most likely unit cell solution, together with cell contents for twenty individual trial solutions was fed into a program implementing a genetic algorithm. 1500 generations were produced, with breeding consisting of "crossover" of information between individuals, "mutation" by some random variation in the cell information, and "carry-over" of the more successful solutions (as judged by an imposed "cost" or penalty function). The ten best solutions of the final population (which had been allowed to grow to one hundred), were selected and energy minimised. The structures obtained fell into two groups, one group in excellent agreement with the experimental data, sufficient to assign a space group. The energy-minimised structural model was finally refined against the X-ray data by a Rietveld method, to yield a satisfactory structural description.

CONCLUSIONS

It has been shown that present computational procedures are sufficiently powerful and refined to produce crystal structures, of both small organic molecules and inorganic materials, which are representative of the structures of Nature, although they cannot yet be relied upon to yield a definitive structure. This is sufficient to provide guidance in refinement of X-ray structures and act as information for the design of materials for successful synthesis.

Clearly, refinements of the force fields are needed, to provide reliable packing descriptions. More importantly, optimization methods need upgrading[14]. Unfortunately, optimization is an exceptionally difficult task when working with a complicated potential energy hypersurface[18], and no early breakthrough in optimization should be anticipated.

REFERENCES

1. R.H. Haüy, "Essai d'une Théorie sur la Structure de Crystaux", Paris (1784); see C. Bunn, "Crystals", Academic Press, New York (1964), p. 7.
2. N. Max, *Nature* 355:115 (1992); I. Stewart, *New Scientist* 131:29 (8 July, 1991).
3. A. Holden, "The Nature of Solids", Columbia Univ. Press, New York (1965), Fig. 10, p. 108.
4. A.I. Kitaigorodskii, "Organic Chemical Crystallography", Consultants Bureau, New York (1961).
5. C.H. MacGillivray, "Symmetry Aspects of M.C. Escher's Periodic Drawings", A. Oosthoek, Utrecht (1965).
6. H.A. Scheraga, *Polish J. Chem.* 68:889 (1994).
7. G. Brink and L. Glasser, *J. Phys. Chem.* 94:981 (1990).
8. A.T. Hagler and L. Leiserowitz, *J. Amer. Chem. Soc.* 100:5879 (1978).
9. L. Glasser and H.A. Scheraga, *J. Mol. Biol.* 199:513 (1988).
10. G. Vanderkooi, *J. Phys. Chem.* 94:4366 (1990).
11. C.J. Camacho and D. Thirumalai, *Proteins: Structure, Function, Genetics* 22:27 (1995).
12. J. Perlstein, *J. Amer. Chem. Soc.* 116:455 and 11420 (1994).
13. A. Gavezotti. *J. Amer. Chem. Soc.* 113:4622 (1991). Program PROMET, Mark I, Milan (1993). Applications are described by D. Braga, F. Grepioni, E. Tedesco and A.G. Orpen, *J. Chem. Soc., Dalton Trans.* 1215 (1995).
14. K.D. Gibson and H.A. Scheraga, *J. Phys. Chem.* 99:3752 and 3765 (1995). Program LMIN, QCPE 664, Quantum Chemistry Program Exchange, Bloomington IN (1995).
15. H.R. Karfunkel and R.J. Gdanitz, *J. Comp. Chem.* 13:1171 (1992).
16. C.M. Freeman and C.R.A. Catlow, *J. Chem. Soc., Chem. Commun.* 89 (1992).
17. T.S. Bush, C.R.A. Catlow and P.D. Battle, *J. Mater. Chem.* 5:1269 (1995).
18. J. Pillardy and L. Piela, *J. Phys. Chem.* 99:11805 (1995)

ON THE ISOSTRUCTURALITY OF SUPRAMOLECULES: PACKING
SIMILARITIES GOVERNED BY MOLECULAR COMPLEMENTARITY

Alajos Kálmán

Central Research Institute for Chemistry
Hungarian Academy of Sciences
Budapest 114, P.O.Box 17, H–1525, Hungary

INTRODUCTION

In general, the term molecular association is used whenever two or more molecules via co-crystallization build crystal lattice together and the asymmetric unit of the new lattice gives room for, at least, one molecule of each component. The "watershed" is that there cannot be covalent bonding between the two compounds, while ionic and dipole-dipole interactions are allowed. When in the new phases the properties of at least two individual components are basically preserved then they are called "binary adducts." *Mutatis mutandis,* the principles governing binary adducts can be carried over without much change to ternary and higher adducts. In the opposite direction, crystal structures with more than one molecule in their asymmetric unit which differ only in their conformations can be termed as quasi-heteromolecular associates[1]. The further (final) step is when the homomolecular associates forming crystals are regarded as zero order supramolecules. In principle, this idea originates from Kitaigorodskii[2] who in 1955 wrote the following: *"Another aspect of organic structures neglected by many workers appears to me to deserve equal attention. Organic substances are built from molecules; many properties of a material depend but slightly on the structure of the molecule and the intramolecular forces, but decisively on the molecular structure of the crystal — that is on the intermolecular forces ....These reasons alone compel us to regard the molecular basis of organic crystal structures as an important branch of sciences"*. By this statement he was the first who outlined the concept of supramolecular chemistry, however, this term was introduced only later by, but widely used now after Lehn. According to Lehn[3] supramolecular chemistry *"is the study of the structures and functions of the supermolecules that result from binding substrates to molecular receptors .... Beyond molecular chemistry based on the covalent bonds lies supramolecular chemistry based on molecular interactions ... "*

Attempting to rationalize these, basically binary adducts of Lehn, we must bear in mind that there are several other proposals for the classification and nomenclature of the heteromolecular associations, e.g. the terms *host, guest, complex* and their binding forces were defined by Cram and coworkers[4]. Weber and Czugler[5] have specified this classification even further. By defining the coordinatoclathrate concept they aim to design the best host molecules and their specific adjustment with guest compounds. They start from the commonly accepted rule: molecular bulkiness and crystal inclusion are closely related. Furthermore, a host designed to the formation of coordinatoclathrates advisably should possess "scissor like" bisection in the molecule. On such kind of bulky molecules there are appended functional (sensor) groups which manage the coordination to the included guest substrate.

To distinguish the archetypes of the most frequently occurring binary adducts, i.e. *molecular compounds and complexes* of A_xB_y type, purportedly the use of Herbstein's classification[6] is recommended. It classifies them in term of interactions $A \cdots A$, $B \cdots B$ and $A \cdots B$ which primarily determine the component arrangement in crystal, as follows: $A \cdots A$ interactions dominate in inclusion complexes,

They can also be divided into three subgroups: (a) zero dimensional *clathrates*, (b) one dimensional *channels* (or *tunnels*), and (c) two dimensional, *lamellar complexes*. $A \cdots A$ and $B \cdots B$ interactions are equally important in *segregated stack transfer* complexes.

All interactions are of roughly equal importance in *packing complexes*, and $A \cdots B$ interactions dominate in *molecular compounds*. Within them Herbstein distinguishes three types of interactions: hydrogen bonded, localized charge transfer and delocalized charge transfer interactions.

From topological point of view, the crystal close packing is governed by the *complementary* surfaces[7] of the host and guest molecules. Consequently, the structures built up by identical (even homochiral) molecules (zero order supramolecules) are also governed by the self-*complementary* surfaces. If self-complementarity is great (i.e. the bumps of the bulky molecules fit perfectly into the hollows formed by themselves[2]) then — as revealed by our studies[8,9] on the *isostructurality* of related organic crystals — these associations tolerate atomic replacements even substitutions or a change in chirality (epimerization) without visible decrease of the already established close packing. The *preserved* or just slightly modified molecular *harmony* indicates a certain ability by which the ca. 30 % empty space (voids os small cavities) in the crystal lattices is exploited without any *morphotropic* phase transitions[2]. Consequently, the study of *isostructurality* is not a goal *per se*, but a good tool of the better understanding of both homo- and heteromolecular close packing. Present work attempts to review the conditions of isostructurality exhibited by the first order supramolecules: the binary adducts.

There is another aspect of the studies on isostructurality, that is to shed more light upon the inverse phenomenon: *polymorphism*. It is hardly to say that polymorphism is of great industrial, in particular pharmaceutical importance[10] e.g. the tableting behavior of powders, physically stable dosage forms and chemical stability are equally dependent on it. The differences between the various polymorphs (at least dimorphs) of a compound manifest themselves as differences in solubility, rate of dissolution, and vapour pressure. Just as different chemical compounds can have different polymorphs, solvates (many of them are in fact inclusion compounds[5]) of different compounds can exhibit polymorphism, too. Our attention to the connection between polymorphism and the less common isostructurality was called first by the study of paradisubstituted benzylideneanilines[11] which are mutually crosslinked by these contrasting phenomena.

ISOSTRUCTURALITY OF ORGANIC CRYSTAL STRUCTURES

A. Isomorphism (historical background)

The first step towards the knowledge of this very important phenomenon widely encountered in nature was made by Romé de l'Isle in 1772. He noticed that by placing potassium alum crystals in saturated ammonium alum solution the crystals start growing and become coated with an ammonium alum layer. After this and other early observations (e.g. Wollaston (1809) recognized that calcite, magnesite and siderite (Ca, Mg, $FeCO_3$), crystallize all as rhombohedra with slightly different faces). Mitscherlich in 1819 found that certain pairs (e.g. KH_2PO_4, KH_2AsO_4, $NH_4H_2PO_4$ and $NH_4H_2AsO_4$) of salts developed the same crystalline form and such salts had similar chemical formulae; one kind of atom in one compound being replaced by another kind in the "related" salt. These pairs were called *isomorphous*. Soon it became clear that isomorphism plays an important role in chemistry. Namely, analyzing such a pair of salts immediately gave the relative atomic weights of these atoms. Thus Berzelius (the teacher of Mitscherlich) established the atomic weight of selenium in 1928, following the discovery of the isomorphism of Na_2SO_4, Ag_2SO_4, Na_2SeO_4 and Ag_2SeO_4. The characterization of the phenomenon has been developed further by von Groth (1874), Náray-Szabó (1969) and others. According to them the basic condition for isomorphism of compounds is that their crystals are closely similar in shape and one of the crystals continues growing in the solution of the other. In other words, their lattice types should be identical and the formation of solid solution (mixed crystals) is guaranteed. However, to define isomorphism precisely has remained impossible, which can be attributed to the fact that this word refers only to the external similarity of crystalline substances. Therefore some authors (e.g. Wells[12] and Bloss[13]) have introduced the terms *isostructural* and *isostructurality* (or isotypism). This controversial situation forced the IUCr Subcommittee on the Inorganic Structure Types[14] to recommend the use of the terms *isotypism* (and *homeotypism*) but exclusively for the inorganic compounds. As far as the organic compounds are concerned, Kitaigorodskii in his fundamental work[2] where he summarized the temporary knowledge on the phenomenon (*cf.* ref. 9) simply used the obsolete term isomorphism.

B. Isostructurality of homomolecular crystals

Since 1984 Kálmán and coworkers step by step have been acquainted with a recurrent phenomenon shown frequently by cardenolides and analogous bufadienolides and which was termed finally as the "main part" isostructuralism of these substances.[15] Crystal structure determinations of novel derivatives (i.e. other steroids) and their solid solutions[16] resulted in the first articulation of the conditions and limits and of the phenomenon together with the introduction of three descriptors[*]:

a) packing coefficient increment: $\Delta(pc)$

b) unit cell similarity index: Π

c) isostructurality index: $I_i(n)$

By the use of these early observations on steroids, furthermore, the crystal structure analyses of the paradisubstituted benzylideneanilines[11] (and references therein) and the packing similarities of $Ph_3Si-SiMe_3$ with its germanium and tin analogues[17] (and references therein) a classification of the forms and degree of isostructurality exhibited by organic molecules has been established[8]. A survey of the literature revealed 45 isostructural pairs formed by more than 60 structures. It was recognized that in highly isostructural pairs (Figs. 1a and b) only *one and the same* atom undergoes either a replacement or substitution or a change

---

[*]Definition and calculation of these descriptors are described in refs. 8, 15 and 16. $I_i(n)$ was first denoted as I_D^n. The calculation of $\Delta(pc)$ was not recommended any more in ref. 8, but it seems to be useful in the case of supramolecules

in chirality[†]. Introduction[8] of isostructurality and molecular isometricity indices ($I_i(n)$ and $I_i(n^*)$, respectively) shed light to the molecular *translations* and *rotations* which aim to maintain packing motifs against alterations suffered by the epimerization (cf. digitoxigenin and 3-epidigitoxigenin depicted in Figs. 1a and c) or isomerization of the related molecules.

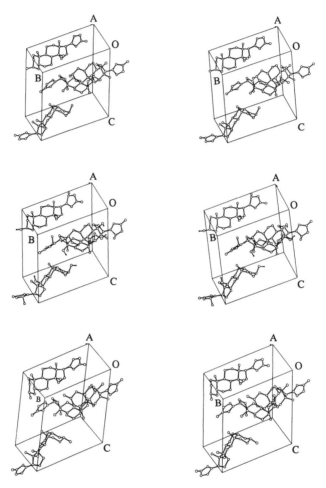

Figure 1. Stereoscopic view of three isostructurally related cardenolides: a) digitoxigenin, b) 21S-methyldigitoxigenin, c) 3-epi-digitoxigenin.

Even the crystals of 1-methyl-5-nitro-2-phenylimidazole and its 2-p-amino derivative[19] exhibit high degree of isostructurality: $I_i(15) = 90\,\%$, although in the latter there are two novel hydrogen bonds formed by the entering $-NH_2$ groups with the $-NO_2$ moieties. The complementary site and distribution of the entering active group(s) is shown by the low increase of the asymmetric unit volume of $\Delta V^* = 13$ Å.[3] In special cases the replacement of even two or more heavy atoms is also tolerated by the existing molecular self-complementarity. E.g. the replacement of the heavy atoms of 5-chloro-7-iodo-8- quinolinol molecule[20] by two bromines does not alter their isostructurality.

[†]It is worth noting that these forms are still equally regarded as isostructural and isomorphous by Glusker, Lewis and Rossi in their recently published monograph[18] and in protein crystallography the isomorphous replacement is a generally accepted term.

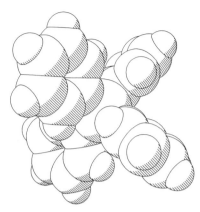

Figure 2. The tetrahedron of tetraphenylmethene calculated from the atomic coordinates.

This can be attributed to the fact that the atomic volume (and mass) goes up at one site, while at the other it goes down, which results in a slight decrease ($\Delta V^* = 13$ Å) of the asymmetric unit volume accompanied by 0.1 g cm^{-3} increase of the crystal density. Similarly, $Ph_3Si–SiMe_3$ and $Ph_3Ge–GeMe_3$ are isostructural[17] since the replacement takes place in the cores of the molecules which has no influence on the complementary surfaces. In addition, the atomic radii of Si and Ge do not differ significantly. However, when they are replaced by the substantially larger tin atom,[21] the molecules are no longer isometric. With similar space group ($P\bar{3}$) and packing array, the isostructurality index is rather low. To distinguish from the high degree forms of isostructurality they are termed *homeostructural*. The stepwise motion from iso- towards homeostructurality is well represented by the "group 14" homologues of tetraphenylmethane,[22] -silane,[23] -germane,[24] -tin[25] and -lead.[26] With a visible distorsion of the Ph_4C tetrahedron (Fig. 2) towards the bisphenoids for Ph_4Sn and Ph_4Pb sitting on axis $\bar{4}$ of space group $P\bar{4}2_1c$ the unit cell vectors **a** *vs* **c**, in accordance with the unit cell similarity indices (Π), are subjected to an inversely related motion summarized in Table 1.

Table 1. Lattice parameters and density of the isostructural the tetraphenyl derivatives of "group 14" elements

Compound	**a** (Å)	**c** (Å)	V (Å^3)	d (g cm^{-3})
Ph_4C	10.896(2)	7.280(1)	864.3	1.22
Ph_4Si	11.450(2)	7.063(1)	926.0	1.21
Ph_4Ge	11.656(11)	6.928(7)	941.3	1.31
Ph_4Sn	12.058(1)	6.581(1)	956.8	1.48
Ph_4Pb	12.092(3)	6.589(2)	963.4	1.75

It looks like as if the unit cell of Ph_4C were squeezed along the **c** axis in stepwise mode through Ph_4Si and Ph_4Ge towards Ph_4Sn and Ph_4Pb. By neighbouring pairs, they can be regarded isostructural, but the remote compounds like Ph4C and Ph4Pb due to the substantial difference in the atomic radii of the core atoms are only homeostructural. In this series the governing role of self-complementary surfaces is also shown by the ortho-methyl derivative of tetraphenyltin[25] which remains also isostructural with the parent compound.

In general, whenever the strict rules of isostructurality are relaxed to some extent, the related pairs are termed as *homeostructural*, which naturally implies a greater variety of packing arrays. A *par excellence* case of homeostructurality was found recently among the isomeric pair of 3-cyano- and 4-cyano-cinnamic acids where the packing arrays formed by slightly different dimers are quite similar. However, their solid state reactivities differ.[27] The proper use of this term is particularly seems to be useful in the case of supramolecules where the guest or even the host molecules are somewhat different within the related pairs.

ISOSTRUCTURALITY OF SUPRAMOLECULES

A. Clathrates with slight differences in the host molecules

The Me_2SO clathrates[28] of two isomeric host molecules: thieno[3,2-b]thiophene and thieno[2,3-b]thiophene differing only in the position of one of S atoms are almost perfectly isostructural in triclinic unit cells. The complementarities of the isomers is shown

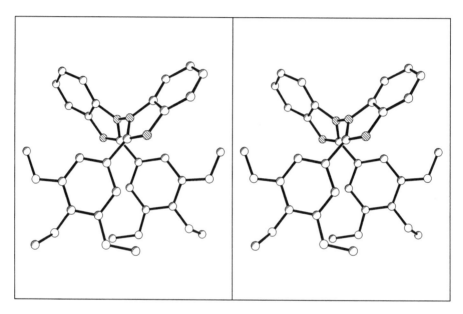

Figure 3. Perspectivic view of the same host molecules 2,2'-bis(3,4,5-trimethoxypheny)-1.1'-bibenzimidazol showing identical conformation in both (2:1) clathrates with acetic and propionic acids.

by the packing coefficient increment $\Delta(pc) = 0.5$ %. Rutherford who is also studying isostructurality,[29] called our attention to their work[30] which reports on the isostructural clathrates of adamantane[29] with thiourea and selenourea with $\Pi = 0.02$ for trigonal unit cells (space group $R\bar{3}c$). They can be regarded as the homologues of the "16 group" elements.

1,1'-binaphthyl-2,2'dicarboxylic acid forms isostructural (1:2) inclusion complexes[31] in the monoclinic space group $C2/c$ with ethanol and propanol (differing in one CH_2 moiety). In the voids of the clathrates both relatively small guest molecules exhibit positional disorder indicating the predominance of the $A \cdots A$ interactions. The host lattices exhibit high degree isostructurality ($I_i(13) = 88$ % showing similar self-complementarity of the bulky scissor–like[5] host molecules. The relatively similar positions of the disordered alcohol molecules is shown by the isostructurality index $I_i(16) = 70$ % calculated with the participation of the alcohol atoms.

Quite recently, my coworkers Czugler and Bombicz[32] succeeded to cocrystallize the achiral acetic and propionic acids with a bulky host molecule: 2,2'-bis(3,4,5-trimethoxyphenyl)-1,1,-bibenzimidazole (2:1 adducts) forming — via spontaneous resolution — isostructural conglomerates in the enantiomorphic hexagonal space group(s) $P3_121/P3_221$. In these supramolecules the host unit preserves its C_2 molecular symmetry which is now perpendicular to the chiral 3_1 or 3_2 axes. With $\Pi = 0.02$ the host molecules have very high isostructurality index of $I_i(21) = 96$ % as depicted in Fig. 3: they cannot be distinguished from each other. However, the positionally disordered guest molecules occupy visibly different positions in the large voids of the crystal lattices. Accordingly, these clathrates basically are governed by $\mathbf{A}\cdots\mathbf{A}$ interactions.

Interestingly enough, Czugler and Bombicz revealed also that the smallest (formic) acid forms a completely different adduct with 2,2'-bis(3,4,5-trimethoxyphenyl)-1,1,-bibenzimidazole. It crystallizes in the monoclinic space group $C2/c$ and remains homeostructural ($\Pi = 0.04$) with the parent compound the structure of which had been reported[33] earlier also from our laboratory. Although their monoclinic angles (β) differ (96.33(1)^0 for the host and 111.15(2))^0 for the adduct crystal) the isostructurality index for the host molecules $I_i(21) = 76$ % (without the three terminal methyl groups $I_i(19) = 78$ %) is high, especially if we take into account that the asymmetric unit of newly formed clathrate incorporates an additional HCOOH molecule. The high self-complementarity of the host molecules is also indicated by the moderate increase of the asymmetric unit volume ($\Delta V^* = 42$ Å^3, i.e. 14 Å^3 per an entering heavy atom). This suggests that the absolute dimension of the guest molecules via weak $\mathbf{A}\cdots\mathbf{B}$ interactions to some extent control directly the formation of the crystal lattices. Neither we can exclude the possibility that these binary adducts, under proper conditions, may form polymorphs.

Finally, in the course of studying the 'main part' isostructuralism of steroids[16] we observed a high degree isostructurality ($I_i(19) = 78$ % and $I_i(21) = 76$ % with $\Pi = 0.004$) between the monohydrates of 5-androstene-3β,17β-diol and 5α-androstane-3β,17β-diol. However, in these binary adducts the guest (water) is common and their host differ, but only in a double bond (C4=C5) of the 5-androstene- which is saturated in 5α-androstane-3β,17β-diol. They cocrystallized readily forming thus a ternary adduct.

B. Three homeostructural adducts formed by different guest molecules

Caira and Mohamed[34] reported on three solvates of 5-methoxysulfadiazine (S) formed with (a) dioxane, (b) tetrahydrofuran and (c) chloroform. These clathrates are based on the common *isostructural* sulfamide host framework with the solvent molecules occupying the framework cavities. As shown by the stereoscopic views of these binary adducts depicted in Fig. 4 the guest molecules occupy the same relative position in the infinite channels formed by the host molecules along the c axis of the common space group $P2_1/c$. The great similarity in the cavity occupation by the guest molecules is well presented by Fig. 6 given by Caira and Mohamed. The adducts exhibit by pair the following three relatively high isostructurality indices accompanied by low Π values (Table 2).

Table 2. Isostructurality $I_i(19)$ and unit cell similarity Π indices with volume differences $\Delta V^*(\text{Å}^3)$ for the asymmetric units

Guest molecules	$I_i(19)$ %	Π	ΔV^*
$C_4H_8O_2/CHCl_3$	90	0.011	9.2
$C_4H_8O/CHCl_3$	86	0.0004	2.8
$C_4H_8O_2/C_4H_8O$	83	0.011	12.1

The highest isostructurality index is shown by the dioxane/chloroform pair which suggests that the morphologically and electronically different guest molecules hardly influence the self-complementarity of the host molecules governed by $\mathbf{A}\cdots\mathbf{A}$ interactions.[6] It is dubious as to

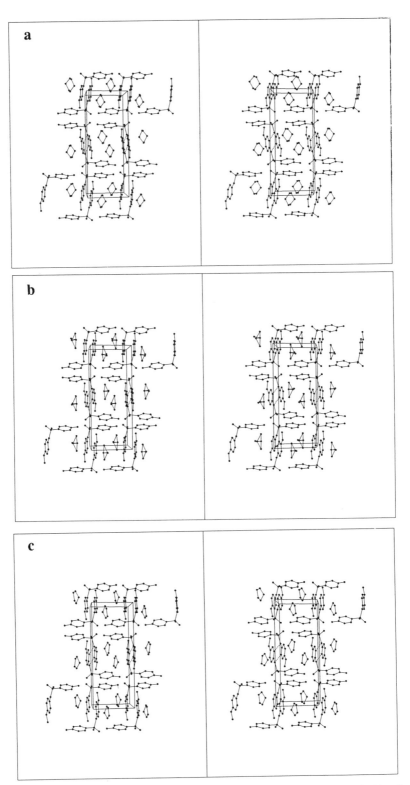

Figure 4. Stereoscopic view of the clathrates of 5-methoxysuphadiazine with a) dioxane, b) chloroform and c) tetrahydrofurane showing the channels built up from the host molecules giving room for the guest unit approximately in the same positions ($CHCl_3$ moelucles exhibit positional disorder).

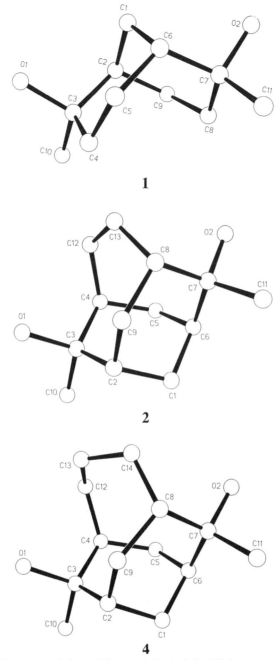

Figure 5. The non-hydrogen atom skeleton of the three helical tubuland diols:
1 (2,6-dimethylbicyclo[3.3.1]nonane-*exo*-2,*exo*-6-diol),
2 (2,7-dimethyltricyclo[4.3.1.1]undecane-*syn*-2,*syn*-7-diol),
4 (2,8-dimethyltricyclo[5.3.1.1]dodecane-*syn*-2,*syn*-8-diol).

whether the lowest isostructurality index shown by the dioxane and tetrahydrofuran clathrates can be attributed to their substantially different conformations, or not. When any conspicuous difference between these pairs are studied, me must bear in mind that the differences in their asymmetric unit volume is rather small, even for the dioxane/tetrahydrofuran pair is less than

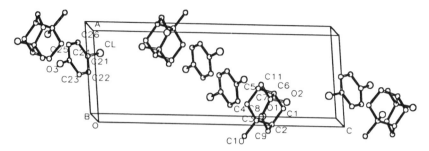

Figure 6. Perspectivic view of the unit cell of the inclusion compound of **1** with p-chlorphenol (**A**).

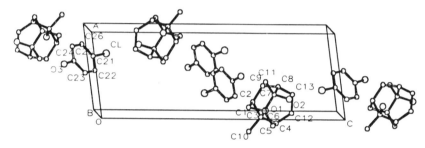

Figure 7. Perspectivic view of the unit cell of the inclusion compound of **2** with p-chlorphenol (**B**).

18 Å^3. At any rate, from the point of view of the guest molecules they have to be regarded homeostructural. The close relationship between isostructuralism and polymorphism (*vide supra*) is shown by the three polymorphs[35,36] of this biologically active sulfonamide prepared and characterized by X-ray crystallography.

C. A group of clathrates formed by different host and guest molecules

A large group of binary adducts formed by closely related host and guest molecules have been prepared and characterized by X-ray structure determinations.[37] The Australian team has cocrystallized three closely related helical tubuland diols labelled hereinafter as **1**, **2** and **4** (Fig. 5) with phenol and five phenol derivatives as p-chlorophenol, hydrochinone, p-methoxyphenol, p-hydroxythiophenol and phloroglucinol (1,3,5-trihydroxybenzene). Since phloroglucinol substantially differs from the other guest molecules its adduct with host **4** is orthorhombic with the polar space group *Fdd*2, while the others crystallize in space group $P2_1/c$ with similar unit cell parameters (*cf.* Table 1 in ref. 37). Only hydrochinone possessing a molecular centre of symmetry crystallizes with a 2:1 ratio of the host and guest molecules in a smaller unit cell ($V = 1370$ Å^3) with respect to the mean of the other five of 1792 Å^3. The other five adducts form a cluster of homeostructural crystals as depicted in Figs. 6-10. The similar complementarity of the helical tubuland diols can be estimated by their superposition, but to quantify their degree is difficult. Nevertheless, in the canals formed by these diols **1**, **2**, and **4**, (cf. Figure 1 in ref. 35) the guest molecules assume approximately the same positions, even if they are disordered (2:p-methoxyphenol). In these channels besides the **A**···**A**

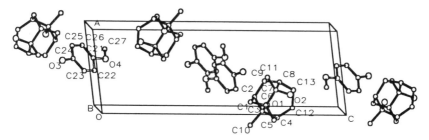

Figure 8. Perspectivic view of the unit cell of the inclusion compound of **2** with p-methoxyphenol (**C**).

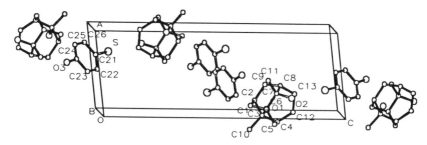

Figure 9. Perspectivic view of the unit cell of the inclusion compound of **2** with p-hydroxythiophenol (**D**).

interactions the effect of the **B**$\cdots$**A** contacts has also to be taken into account. The highest degree of isostructurality could be expected between structures **B** and **D** ($\Pi = 0.005$) where the same **2** is cocrystallized with p-Cl-phenol (**B**) and p-hydroxythiophenol (**D**). Indeed, they show very high isostructurality index for the guest molecules $I_i(8) = 93\ \%$ and somewhat lower for the host molecules $I_i(15) = 67\ \%$. Naturally, their common isostructurality index is still high: $I_i(23) = 73\ \%$. In contrast, for the **A/E** pair ($\Pi = 0.023$) where both hosts (**1** vs. **4**) and guests (p-Cl-phenol vs. phenol) are different only the guest molecules show high isostructurality $I_i(7) = 86\ \%$. The fit of the host molecules has not been satisfactory so far. Similarly, the proper fit of host (**1** and **2**) skeletons for the **A/B** pair should be solved. At any rate, the isostructurality of the guest molecules is high $I_i(8) = 84\ \%$. The corresponding isometricity index: $I_i(8^*) = 99\ \%$. The **B/C** pair has the same host **2** with the lowest isometricity of the guest molecules. Their $I_i(8)$ is only 48 % disturbed by the positional disorder of the p-methoxyphenol molecules. In spite of this, the self-complementarity of the host molecules are similar, consequently their isostructurality is high: $I_i(15) = 74\ \%$. The common isostructurality index of the host and= guest molecules is still: $I_i(22) = 63\ \%$.

The highest unit cell similarity index ($\Pi = 0.04$) is shown by the adducts **A** and **C** which can be attributed to the substantial differences between the both host and guest molecules, respectively. However, without the appropriate calculation of isostructurality and isometricity indices for the different host molecules only their homeostructurality can be established as depicted in Figs 6-10. The appropriate fit of these host molecules depends on the formation of the distance differences (ΔR_i) between the crystal coordinates of the *positionally similar* non-H atoms within the same section of the asymmetric units (*cf.* the discussion of eq. 2 in ref. 8).

219

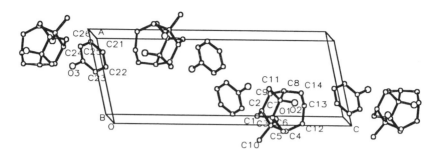

Figure 10. Perspectivic view of the unit cell of the inclusion compound of **2** with phenol (**E**).

Further on, it seems to be advisable if these calculations for supramolecules were extended to the whole unit cell, instead of the present forms restricted only to one the asymmetric units.

D. Concluding remarks

The meticulous studies on the iso- or homeostructurality of supramolecules seems to be useful to a deeper understanding of the substrate and molecular receptor interactions. While this manuscript has been prepared my coworkers (Drs. Bombicz and Czugler) have found further binary adducts of 2,2'-bis(3,4,5-trimethoxyphenyl)-1,1,-bibenzimidazole with novel and larger guest acid molecules (e.g. trimethylacetic acid, etc.) which are also homeostructural with the host compound.[36] In addition to these novel binary systems even ternary adducts (e.g. nitrate and perchlorate salts of the double betain (L = $cis-pMe_2N^+C_5H_4N)C_2(COO^-)_2]$ monohydrate) related by isostructurality[38] were found ($I_i(28) = 88$ %). Of course, to a better characterization of such systems the calculation of the isostructurality indices etc. should be improved which is now under consideration.

ACKNOWLEDGMENT

I wish to express my sincere thanks to my colleagues Drs. Gyula Argay, Petra Bombicz, Mátyás Czugler and László Párkányi (Budapest) for their invaluable help in data collections, calculations and brainstorming discussions. Thanks are also due to Professors Mino Caira (Cape Town, SA), Roger Bishop (Kensington, Australia) and John Rutherford (Umtata, Transkei) for their cooperation and useful information concerning their own works. I do thank Professor Jan C. A. Boeyens (Johannesburg, SA) for the kind invitation to serve the first *Small molecules Indaba* (held in the Skukuza camp of the Kruger National Park, between 20-25 August 1995) as an invited speaker. I am also very grateful to my secretary Mrs. Györgyi Csákvári for her assistance in preparing the manuscript. This work has been sponsored by the Hungarian Research Fund, Grant No. OTKA T014539.

REFERENCES

1. P.M. Zorky, A.E. Razumaeva and V.K. Belskii, *Acta Cryst.* A33:1001 (1977).
2. A.J. Kitaigorodskii, "Organic Chemical Crystallography," pp 222-231, Consultants Bureau, New York (1961).
3. J.-M. Lehn, *Science* 227:849-856 (1985).

4. E.P. Kyba, H.C. Helgeson, K. Madan, G.W. Gokel, T.L. Tarnowski, S.S. Moore, D. J. Cram, *J. Am. Chem. Soc.* 99:2564 (1977).

5. E. Weber, M. Czugler, *Topics in Current Chemistry* 149:45, Springer Verlag, Berlin-Heidelberg (1988).

6. F. H. Herbstein, *Acta Chim. Hung.* 130:377 (1993).

7. C. Pratt Brock and J. D. Dunitz, *Chem. Mater.* 6:1118 (1994).

8. A. Kálmán, L. Párkányi, and Gy. Argay, *Acta Cryst.* B49:1039 (1993).

9. A. Kálmán, L. Párkányi, and Gy. Argay, *Acta Chim. Hung.* 130:279 (1993).

10. J. K. Haleblian, *J. Pharm. Sci.* 64:1269-1288 (1975).

11. I. Bar and J. Bernstein, *Tetrahedron* 43:1299-1305 (1987).

12. A.F. Wells, "Structural Inorganic Chemistry," pp 182-186; Clarendon Press, Oxford (1962)

13. F.D. Bloss "Crystallography and Crystal Chemistry," pp 249-250; Holt, Rinehart and Winston, Inc., New York (1971).

14. J. Lima-de-Faria, E. Heller, F. Liebau, E. Makovicky and E. Parthé, *Acta Cryst.* A46:1 (1990).

15. A. Kálmán, Gy. Argay, D. Scharfenberg-Pfeiffer, E. Höhne, B. Ribár, *Acta Cryst.* B47:68 (1991)

16. A. Kálmán, Gy. Argay, D. Zivanov-Stakić, S. Vladimirov, B. Ribár, *Acta Cryst.* B48:812 (1992)

17. L. Párkányi, A. Kálmán, S. Sharma, D. M. Nolen and K.H. Pannell, *Inorg. Chem.* 33:180 (1994).

18. J.P. Glusker, M. Lewis and M. Rossi, "Crystal Structure Analysis for Chemists and Biologists," p. 284, VCH Publ., Inc., New York (1994).

19. T.A. Olszak, O.M. Peeters, N.M. Blaton and C.J. de Ranter, *Acta Cryst.* C50:761 (1994).

20. S. Kashino and M. Haisa, *Bull. Chem. Soc. Jpn.* 46:1094 (1973).

21. L. Párkányi, A. Kálmán, K.H. Pannell, F. Carvantes-Lee, and R.N. Kapoor, *Inorg. Chem.*, submitted for publication (1994). 22. A. Robbins, G.A. Jeffrey, J.P. Chesick, J. Donohue, F.A. Cotton, B.A. Frenz and C. A. Murillo, *Acta Cryst.* B31:2395 (1975)

23. V. Gruhnert, A. Kirfel, G. Will, F. Wallrafen and Recker, Z. *Krist.* 163:53 (1983).

24. A. Karipides and D.A. Haller, *Acta. Cryst.* B28:2889 (1972).

25. V.K. Belsky, A.A. Simonenko, V.O. Reikhsfeld, and I.E. Saratov, *J. Organomet. Chem.* 244:125 (1983).

26. V. Busetti, M. Mammi, A. Signor, and A. Del Pra, *Inorg. Chim. Acta* 1:424 (1967).

27. M.S.K. Dhurjati, J.A.R.P. Sarma and G.R. Desiraju, *J. Chem. Soc. Chem. Commun.* 1702 (1991).

28. N. Hayashi, Y. Mazaki and K. Kobayashi, *J. Chem. Soc. Chem. Commun.* 2351 (1994).

29. J.S. Rutherford, On comparing lattice parameters among isostructural molecular crystals, to be published.

30. R. Gopal, B.E. Robertson and J.S. Rutherford, *Acta. Cryst.* C45:257 (1989).

31. E. Weber, I. Csöregh, B. Stensland and M. Czugler, *J. Am. Chem. Soc.* 106:3297 (1984).

32. M. Czugler and P. Bombicz, *J. Chem. Soc. Chem. Commun.*, submitted for publication (1995).

33. G. Speier and L. Párkányi, *J. Org. Chem.* 51:218 (1986).

34. M.R. Caira and R. Mohamed, *Supramolec. Chem.* 2:201 (1993).

35. G. Giuseppetti, C. Tadini, G.P. Bettinetti and F. Giordano, *Cryst. Struct. Commun.* 6:263 (1977).

36. M.R. Caira, *J. Chem. Cryst.* 24:695 (1994).

37. A.T. Ung, R. Bishop, D.C. Craig, I.G. Dance and M.L. Scudder, *Chem. Mater.* 6:1269 (1994).

38. D-D. Wu and T.C.W. Mak, *J. Chem. Cryst.* 24:689 (1994).

THE PHENOMENON OF CONGLOMERATE CRYSTALLIZATION. PART 44. COUNTERION CONTROL OF CRYSTALLIZATION PATHWAY SELECTION. PART VI.
THE CRYSTALLIZATION BEHAVIOR OF [Co(abap)(NO$_2$)$_2$]Cl (I), [Co(abap)(NO$_2$)$_2$]ClO$_4$ (II), [Co(abap)(NO$_2$)$_2$]PF$_6$·H$_2$O (III) AND [Co(abap)(NO$_2$)$_2$]I·H$_2$O (IV)

Ivan Bernal*, Xubin Xia⁺, and Fernando Somoza⁺

Department of Chemistry
University of Houston
Houston, Texas 77204-5641
USA

Abstract

Coordination compounds of Co(III) and the dissymmetric tripodal ligand [N-(2-aminoethyl)-N,N'-bis(3-aminopropyl)-amine] produces a series of anionic derivatives of the cation [Co(abap)(NO$_2$)$_2$]⁺ whose single crystal structures have been determined here in order to document the crystallization pathway selected by these species and to further explore the effect the counter ion has in the selection of the crystallization pathway of racemic solutions.

[Co(abap)(NO$_2$)$_2$]Cl and [Co(abap)(NO$_2$)$_2$]ClO$_4$·H$_2$O crystallize as conglomerates in space group $P2_1$ and $P2_12_12_1$, respectively, while [Co(abap)(NO$_2$)$_2$]PF$_6$·H$_2$O and [Co(abap)(NO$_2$)$_2$]I·H$_2$O crystallize as racemates. The conformations of the two six-membered rings in these complexes are chairs which are the expected, stable conformations. In the chloride and perchlorate complexes there are strong hydrogen bonds between anions and hydrogens of terminal nitrogens and also between the the -NO$_2$ oxygens and those hydrogens, interactions which lock the cations into a specific dissymmetric conformation. Moreover, inter-cationic hydrogen bonds lead to the formation of spiral strings, adjacent strings being of the same helicity in the case of the conglomerates. Such strong hydrogen bonds and helical strings are noticeably absent in the case of the racemic crystals of [Co(abap)(NO$_2$)$_2$]PF$_6$·H$_2$O and [Co(abap)(NO$_2$)$_2$]I·H$_2$O.

INTRODUCTION

Definitions: If we limit ourselves exclusively to ordered crystals, then a centrosymmetric lattice is one which is compatible with the presence of inversion centers; if asymmetric or dissymmetric moieties are present in such lattice, they will be in enantiomorphic pairs related by the inversion centers. A polar lattice is one which contains mirror planes but no inversion centers (a sharpened pencil is a good example of a polar object). If asymmetric or dissymetric moieties are present in such lattice, they will be in enantiomorphic pairs related by mirror planes. An enantiomorphic lattice is one whose point group operations are all proper or

Table 1. Space Groups of Various *cis*–dinitrocobalt(III) Amines

Compound	Space Group	Cryst. Mode	Ref.
[*cis*–Co(en)$_2$(NO$_2$)$_2$]Cl[a]	$P2_1$	Conglomerate	1
[*cis*–Co(en)$_2$(NO$_2$)$_2$]Br	$P2_1$	Conglomerate	23
[*cis*–Co(en)$_2$(NO$_2$)$_2$]I	$P2_1$	Conglomerate	10
[*cis*–Co(en)$_2$(NO$_2$)$_2$]NO$_3$	$P2_1/c$	Racemic	21
[*cis*–Co(en)$_2$(NO$_2$)$_2$]NO$_2$	Cc	Racemic	19
[*cis*–α-Co(trien)$_2$(NO$_2$)$_2$]Cl·H$_2$O[b]	$P2_12_12_1$	Conglomerate	1
[*cis*–α-Co(trien)$_2$(NO$_2$)$_2$]I·H$_2$O	$P2_12_12_1$	Conglomerate	10
[*cis*–β-Co(trien)$_2$(NO$_2$)$_2$]I	Cc	Racemic	4
[*cis*–β-Co(trien)$_2$(NO$_2$)$_2$]NO$_3$	$P2_1/n$	Racemic	4
[*cis*–Co(tren)(NO$_2$)$_2$]Cl[c]	$Pcmb$	Racemic	25
[*cis*–Co(tren)(NO$_2$)$_2$]Br	$Pca2_1$	Racemic	22
[*cis*–Co(tren)(NO$_2$)$_2$]ClO$_4$	$Pna2_1$	Racemic	22
[*cis*–Co(tren)(NO$_2$)$_2$]$_2$Br·ClO$_4$	$P2_12_12_1$	Kryptoracemic	22
[*cis*–Co(tren)(NO$_2$)$_2$]NO$_3$	$Pca2_1$	Racemic	22

[a]en = ethylenediamine (H$_2$NCH$_2$CH$_2$NH$_2$); [b]trien = 1,8-diamino-3,6-diaza-octane
(H$_2$NCH$_2$CH$_2$NHCH$_2$CH$_2$NHCH$_2$CH$_2$NH$_2$); [c]tren = tris(2-aminoethyl)amine (N(CH$_2$CH$_2$NH$_2$)$_3$).

improper axes of symmetry; in the case when asymmetric or dissymmetric moieties are the asymmetric unit of such a crystal, they will all be homochiral. A kryptoracemic crystal is one whose enantiomorphic lattice bears an asymmetric unit which is a racemic pair. Appropriate examples of these are given in the text.

Conglomerate Behavior of Cobalt(III) Amine Nitrite Complexes

The conglomerate crystallization of Werner coordination compounds of cobalt amines belonging to a class characterized by having ligands, such as -NO$_2$, which can be anchored into specific dissymmetric arrangements by intramolecular hydrogen bonds, steric hindrance, or both have been investigated by Bernal and co-workers.[1-24] The information is summarized they obtained in Table 1.

More than one third of the compounds they investigated are conglomerates. Therefore, is this just a coincidence, or do these compounds bear an inner similarity which drives them to crystallize as conglomerates? They summarized the results of many experiments by some general observations.

1. In substituted cobalt amine series, hydrogen bonding plays an important role in selecting the crystallization pathway; in cis-dinitro cobalt amines series, the [*cis*–α-Co(trien)(NO$_2$)$_2$]X·H$_2$O (X = Cl[1], I[10]), [*cis*–Co(en)$_2$(NO$_2$)$_2$]X (X = Cl[1], Br[23], I[20]), strong intramolecular bonds are formed by the axial amine -NH$_2$ hydrogens with the -NO$_2$ oxygens, which locks the molecules (or ions) into a specific dissymmetric conformation. I At the same time, these oxygens form intermolecular hydrogen bond with amine hydrogens of another cation. The halides link the pair of the basal -NH hydrogens, and together with the water of crystallization, help to hold the cations together and result in a conglomerate crystal. Thus, it seems that for this class of compounds, the formation of such bonds are a necessary condition for conglomerate crystallization.

2. When the charge-compensating anion is a halide, or some other relatively poor hydrogen bonding species, cations such as [*cis*–Co(en)$_2$(NO$_2$)$_2$]$^+$ form conglomerates; however, if the compensating anion is a powerful hydrogen bonding species, such as NO$_3^-$ or NO$_2^-$, they favorably compete for the formation of hydrogen bonds with the cation's -NH$_2$ hydrogens and thus block the formation of intramolecular ones. When this hap-

pens, cations that crystallize as conglomerates with halide counter ions now crystallize as racemates.

3. All the conglomerates studied have (a) the exact same pattern of hydrogen bonded attachments forming between their transition metal ions, leading to the formation of polymeric, spiral strings resembling helical polypeptides (b) concerning the interactions between the strings: these interactions can be of two types (i) direct interactions between them, as in the two halves of a zipper, or protein strands, examples of which are the structures of mer-$[Co(NH_3)_3(NO_2)_3]$ (space group $P2_12_12_1$)[26] and of fac-$[Co(NH_3)_3(NO_2)_3]$ (space group $P2_12_12_1$)[27] (c) counter ions (and/or waters of crystallization, where relevant) are present, as is the case of compound [cis–α-Co(trien)(NO_2)_2]Cl·H_2O, (Fig. 2) in such case, the strings are joined by the agency of the halides and the waters of crystallization. Linkages must have a clear preference for joining strings of the same chirality.

One way of altering this hydrogen bonding pattern is to modify the amine system of [cis–α-Co(trien)(NO_2)_2]^+ such that, while as much of the system is kept intact, there is either no equatorial -NH hydrogens left in the modified cation, or one, or both, of the equatorial positions contains a quaternary nitrogen. The simplest way to block both of the amine equatorial sites of trien is to synthesize a 4,7-N,N'-dimethyltrien ligand, which has been prepared.[24] Briefly, the resulting (+/–)-[cis–α-Co(4,7-di-N,N'-methyltrien)(NO_2)_2]Cl·H_2O (1) crystals are also examples of kryptoracemates crystallizing in space group C_2 ($z = 8$) while those of the iodide (+/–)-[cis–α-Co(4,7-di-N,N'-methyltrien)(NO_2)_2]I (2) are true racemates, space group Pccn ($z = 8$), as expected. Thus, the validity of our proposal is demonstrated with the results obtained for (2) while (1) represents a highly welcomed, but unexpected, variant of our proposed racemate crystallization pathway selection.

Methylating only one of the trien secondary nitrogen is not a simple chore, as we learned by methylating both of them, which is a simpler task. An equally effective way of eliminating only one of the equatorial -NH moieties present in the basal plane of the metal, yet making use of the asymmetric tripodal ligand abap [N-(2-aminoethyl)-N,N'-bis(3-aminopropyl)-amine] which was prepared according to the literature[28] (see Scheme 1, below).

EXPERIMENTAL

Melting points were uncorrected. Proton and carbon-13 NMR were recorded on a General Electric QE-300 spectrometer at 300 MHz for ^{1}H NMR and 75 MHz for ^{13}C NMR. Elemental analyses were performed by Galbraith Laboratory, Inc. (2323 Sycamore Dr., Knoxville, Tenn., 37921). All commercial chemicals were of reagent grade.

Diphthalimidodipropylamine (3)[28]

Phthalic anhydride (5.9 g, 40 mmol) was melted at 180 ^{0}C with an oil bath. 3,3'-Iminobispropylamine (2.6 g, 20 mmol) was added dropwise with vigorous stirring. The glassy solid which formed on cooling was extracted with hot absolute ethanol. The products precipitated on cooling (5.8 g, 72 %): mp 140-143 ^{0}C (lit.[29] mp 136-137 ^{0}C); ^{1}H NMR (CDCl_3): d 3.53 (m, 4H), 3.37 (m, 4H), 3.12 (t, 4H), 2.16 (m, 4H). ^{13}C NMR (CDCl_3): δ 51.7, 50.5, 37.7, 35.0, 22.9.

(2-Aminoethyl)-N,N-bis(3-aminopropyl)-amine·4HCl·2.5H$_2$O (Abap) (6)[28]

To a melt of 3,3'-Diphthalimidodipropylamine (3) (2.77 g, 7.1 mmol) at 155 ^{0}C was added solid (2-bromoethyl)phthalimide (1.8 g, 7.1 mmol) in portions over 10 min. The mixture was stirred at 160-170 ^{0}C for 45 min and allowed to cool. The solid mass was ground to a powder and refluxed in 8 M HCl (30 mL) for 11 h. After cooling in ice and removal of phthalic acid, the filtrate was reduced to dryness (rotavap). The crude product was dissolved in water (8 mL), the mixture filtered (Celite), and the filtrate added dropwise with stirring to EtOH (50 mL). The resulting precipitate was removed by filtration and suspended in hot EtOH (ca. 27 mL). Water (6.5 mL) was added followed by another portion of EtOH (25 mL). Cooling in ice gave the pure product as white crystals (1.0 g, 40 %): ^{1}H NMR (D$_2$O): d 3.53 (m, 4H), 3.37 (m, 4H), 3.12 (t, 4H), 2.16 (m, 4H). ^{13}C NMR (D$_2$O): δ 51.7, 50.5, 37.7, 35.0, 22.9.

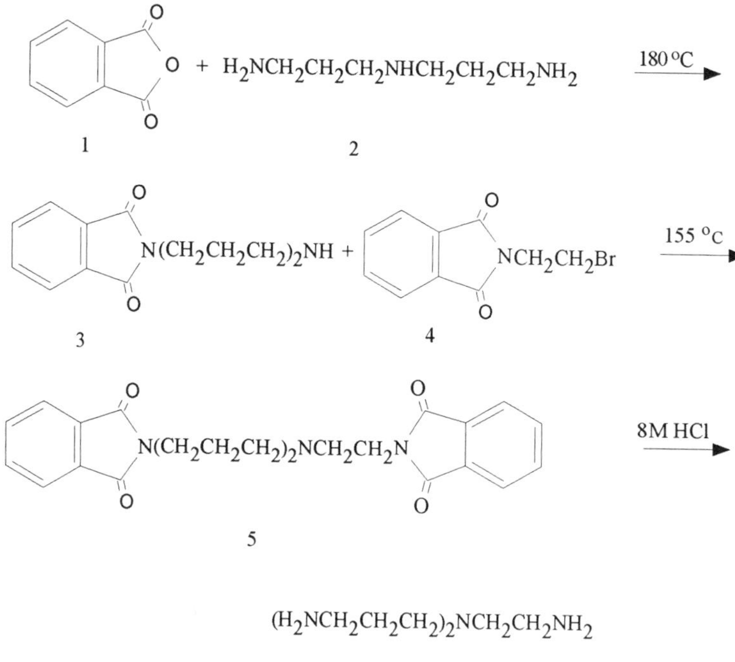

Scheme 1. The synthesis of the ligand abap

[Co(abap)(NO$_2$)$_2$]Cl (I)

CoCl$_2$·6 H$_2$O (0.74 g, 3.1 mmol) were dissolved in 15 ml H$_2$O and 1 ml 30 % H$_2$O$_2$ was added drop by drop. Abap salt (1.13 g, 3.1 mmol) and 0.5 g NaOH were dissolved in another 15 ml H$_2$O. The two solutions were mixed and treated with NaNO$_2$ (0.45 g, 6.5 mmol). The resulting solution was heated to 60 ^{0}C for 3 h and then cooled to room temperature (20 ^{0}C). After several days, red-orange single crystals were obtained which can be used for the X-ray structure determination. Yield 72 %. Elemental Analysis Calcd (found) for C$_8$H$_{22}$N$_6$O$_4$CoCl: C, 26.61 (26.38), H, 6.10 (5.96), N, 23.28 (22.97).

[Co(abap)(NO$_2$)$_2$]ClO$_4$·H$_2$O (II)

[Co(abap)(NO$_2$)$_2$]Cl was converted to [Co(abap)(NO$_2$)$_2$]ClO$_4$·H$_2$O by treatment with an excess of NaClO$_4$ in deionized water. A precipitate was obtained, filtered and washed with cold water, ethanol, dried and redissolved in deionized water. On long standing at room temperature (22 ^{0}C), brown-yellow crystals were obtained for X-ray analysis. Elemental Analysis Calcd (found) for $C_8H_{22}N_6O_8CoCl \cdot H_2O$: C, 21.72 (21.80), H, 5.42 (5.47), N, 18.97 (18.92).

$[Co(abap)(NO_2)_2]PF_6 \cdot H_2O$ (III)

$[Co(abap)(NO_2)_2]Cl$ was converted to $[Co(abap)(NO_2)_2]PF_6 \cdot H_2O$ by treatment with an excess of NH_4PF_6 in deionized water. A precipitate was obtained, filtered and washed with cold water, ethanol, dried and redissolved in deionized water. On long standing at room temperature (22 ^{0}C), the sample crystals were obtained for X-ray analysis. Elemental Analysis Calcd (found) for $C_8H_{22}N_6O_4CoPF_6 \cdot H_2O$: C, 20.43 (19.73), H, 4.68 (5.02), N, 17.87 (17.16).

$[Co(abap)(NO_2)_2]I \cdot H_2O$ (IV)

$[Co(abap)(NO_2)_2]Cl$ was converted to $[Co(abap)(NO_2)_2]I \cdot H_2O$ by treatment with an excess of NH_4I in deionized water. A precipitate was obtained, filtered and washed with cold water, ethanol, dried and redissolved in deionized water. Since, in solution, the iodide was photochemically decomposed by light, the brown-yellow crystals were obtained for X-ray analysis on long standing at room temperature (22 ^{0}C) in total darkness. Elemental Analysis Calcd (found) for $C_8H_{22}N_6O_4CoI \cdot H_2O$: C, 20.51 (20.62), H, 4.70 (4.95), N, 17.94 (17.66).

X-Ray Crystallography

Data were collected with an Enraf-Nonius CAD-4 diffractometer operating with a Molecular Structure Corporation TEXRAY-230 modification[30] of the SDP-Plus software package.[31] The procedure used for crystal alignment, cell constant determination, space group determination and data collection were uniform for all four crystals. Crystals were centered with data in the $20 < 2\Theta < 40$ range and examination of the cell constants, absences, and Niggli matrix.[32] clearly showed (**I**) to crystallize in space group $P2_12_12_1$. (**II**) crystallizes in space group $P2_1$ or $P2_1/m$. (**III**) crystallizes in space group $P2_1/c$. (**IV**) crystallizes in space group $P2_1/n$. The details of data collection are summarized in Tables 2, 3, 4 and 5, respectively, for compounds (**I**), (**II**), (**III**) and (**IV**).

The intensity data sets were corrected for absorption using empirical curves derived from Psi scans[51,52] of suitable reflections. The scattering curves were taken from Cromer and Waber's compilation.[54] During data collection, intensity and orientation standards were monitored and showed no significant deviations from the initial values. Processing of the data were carried out with the PC version of the NRCVAX package.[55] All four structures were solved from their Patterson maps using their heaviest atom as the initial phasing species for a difference Fourier map. All non-hydrogen atoms were readily found and refined isotropically, whereupon the hydrogens of the cations were added at their idealized positions (C-H = N-H = 0.95 Å). Water hydrogens, when relevant, were found experimentally and fixed at those positions, if they made reasonable stereochemical sense; otherwise, they were ignored. Conversion of the heavy atoms to anisotropic motion resulted in refinement of the overall structure to final $R(F)$ and $R_w(F)$ factors listed in Tables 2, 3, 4 and 5. In the case of (**I**) and (**III**), the absolute configurations were determined by use of the Flack test[34] which for (**I**) was 0.00 and for (**III**) it was 0.06; thus, the initial solutions were accidentally that of the correct enantiomorph in both cases.

Table 2. Summary of Data Collection and Processing Parameters for $[Co(abap)(NO_2)_2]Cl$ **(I)**

Space Group	$P2_12_12_1$ (No. 19)
Cell Constants	$a = 8.560(3)$ Å
	$b = 9.717(2)$
	$c = 17.363(6)$
Cell Volume	$V = 1444.27$Å^3
Molecular Formula	$CoClO_4N_6C_8H_{22}$
Molecular Weight	360.68 g mole^{-1}
Density (calc; $z = 4$ mol/cell)	1.659 g cm^{-3}
Radiation Employed	MoK$_\alpha$ ($\lambda = 0.71073$ Å)
Absorption Coefficient	$\mu = 13.90$ cm^{-1}
Relative Transmission Coefficients	0.6444 to 0.7334
Data Collection Range	$4^0 \leq 2\Theta \leq 50^0$
Scan Width	$\Delta\Theta = 0.95 + 0.35\tan\Theta$
Total Unique Data Collected	3217
Data Used In Refinement[a]	2393
$R = \Sigma||F_o| - |F_c||/\Sigma|F_o|$	0.032
$R_w = [\Sigma w(|F_o| - |F_c|)^2/\Sigma|F_o|^2]^{1/2}$	0.035
Weights Used	$w = [\sigma(F_o)]^{-2}$

[a]The difference between this number and the total is due to subtraction of 824 that were standards, symmetry related or did not meet the criterion that $I \geq 2.5\sigma(I)$.

Table 3. Summary of Data Collection and Processing Parameters for $[Co(abap)(NO_2)_2]ClO_4 \cdot H_2O$ **(II)**

Space Group	$P2_1$ (No. 4)
Cell Constants	$a = 7.926(1)$ Å
	$b = 11.436(2)$
	$c = 9.307(1)$
	$\beta = 95.32(1)^0$
Cell Volume	$V = 839.9$ Å^3
Molecular Formula	$CoClO_9N_6C_8H_{24}$
Molecular Weight	442.70 g mole^{-1}
Density (calc; $z = 2$ mol/cell)	1.756 g cm^{-3}
Radiation Employed	MoK$_\alpha$ ($\lambda = 0.71073$ Å)
Absorption Coefficient	$\mu = 12.30$ cm^{-1}
Relative Transmission Coefficients	0.6949 to 0.7593
Data Collection Range	$4^0 \leq 2\Theta \leq 60^0$
Scan Width	$\Delta\Theta = 1.00 + 0.35\tan\Theta$
Total Unique Data Collected	3323
Data Used In Refinement[a]	2967
$R = \Sigma||F_o| - |F_c||/\Sigma|F_o|$	0.025
$R_w = [\Sigma w(|F_o| - |F_c|)^2/\Sigma|F_o|^2]^{1/2}$	0.029
Weights Used	$w = [\sigma(F_0)]^{-2}$

[a]The difference between this number and the total is due to subtraction of 356 that were standards, symmetry related or did not meet the criterion that $I \geq 2.5\sigma(I)$.

Table 4. Summary of Data Collection and Processing Parameters for
[Co(abap)(NO$_2$)$_2$]PF$_6$·H$_2$O (**III**)

Space Group	$P2_1/c$ (No. 14)
Cell Constants	$a = 7.971(1)$ Å
	$b = 11.480(2)$
	$c = 19.563(5)$
	$\beta = 95.84(3)^0$
Cell Volume	$V = 1780.7$ Å^3
Molecular Formula	CoPF$_6$O$_5$N$_6$C$_8$H$_{24}$
Molecular Weight	488.21 g mole^{-1}
Density (calc; $z = 4$mol/cell)	1.821 g cm^{-3}
Radiation Employed	MoK$_\alpha$ ($\lambda = 0.71073$ Å)
Absorption Coefficient	$\mu = 11.40$ cm^{-1}
Relative Transmission Coefficients	0.7683 to 0.8349.
Data Collection Range	$4^0 \leq 2\Theta \leq 65^0$
Scan Width	$\Delta\Theta = 0.90 + 0.35\tan\Theta$
Total Unique Data Collected	5106
Data Used In Refinement[a]	3030
$R = \Sigma ||F_0| - |F_c||/\Sigma|F_0|$	0.034
$R_w = [\Sigma w(|F_0| - |F_c|)^2/\Sigma|F_0|^2]^{1/2}$	0.035
Weights Used	$w = [\sigma(F_0)]^{-2}$

[a]The difference between this number and the total is due to subtraction of 2076 that were standards, symmetry related or did not meet the criterion that $I \geq 2.5i\sigma(I)$.

Table 5. Summary of Data Collection and Processing Parameters for
[Co(abap)(NO$_2$)$_2$]I·H$_2$O (**IV**)

Space Group	$P2_1/n$ (No. 4)
Cell Constants	$a = 8.031(3)$ Å
	$b = 17.896(7)$
	$c = 11.237(4)$
	$\beta = 102.02(5)^0$
Cell Volume	$V = 1579.64$ Å^3
Molecular Formula	ICoO$_5$N$_6$C$_8$H$_{24}$
Molecular Weight	468.13 g mole^{-1}
Density (calc; $z = 2$ mol/cell)	1.969 g cm^{-3}
Radiation Employed	MoK$_\alpha$ ($\lambda = 0.71073$ Å)
Absorption Coefficient	$\mu = 30.50$ cm^{-1}
Relative Transmission Coefficients	0.4464 to 0.5148
Data Collection Range	$4^0 \leq 2\Theta \leq 50^0$
Scan Width	$\Delta\Theta = 0.90 + 0.35\tan\Theta$
Total Unique Data Collected	2828
Data Used In Refinement[a]	1597
$R = \Sigma ||F_0| - |F_c||/\Sigma|F_0|$	0.030
$R_w = [\Sigma w(|F_0| - |F_c|)^2/\Sigma|F_0|^2]^{1/2}$	0.032
Weights Used	$w = [\sigma(F_0)]^{-2}$

[a]The difference between this number and the total is due to subtraction of 1231 that were standards, symmetry related or did not meet the criterion that $I \geq 2.5\sigma(I)$.

Table 6. Positional Parameters and Their esd's for (**I**)

Atom	x	y	z	$B(\text{Å})$
Co	0.34753(11)	0.47389(10)	0.10457(5)	1.66(3)
Cl	0.29491(19)	0.82598(19)	0.25080(11)	2.60(7)
O1	0.2358(8)	0.3472(7)	−0.0279(3)	5.1(3)
O2	0.3519(8)	0.5353(6)	−0.0503(3)	4.5(3)
O3	0.6436(6)	0.4712(6)	0.0377(3)	3.82(24)
O4	0.6007(5)	0.6471(6)	0.1090(4)	3.55(25)
N1	0.1325(7)	0.4119(6)	0.1447(3)	2.42(25)
N2	0.3972(6)	0.5074(6)	0.2141(3)	2.29(23)
N3	0.4434(7)	0.2898(6)	0.1121(4)	2.7(3)
N4	0.2799(7)	0.6656(6)	0.0895(3)	2.39(25)
N5	0.3052(7)	0.4466(6)	−0.0037(3)	2.53(24)
N6	0.5552(7)	0.5382(7)	0.0804(3)	2.7(3)
C1	0.1305(8)	0.4434(8)	0.2305(4)	3.1(3)
C2	0.2913(9)	0.4316(9)	0.2646(4)	3.2(4)
C3	0.0926(9)	0.2632(9)	0.1313(5)	3.6(4)
C4	0.2118(11)	0.1563(8)	0.1566(5)	4.1(4)
C5	0.3542(10)	0.1584(7)	0.1071(5)	3.4(3)
C6	−0.0048(8)	0.4866(9)	0.1098(6)	3.5(4)
C7	−0.0024(9)	0.6420(9)	0.1103(6)	3.8(4)
C8	0.1218(9)	0.7025(8)	0.0617(4)	3.2(3)

Anisotropically refined atoms are given in the form of the isotropic equivalent thermal parameter defined as $(4/3)\cdot[a^2\cdot\beta_{11}+b^2\cdot\beta_{22}+c^2\cdot\beta_{33}+ab(\cos\gamma)\cdot\beta_{12}+ac(\cos\beta)\cdot\beta_{13}+bc(\cos\alpha)\cdot\beta_{23}]$.

Table 7. Positional Parameters and Their esd's for (**II**)

Atom	x	y	z	$B(\text{Å})$
Co	0.07942(3)	0.50000	0.15710(3)	1.501(10)
Cl	0.27634(8)	0.98518(10)	0.37099(7)	2.291(22)
O1	−0.1572(3)	0.6535(3)	0.2486(3)	3.60(10)
O2	−0.2344(3)	0.5826(3)	0.0394(3)	3.30(9)
O3	−0.0046(3)	0.29060(24)	0.0215(3)	3.59(10)
O4	−0.1786(3)	0.3378(3)	0.1737(3)	3.74(11)
O5	0.2841(3)	0.9238(3)	0.5071(3)	3.77(11)
O6	0.2495(4)	0.9047(3)	0.2539(3)	4.61(13)
O7	0.1370(3)	1.0659(3)	0.3639(3)	4.12(12)
O8	0.4331(3)	1.0472(3)	0.3620(3)	4.03(11)
N1	0.3167(3)	0.42210(23)	0.17500(25)	1.85(8)
N2	0.2009(3)	0.64739(23)	0.2022(3)	2.14(8)
N3	0.0505(3)	0.47339(22)	0.36294(24)	2.09(8)
N4	0.0796(3)	0.53403(25)	−0.05213(24)	2.23(8)
N5	−0.0492(3)	0.35954(24)	0.1129(3)	2.15(8)
N6	−0.1313(3)	0.5890(3)	0.1463(3)	2.28(9)
C1	0.4383(3)	0.5162(3)	0.2306(3)	2.50(10)
C2	0.3811(4)	0.6357(3)	0.1767(4)	2.72(11)
C3	0.3373(3)	0.3202(3)	0.2776(3)	2.40(9)
C4	0.2896(4)	0.3390(3)	0.4288(3)	2.67(10)
C5	0.1026(4)	0.3616(3)	0.4355(3)	2.49(10)
C6	0.3725(4)	0.3719(3)	0.0371(3)	2.73(11)
C7	0.3610(4)	0.4485(4)	−0.0951(4)	3.24(13)
C8	0.1802(4)	0.4703(3)	−0.1547(3)	2.91(11)
Ow1	0.2662(4)	0.6668(3)	0.5250(3)	4.15(11)
Hw1	0.3167(9)	0.7364(7)	0.4875(8)	1.7(3)
Hw2	0.3330(9)	0.6444(7)	0.5762(8)	1.7(3)

Anisotropically refined atoms are given in the form of the isotropic equivalent thermal parameter defined as $(4/3)\cdot[a^2\cdot\beta_{11}+b^2\cdot\beta_{22}+c^2\cdot\beta_{33}+ab(\cos\gamma)\cdot\beta_{12}+ac(\cos\beta)\cdot\beta_{13}+bc(\cos\alpha)\cdot\beta_{23}]$.

Table 8. Positional Parameters and Their esd's for (**III**)

Atom	x	y	z	$B(\text{Å})$
Co	0.08064(4)	0.27451(3)	0.325029(19)	1.576(10)
P	0.72608(10)	0.23779(7)	0.06426(4)	2.54(3)
F1	0.7157(3)	0.29305(22)	−0.01097(12)	4.23(10)
F2	0.8790(3)	0.15626(22)	0.04686(16)	4.45(11)
F3	0.8559(3)	0.33638(18)	0.09185(14)	4.08(10)
F4	0.5960(4)	0.13826(21)	0.03615(17)	5.00(13)
F5	0.7365(4)	0.18221(20)	0.13911(13)	4.28(10)
F6	0.5729(3)	0.31832(21)	0.08106(14)	4.14(10)
O1	−0.0005(4)	0.07503(21)	0.24944(17)	3.66(10)
O2	−0.1749(4)	0.11025(24)	0.32332(18)	3.98(12)
O3	−0.2305(3)	0.36144(22)	0.27003(14)	3.39(10)
O4	−0.1581(4)	0.42097(23)	0.37262(16)	3.84(11)
Ow1	0.2376(5)	0.4047(3)	0.51386(17)	4.38(13)
N1	0.3171(3)	0.19723(19)	0.33235(13)	2.09(8)
N2	0.2009(3)	0.41956(19)	0.35467(14)	2.26(8)
N3	0.0460(3)	0.23522(21)	0.42085(13)	2.42(8)
N4	0.0861(3)	0.32165(21)	0.22727(13)	2.27(8)
N5	−0.0464(3)	0.13641(20)	0.29620(15)	2.43(9)
N6	−0.1286(3)	0.36270(20)	0.32160(15)	2.35(9)
C1	0.4373(4)	0.2892(3)	0.36290(19)	2.85(11)
C2	0.3810(4)	0.4100(3)	0.34231(20)	2.82(11)
C3	0.3370(4)	0.0910(3)	0.37786(19)	2.84(11)
C4	0.2862(5)	0.1018(3)	0.44907(20)	3.17(12)
C5	0.0978(5)	0.1211(3)	0.45200(19)	3.05(12)
C6	0.3748(4)	0.1546(3)	0.26542(18)	2.87(11)
C7	0.3654(4)	0.2379(3)	0.20593(19)	3.33(12)
C8	0.1863(5)	0.2649(3)	0.17722(17)	3.21(12)
Hw1	0.171	0.399	0.561	1.9
Hw2	0.250	0.494	0.506	1.9

Anisotropically refined atoms are given in the form of the isotropic equivalent thermal parameter defined as $(4/3) \cdot [a^2 \cdot \beta_{11} + b^2 \cdot \beta_{22} + c^2 \cdot \beta_{33} + ab(\cos\gamma) \cdot \beta_{12} + ac(\cos\beta) \cdot \beta_{13} + bc(\cos\alpha) \cdot \beta_{23}]$.

Table 9. Positional Parameters and Their esd's for (**IV**)

Atom	x	y	z	$B(\text{Å})$
I	0.17847(6)	0.56312(3)	0.76347(5)	4.29(3)
Co	0.06939(9)	0.67600(4)	0.33660(7)	1.43(3)
O1	−0.2273(6)	0.7229(3)	0.4088(5)	4.2(3)
O2	−0.1720(7)	0.6102(3)	0.4469(6)	5.1(3)
O3	0.1384(6)	0.7638(3)	0.5435(4)	3.46(23)
O4	0.3537(6)	0.7025(3)	0.5166(4)	3.83(23)
N1	0.2596(6)	0.6805(3)	0.2372(5)	2.24(21)
N2	−0.0771(6)	0.6324(3)	0.1900(4)	2.12(21)
N3	−0.0053(6)	0.7775(3)	0.2829(5)	2.22(21)
N4	0.1301(6)	0.5748(3)	0.4043(5)	2.35(22)
N5	−0.1320(6)	0.6697(3)	0.4092(4)	2.23(23)
N6	0.2014(7)	0.7190(3)	0.4824(5)	2.27(23)
C1	0.1685(8)	0.6642(4)	0.1072(5)	2.9(3)
C2	0.0261(9)	0.6092(4)	0.1008(6)	3.1(3)
C3	0.3475(8)	0.7557(4)	0.2393(6)	2.7(3)
C4	0.2371(8)	0.8231(4)	0.2027(6)	3.1(3)
C5	0.1174(8)	0.8390(4)	0.2856(6)	2.9(3)
C6	0.4060(8)	0.6270(4)	0.2746(7)	3.1(3)
C7	0.3665(9)	0.5470(4)	0.3036(7)	3.8(3)
C8	0.3007(8)	0.5399(4)	0.4163(7)	3.4(3)
Ow1	0.406(3)	0.4592(15)	0.9955(22)	37.1(25)

Anisotropically refined atoms are given in the form of the isotropic equivalent thermal parameter defined as $(4/3) \cdot [a^2 \cdot \beta_{11} + b^2 \cdot \beta_{22} + c^2 \cdot \beta_{33} + ab(\cos\gamma) \cdot \beta_{12} + ac(\cos\beta) \cdot \beta_{13} + bc(\cos\alpha) \cdot \beta_{23}]$.

RESULTS

Racemic [Co(abap)(NO$_2$)$_2$]Cl (**I**), CoClO$_4$N$_6$C$_8$H$_{22}$, crystallizes as a conglomerate, at 22 ^{0}C, from a deionized water solution in space group $P2_12_12_1$ (No. 19). The five-membered ring for the molecule described is δ (N1-C1-C2-N2 = +47.7^0). The Cl$^-$ anion forms Cl....H bonds with a hydrogen of each of a terminal -NH$_2$, thus rendering those hydrogens different and the terminal nitrogens chiral. In the cation of the crystal we used both are R.

Racemic [Co(abap)(NO$_2$)$_2$]ClO$_4 \cdot$H$_2$O (**II**), CoClO$_9$N$_6$C$_8$H$_{24}$, crystallizes as a conglomerate, at 22 ^{0}C, from a deionized water solution in space group $P2_1$ (No. 4), The five-membered ring for the molecule described here is δ (+48.6^0). The water of crystallization forms hydrogen bonds with a terminal NH$_2$ hydrogen rendering its nitrogen chiral in the solid state. The ClO$_4^-$ anion forms three O....H bonds with hydrogens of the water, as well as as six with amine hydrogens. In turn, a non-bonded pair on the water oxygen forms a hydrogen bond with an amine hydrogen.

Racemic [Co(abap)(NO$_2$)$_2$]PF$_6 \cdot$H$_2$O (**III**), CoPF$_6$O$_5$N$_6$C$_8$H$_{24}$, crystallizes, at 22 ^{0}C, from a deionized water solution in space group $P2_1/c$ (No. 14). The five-membered ring for the cation described here is δ (N1-C1-C2-N2 = +48.9^0). Finally, racemic [Co(abap)(NO$_2$)$_2$]I·H$_2$O (**IV**), ICoO$_5$N$_6$C$_8$H$_{24}$, crystallizes, at 22 ^{0}C, from a deionized water solution in space group $P2_1/n$. The five-membered ring for the cation described here is δ (N1-C1-C2-N2 = +47.7^0). The I$^-$ anion and the water of crystallization form hydrogen bonds with terminal nitrogen hydrogens and the terminal nitrogens are chiral.

The space groups of compounds (**I**), (**III**) and (**IV**) are unambiguously determined by their systematic absences. The space group of (**II**) is ambiguous inasmuch as its systematic absences are shared by the centrosymmetric space group $P2_1/m$; however, we attempted to solve the structure in the centric choice and the results were very poor (heavily distorted cations and anions; unacceptable thermal parameters and R factors). On the other hand, solution and refinement proceeded smoothly in the enantiomorphic space group and with a very attractive

Table 10. Bond Distances (Å) and Angles (^{0}) and Selected H-Bonding for (**I**)

(1) Bond Distances (Å)

Co-N1	2.059(6)	N1-C3	1.503(10)	Co-N2	1.975(5)
N1-C6	1.509(10)	Co-N3	1.973(6)	N2-C2	1.461(9)
Co-N4	1.969(6)	N3-C5	1.490(9)	Co-N5	1.933(6)
N4-C8	1.480(9)	Co-N6	1.930(6)	C1-C2	1.502(11)
O1-N5	1.209(8)	C3-C4	1.522(13)	O2-N5	1.247(8)
C4-C5	1.492(12)	O3-N6	1.244(8)	C6-C7	1.510(13)
O4-N6	1.231(9)	C7-C8	1.479(12)	N1-C1	1.522(9)

(2) Bond Angles (^{0})

N1-Co-N2	85.12(23)	C1-N1-C6	106.8(6)
N1-Co-N3	94.81(24)	C3-N1-C6	102.9(6)
N1-Co-N4	93.39(24)	Co-N2-C2	111.2(4)
N1-Co-N5	96.97(24)	Co-N3-C5	124.1(5)
N1-Co-N6	172.66(24)	Co-N4-C8	122.7(5)
N2-Co-N3	89.78(25)	Co-N5-O1	122.7(5)
N2-Co-N4	92.02(24)	Co-N5-O2	118.4(5)
N2-Co-N5	177.7(3)	O1-N5-O2	118.9(6)
N2-Co-N6	87.57(23)	Co-N6-O3	121.3(5)
N3-Co-N4	171.73(25)	Co-N6-O4	118.8(5)
N3-Co-N5	91.0(3)	O3-N6-O4	119.8(6)
N3-Co-N6	85.7(3)	N1-C1-C2	111.1(6)
N4-Co-N5	86.88(25)	N2-C2-C1	107.1(5)
N4-Co-N6	86.3(3)	N1-C3-C4	117.3(7)
N5-Co-N6	90.33(25)	C3-C4-C5	111.8(7)
Co-N1-C1	106.4(4)	N3-C5-C4	113.4(6)
Co-N1-C3	115.6(5)	N1-C6-C7	118.0(6)
Co-N1-C6	114.8(5)	C6-C7-C8	113.8(7)
C1-N1-C3	110.0(6)	N4-C8-C7	112.1(6)

(3) Selected H-Bonding (Å)

	Å	^{0}
N2 - H5.....O4	2.73	88.04
N2 - H5.....Cl	2.34	43.65
N2 - H6.....O4	2.72	89.205
N2 - H6.....Cl[a]	2.34	157.25
N3 - H13...Cl[b]	2.39	155.8
N3 - H14...O1[c]	2.39	82.39
N3 - H14...O3	2.17	120.12
N4 - H21...O2	2.45	102.19
N4 - H21...O4	2.41	102.58
N4 - H22...O4	2.76	80.93
N4 - H22...Cl	2.26	173.73

[a]Cl at $-x+1, y-1/2, -z+1/2$
[b]Cl at $-x+1, y-1/2, -z+1/2$
[c]O1 at $x+1/2, -y+1/2, -z$

Table 11. Bond Distances(Å) and Angles(^{0}) and Selected H-Bonding for (**II**)

(1) Bond Distances (Å)

Co-N1	2.0740(22)	O4-N5	1.241(3)	Co-N2	1.967(3)
N1-C1	1.504(4)	Co-N3	1.974(4)	N1-C3	1.506(4)
Co-N4	1.986(4)	N1-C6	1.509(4)	Co-N5	1.927(3)
N2-C2	1.475(4)	Co-N6	1.9504(24)	N3-C5	1.487(4)
Cl-O5	1.444(3)	N4-C8	1.490(4)	Cl-O6	1.428(3)
C1-C2	1.511(5)	Cl-O7	1.436(3)	C3-C4	1.506(5)
Cl-O8	1.4393(25)	C4-C5	1.512(4)	O1-N6	1.236(4)
C6-C7	1.507(5)	O2-N6	1.230(4)	C7-C8	1.509(5)
O3-N5	1.235(4)				

(2) Bond Angles (^{0})

Co-N1-C3	115.30(16)	N1-Co-N3	92.41(9)
Co-N1-C6	115.49(17)	N1-Co-N4	94.59(9)
N1-Co-N2	85.90(10)	C1-N1-C3	107.91(22)
N1-Co-N5	96.73(10)	C1-N1-C6	109.63(22)
N1-Co-N6	173.83(11)	C3-N1-C6	102.80(25)
N2-Co-N3	91.37(11)	Co-N2-C2	110.50(20)
N2-Co-N4	89.79(11)	Co-N3-C5	121.38(19)
N2-Co-N5	177.37(10)	Co-N4-C8	125.54(20)
N2-Co-N6	87.94(11)	Co-N5-O3	119.96(20)
N3-Co-N4	172.97(10)	Co-N5-O4	120.88(23)
N3-Co-N5	88.38(11)	O3-N5-O4	119.2(3)
N3-Co-N6	87.39(10)	Co-N6-O1	118.28(21)
N4-Co-N5	90.15(11)	Co-N6-O2	120.81(22)
N4-Co-N6	85.72(10)	O1-N6-O2	120.9(3)
N5-Co-N6	89.43(11)	N1-C1-C2	111.71(22)
O5-Cl-O6	110.32(22)	N2-C2-C1	107.09(24)
O5-Cl-O7	108.84(18)	N1-C3-C4	117.4(3)
O5-Cl-O8	108.87(17)	C3-C4-C5	113.27(24)
O6-Cl-O7	108.64(19)	N3-C5-C4	111.14(24)
O6-Cl-O8	109.98(18)	N1-C6-C7	118.2(3)
O7-Cl-O8	110.18(19)	C6-C7-C8	112.3(3)
Co-N1-C1	105.52(18)	N4-C8-C7	113.1(3)

(3) Selected H-Bonding (Å)

	Å	^{0}		Å	^{0}
N2 - H5......O1	2.78	88.37	N3 - H14....Ow1	2.16	176.14
N2 - H5......O6	2.80	72.96	N4 - H21....O2	2.99	66.67
N2 - H5......Ow1	2.11	156.16	N4 - H21....O3[d]	2.24	138.35
N2 - H6......O1	2.79	87.58	N4 - H21....O4[e]	2.86	160.28
N2 - H6......O3[a]	2.05	160.65	N4 - H22....O2	2.18	118.45
N2 - H6......O6	2.57	107.71	N4 - H22....O3	2.89	84.69
N3 - H13....O1	2.35	107.08	N4 - H22....O6[f]	2.58	146.24
N3 - H13....O4	2.57	97.87	N4 - H22....O7[g]	2.63	123.20
N3 - H13....O5[b]	2.24	145.58	Ow1 - Hw1...O5	2.17	136.40
N3 - H13....O7[c]	2.73	114.01	Ow1 - Hw1...O6	2.91	139.19
N3 - H14....O1	2.87	75.37	Ow1 - Hw2...O8[h]	2.19	154.08

[a]O3 at $-x, y + 1/2, -z$ [b]O5 at $-x, y - 1/2, -z + 1$ [c]O7 at $-x, y - 1/2, -z + 1$ [d]O3 at $-x, y + 1/2, -z$
[e]O4 at $-x, y + 1/2, -z$ [f]O6 at $-x, y - 1/2, -z$ [g]O7 at $-x, y - 1/2, -z$ [h]O8 at $-x + 1, y - 1/2, -z + 1$

Table 12. Bond Distances(Å) and Angles(^{0}) and Selected H-Bonding for (**III**)

(1) Bond Distances (Å)

Co-N1	2.0748(23)	P-F4	1.602(3)	N1-C6	1.513(5)
Co-N2	1.9784(21)	P-F5	1.592(3)	N2-C2	1.484(5)
Co-N3	1.975(3)	P-F6	1.592(3)	N3-C5	1.486(4)
Co-N4	1.992(3)	O1-N5	1.239(5)	N4-C8	1.476(5)
Co-N5	1.9339(22)	O2-N5	1.238(5)	C1-C2	1.500(4)
Co-N6	1.946(3)	O3-N6	1.229(3)	C3-C4	1.495(6)
P-F1	1.597(3)	O4-N6	1.244(4)	C4-C5	1.525(6)
P-F2	1.601(3)	N1-C1	1.508(3)	C6-C7	1.502(5)
P-F3	1.5905(22)	N1-C3	1.508(4)	C7-C8	1.513(4)

(2) Bond Angles (^{0})

N1-Co-N2	85.99(10)	F1-P-F5	179.78(15)	Co-N2-C2	109.76(19)
N1-Co-N3	92.90(11)	F1-P-F6	89.47(15)	Co-N3-C5	122.05(24)
N1-Co-N4	94.24(11)	F2-P-F3	90.28(13)	Co-N4-C8	125.65(18)
N1-Co-N5	96.57(10)	F2-P-F4	89.44(14)	Co-N5-O1	119.8(3)
N1-Co-N6	173.70(9)	F2-P-F5	89.52(16)	Co-N5-O2	120.37(25)
N2-Co-N3	91.40(11)	F2-P-F6	179.50(11)	O1-N5-O2	119.8(3)
N2-Co-N4	89.74(11)	F3-P-F4	179.68(21)	Co-N6-O3	120.97(25)
N2-Co-N5	177.43(12)	F3-P-F5	90.10(13)	Co-N6-O4	118.91(17)
N2-Co-N6	87.71(10)	F3-P-F6	90.17(13)	O3-N6-O4	120.1(3)
N3-Co-N4	172.82(11)	F4-P-F5	90.08(15)	N1-C1-C2	112.34(19)
N3-Co-N5	88.27(12)	F4-P-F6	90.11(14)	N2-C2-C1	107.0(3)
N3-Co-N6	87.10(12)	F5-P-F6	90.70(16)	N1-C3-C4	117.5(3)
N4-Co-N5	90.27(12)	Co-N1-C1	105.31(17)	C3-C4-C5	114.07(25)
N4-Co-N6	85.87(12)	Co-N1-C3	115.24(22)	N3-C5-C4	110.3(3)
N5-Co-N6	89.72(11)	Co-N1-C6	115.63(15)	N1-C6-C7	118.0(3)
F1-P-F2	90.31(15)	C1-N1-C3	107.92(19)	C6-C7-C8	112.9(3)
F1-P-F3	90.06(13)	C1-N1-C6	109.4(3)	N4-C8-C7	113.4(3)
F1-P-F4	89.76(14)	C3-N1-C6	103.09(24)		

(3) Selected H-Bonding (Å)

	Å	^{0}	
N2 - H6....... O1	2.13	159.00	O1 at $-x, y-1/2, -z+1/2$
N4 - H21......O1	2.28	135.52	O1 at $-x, y-1/2, -z+1/2$
N4 - H22......O1	2.90	83.99	
N3 - H13......O2	2.55	98.21	
N4 - H21.....O2	2.66	155.30	O2 at $-x, y+1/2, -z+1/2$
N4 - H22......O3	2.18	119.28	
Ow1 - Hw1...O4	2.45	116.02	O4 at $-x, -y+1, -z+1$
N2 - H6........O4	2.78	88.75	
N3 - H13......O4	2.36	106.81	
N3 - H14......O4	2.85	76.63	
Ow1- Hw2....O4	2.73	100.89	O4 at $-x, -y+1, -z+1$
N3 - H13....F1	2.25	145.81	F1 at $x-1, -y+1/2, z-1/2$
N2 - H5........F2	2.84	120.78	F2 at $-x+1, y+1/2, -z+1/2$
N3 - H13......F2	2.73	108.93	F2 at $x-1, -y+1/2, z+1/2$
N3 - H14......F2	2.82	103.19	F2 at $x-1, -y+1/2, z+1/2$
Ow1 - Hw1...F2	2.40	121.10	F2 at $x-1, -y+1/2, z+1/2$
Ow1 - Hw2...F2	2.31	143.94	F2 at $-x+1, y+1/2, -z+1/2$
Ow1 - Hw2...F4	2.26	146.32	F4 at $-x+1, y+1/2, -z+1/2$
N3 - H6.......F5	2.51	117.00	F5 at $-x+1, y+1/2, -z+1/2$

Table 13. Bond Distances(Å) and Angles(^{0}) and Selected H-Bonding for (**IV**)

(1) Bond Distances (Å)

Co-N1	2.074(5)	N1-C3	1.518(8)	Co-N2	1.975(5)
N1-C6	1.507(8)	Co-N3	1.968(5)	N2-C2	1.486(8)
Co-N4	1.986(5)	N3-C5	1.474(8)	Co-N5	1.961(5)
N4-C8	1.486(8)	Co-N6	1.918(5)	C1-C2	1.499(10)
O1-N5	1.220(8)	C3-C4	1.501(10)	O2-N5	1.214(7)
C4-C5	1.499(10)	O3-N6	1.230(7)	C6-C7	1.516(10)
O4-N6	1.239(7)	C7-C8	1.475(11)	N1-C1	1.519(8)

(2) Bond Angles (^{0})

N1-Co-N2	86.76(20)	C1-N1-C6	109.8(5)
N1-Co-N3	90.60(20)	C3-N1-C6	103.1(5)
N1-Co-N4	95.30(21)	Co-N2-C2	110.7(4)
N1-Co-N5	172.16(20)	Co-N3-C5	121.6(4)
N1-Co-N6	96.11(21)	Co-N4-C8	124.5(4)
N2-Co-N3	91.04(21)	Co-N5-O1	121.6(4)
N2-Co-N4	90.89(21)	Co-N5-O2	119.8(4)
N2-Co-N5	85.68(20)	O1-N5-O2	118.3(5)
N2-Co-N6	177.06(22)	Co-N6-O3	121.3(4)
N3-Co-N4	173.89(21)	Co-N6-O4	120.2(4)
N3-Co-N5	87.46(22)	O3-N6-O4	118.5(5)
N3-Co-N6	88.29(22)	N1-C1-C2	112.5(5)
N4-Co-N5	86.90(22)	N2-C2-C1	108.1(5)
N4-Co-N6	89.49(21)	N1-C3-C4	117.5(5)
N5-Co-N6	91.43(21)	C3-C4-C5	113.7(5)
Co-N1-C1	104.4(4)	N3-C5-C4	110.9(5)
Co-N1-C3	114.7(4)	N1-C6-C7	118.3(5)
Co-N1-C6	116.4(4)	C6-C7-C8	113.2(6)
C1-N1-C3	108.3(5)	N4-C8-C7	112.0(6)

(3) Selected H-Bonding (Å)

	Å	^{0}	
N2 - H5......O3	2.48	175.32	O3 at $x - 1/2, y, z - 1/2$
N2 - H5......O4	2.78	134.18	O4 at $x - 1/2, -y + 1.5, z - 1/2$
N2 - H6......O2	2.72	108.19	
N2 - H6......Ow1	2.97	110.13	Ow1 at $-x, -y + 1, -z + 1$
N2 - H6......I	2.78	151.58	I at $-x, -y + 1, -z + 1$
N3 - H13....O1	2.02	124.66	
N3 - H13....O3	2.85	89.92	
N3 - H14....O3	2.66	147.95	O3 at $x - 1/2, y, z - 1/2$
N3 - H14....O4	2.07	159.74	O4 at $x - 1/2, -y + 1.5, z - 1/20$
N4 - H21....O2	2.57	83.82	
N4 - H21....I	2.76	173.73	I at $-x, -y + 1, -z + 1$
N4 - H22....O2	2.27	102.25	
N4 - H22....O4	3.00	81.87	

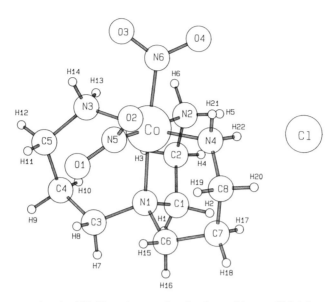

Figure 1. The asymmetric unit of (**I**). The anion was placed at that position at which it forms the shortest hydrogen bond to an amine hydrogen (H22). The ring defined by N1-C3-C4-C5-N3-Co and that defined by Ni-C6-C7-C8-N4-Co are chairs; also the fragment N1-C1-C2-N2 is helical δ (+47.7⁰). The orientation of the two -NO₂ are measured by the torsional angles N6-Co-N5-O2 (−41.2⁰) and N5-Co-N6-O3 (−50.7⁰); thus, the cation is dissymetric, as is the case with (**II**), (**III**), and (**IV**).

geometry, thermal parameters and residuals. Thus, we selected this one. Figures 1, 2, 3 and 4 give labelled views of the asymmetric units found for molecules (**I**), (**II**), (**III**) and (**IV**). Figures 5, 6, 7 and 8 depict the packing of the ions in their respective unit cells. Final positional and equivalent-isotropic thermal parameters are given in Tables 6, 7, 8 and 9 and bond lengths, angles, selected torsional angles and useful hydrogen bonds are listed in Tables 10, 11, 12 and 13.

DISCUSSION

Earlier, we reported[37] the crystallization behavior of a series of compounds with composition [cis–Co(tren)(NO₂)₂]ₙY·mH₂O [(**V**), n = 1, Y = Br⁻, n = 0]; [(**VI**), n = 2, Y = (Br⁻ + ClO₄⁻), m = 1]; [(**VII**), n = 1, Y = ClO₄⁻, m = 0]; and [(**VIII**), n = 1, Y = NO₃⁻, m = 0] crystallize, respectively, a racemate, a kryptoracemate, a racemate and a racemate. Recently, we also discovered[38] that [cis–Co(tren)(NO₂)₂]ₙY : [(**IV**), n = 1, Y = Cl⁻]; [(**I**), n = 1, Y = BF₄⁻]; and [(**II**), n = 2, Y = SiF₆²⁻] crystallize, respectively, as a racemate, a racemate and a conglomerate. [tren = N(CH₂-CH₂-NH₂)₃].

In those studies of the [cis–Co(tren)(NO₂)₂]Y series,[37, 38] we concentrated on recording the changes in crystallization pathway introduced by a change in counter anion (which may also introduce a change in the numbers of waters of crystallization) while retaining the temperature of crystallization and the solvent constant. In both studies, we noted that, indeed, their crystallization behavior appears to be controlled by the nature of the counter anion used. Interestingly, the space groups in which these substances crystallize contain examples of centrosymmetric, polar, enantiomorphic and kryptoracemic[37, 38] crystals. Thus, that class of cations has provided us with examples of crystals displaying all possible modes of crystallization known, except examples of unbalanced crystallization.[39]

In order to further probe the effect of a small change on the ligand (tren to abap) on the

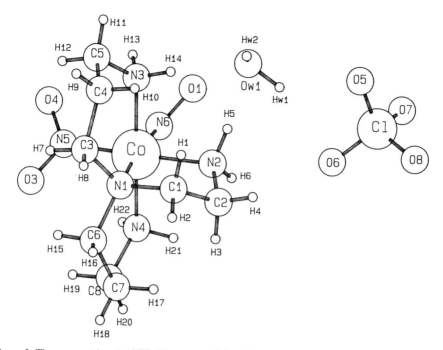

Figure 2. The asymmetric unit of (**II**). The water and the anion were placed at those positions at which they form the shortest hydrogen bonds to an amine hydrogen, both with H5. The ring defined by N1-C3-C4-C5-N3-Co and that defined by Ni-C6-C7-C8- N4-Co are chairs; also the fragment N1-C1-C2-N2 is helical δ (+48.6^0). The orientation of the two -NO$_2$ are measured by the torsional angles N6-Co-N5-O4 (-45.0^0) and N5-Co-N6-O2 (-53.1^0); thus, the cation is also dissymetric.

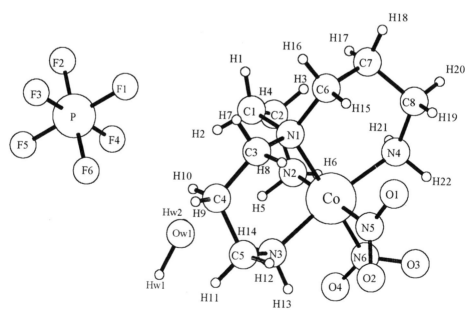

Figure 3. The asymmetric unit of (**III**). The water and the anion were placed at those positions at which they form the shortest hydrogen bonds, the water to F2. The remarks made in the captions of Figs. 1 and 2 concerning the dissymmetry of the five membered ring and the chair conformation of the two six-membered rings apply; however see Table 14 and the Discussion concerning torsional angles of the -NO$_2$'s.

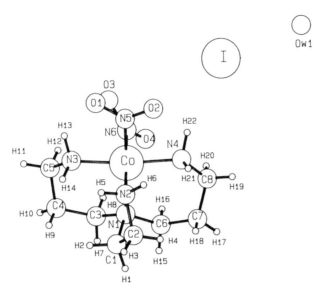

Figure 4. The asymmetric unit of (**IV**). The water and the anion were placed at those positions at which they form the shortest hydrogen bonds, the Iodine to H22 and the water to the iodide. The remarks made in the captions of Figs. 1 and 2 concerning the dissymmetry of the five membered ring and the chair conformation of the two six-membered rings apply; however see Table 14 and the Discussion concerning torsional angles of the -NO$_2$'s.

crystallization pathway of the cations, and hoping to obtain clues as to the reasons for their choice of crystallization pathway, we have prepared the title compounds whose structures and crystallization behavior we describe in this report.

The asymmetric tripodal ligand abap [(N-(2-aminoethyl)-N,N-bis(3-aminopropyl)-amine] was first made by Blackman.[28] The Co(III) complexes of abap were prepared in our laboratory by oxidation of Co(II) species using a stronger oxidant (30 % H$_2$O$_2$) based on the suggestion that the resulting Co(III) abap complexes should be less prone to reduction. This contrasts with reported synthesis of abap complexes, where air oxidation is required. But air oxidation of a solution of Co(ClO$_4$)$_2$·6H$_2$O, abap and NaNO$_2$ gives the peroxo-bridged dimer [(abap)(NO$_2$)CoO$_2$Co(NO$_2$)(abap)]$^{2+}$, which was not expected. Acid hydrolysis of this dimer in dilute aqueous HNO$_3$ and addition of NaClO$_4$ gave [Co(abap)(O$_2$NO)] in good yield.[28]

The compounds (**I**) and (**II**), crystallizing in space group $P2_12_12_1$ and $P2_1$, are conglomerates. While the compounds (**III**) and (**IV**) crystallize in space group $P2_1/c$ and $P2_1/n$, which are true racemates. As we mentioned earlier, the difference between the [Co(tren)(NO$_2$)$_2$]$^{2+}$ series and [Co(abap)(NO$_2$)$_2$]$^{2+}$ series is that, in the former, three five-membered rings form upon complexion while two six-membered rings and one five-membered ring are formed in the abap series. This small change in structure completely changes the crystallization pathway from racemate ([Co(tren)(NO$_2$)$_2$]Y, Y = Cl, ClO$_4$) to conglomerate ([Co(abap)(NO$_2$)$_2$]Y, Y = Cl, ClO$_4$).

As the data in Tables 14 and 15 illustrate, there is a remarkable change in the conformation of the -NO$_2$ groups. For the [Co(tren)(NO$_2$)$_2$]X (X = Cl, ClO$_4$, both are racemates), the torsional angles are clearly different from those present in (**I**) and (**II**) (both are conglomerates). Unlike the results for [Co(tren)(NO$_2$)$_2$]X (X = Cl, ClO$_4$), in which the two -NO$_2$ group planes are oriented almost perpendicularly to each other, the relevant torsional angles are around 90^0. But for (**I**) and (**II**), the two NO$_2$ planes twist a lot, the torsional angles being around -45^0 or 135^0, as shown in Fig. 1 and Fig. 2. The twist appears to prevent the forma-

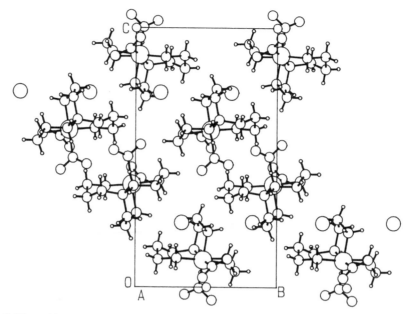

Figure 5. The packing of ions in the unit cell of (**I**), which is an *a*-projection. Note the hydrogen bonded, helical, strings running along the diagonals of the *bc*-plane. For discussion of the importance of these helical strings see the Discussion and the tables of hydrogen bonds.

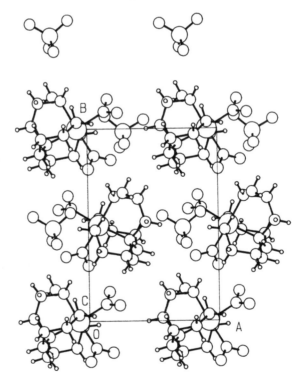

Figure 6. The packing of ions in the unit cell of (**II**), which is a *c*-projection. Note the hydrogen bonded, helical, strings running along the *b*-direction. For discussion of the importance of these helical strings see the Discussion and the tables of hydrogen bonds.

240

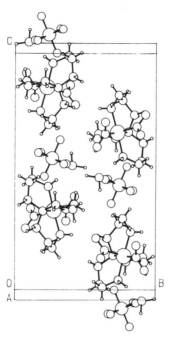

Figure 7. The packing of ions in the unit cell of (**III**), which is an *a*-projection. Note the the absence of hydrogen bonded interaction between cations, which are linked by the agency of hydrogen bond with the anion and the waters of crystallization. The absence of helical strings is also notable. The inversion center of this cell is obvious and located at the center of this centrosymmetric cell.

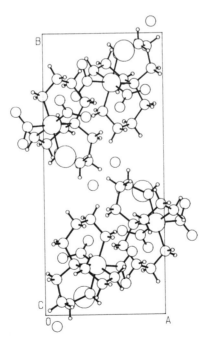

Figure 8. The packing of ions in the unit cell of (**IV**), which is a *c*-projection. Note the the absence of hydrogen bonded interaction between cations, which are linked by the agency of hydrogen bond with the anion and the waters of crystallization. The absence of helical strings is also notable. The inversion center of this cell is obvious and located at the center of this centrosymmetric cell.

Table 14. Selected Torsional Angles for tren series (**V**)[a], (**VI**)[b] and abap series (**I**) and (**II**)

	(**V**)	(**VI**)	(**I**)	(**II**)
N6-Co-N5-O1[c]	89.1	89.2	138.7	133.9
N6-Co-N5-O2	−86.7	−86.6	−41.2	−45.0
N5-Co-N6-O3	−89.1	−89.2	−50.7	−53.1
N5-Co-N6-O4	86.7	86.6	130.3	128.2

[a]The torsional angles data from ref. 25   [b]taken from ref. 40.   [c]In the original, the atoms may have different numbers in the figures for each compound.

Table 15. Torsional Angles for abap series (**I**), (**II**), (**III**) and (**IV**)

		(**I**)	(**II**)	(**III**)	(**IV**)
Ring 1	Co-N1-C6-C7	−51.5	+50.6	+50.9	+43.3
	N1-C6-C7-C8	+66.0	−70.4	−69.4	−69.0
	C6-C7-C8-N4	−63.4	+62.6	+61.8	+67.0
	C7-C8-N4-Co	+55.3	−44.5	−44.5	−47.5
Ring 2	Co-N1-C3-C4	+51.3	−52.6	−51.7	54.0
	N1-C3-C4-C5	−70.1	+65.6	+66.1	+63.5
	C3-C4-C5-N3	+64.2	−65.1	−65.1	−63.4
	C4-C5-N3-Co	−47.8	+59.5	+58.1	+62.6

tion of the centrosymmetric space groups, as expected. ([cis–Co(abap)(NO$_2$)$_2$]Y·mH$_2$O ([(**I**), Y = Cl$^-$, n = 0, space group: $P2_12_12_1$]; [(**II**), Y = ClO$_4^-$, m = 1, space group: $P2_1$]; [(**III**), Y = PF$_6^-$, m = 1, space group: $P2_1/c$]; and [(**IV**), Y = I$^-$, m = 1, space group: $P2_1/n$]). It is believed that these conformational changes for -NO$_2$ groups contribution to the pathway selection may be due to steric hindrance — the formation of the two six-membered rings in stead of five-membered rings and the change in the formation of intra- and inter-molecular hydrogen bonding.

After a careful study of the abap series, we found that the six-membered rings of the cation of all four complexes, (**I**), (**II**), (**III**) and (**IV**) share the same chair conformation. But their torsional angles are of opposite sign in order to make central nitrogen tetrahedral. For example, for the chloride, the torsional angles of ring 1 are −, +, −, + while the torsional angles of ring 2 are opposite, +, −, +, −. The torsional angles of the two six-member rings in different compounds are listed in Table 15.

For six-membered aliphatic rings, the chair conformation is more stable than the boat and twist boat. For example, for cyclohexane, the chair form is converted to the boat form by the flipping of one carbon atom and vice versa. In the process of flipping from chair to boat form, the valence angles are slightly deformed, and the molecules goes through a maximum of potential energy. The height of this energy barrier was estimated at 10-11 kcal/mole with the help of nuclear magnetic resonance spectroscopy.[42] On the basis of calculations and experimental results, it may be stated that the energy difference between the chair form and the twist boat of cyclohexane may be assumed to be between 5 and 6 kcal/mole. Because the twist form is more flexible in comparison to the chair form, the higher entropy values of 5 e.u. must be considered. The difference in free energy (ΔG) is easily determined at 3.5 to 4.5 kcal/mole (at 25 ^{0}C) from this information.[40,42]

In our early studies, we found it possible that the higher energy conformation exists. For example, in [cis–Co(NO$_2$)(SO$_3$)] and [cis–Co(N$_3$)(SO$_3$)],[43] we saw one en ring inverted such that instead of both being λλ or δδ (lowest conformational energy) in the latter one of the rings had the higher energy conformation λδ or δλ. Again, in [Co(tn)$_2$ox]$^+$ and [Co(tn)$_2$CO$_3$]$^+$ cations (tn = 1,3-diaminopropane), the six membered rings are mostly chairs; however, in the case of the oxalato ClO$_4$-derivative one ring was chair and the other one a twist boat.[41] It is a reasonable expectation that this is the outcome of having different numbers of hydrogen bonds which are also of different strengths.

Finally, the strengths of intramolecular hydrogen bonding for the four abap complexes are different. In conglomerates (**I**) and (**II**), they have stronger intramolecular hydrogen bonding between the -NO$_2$ oxygens and hydrogens of terminal amine nitrogens and also between the anions and those hydrogens. It is believed that those intramolecular hydrogen bonds may lock the molecules into specific dissymmetric conformations which influence the crystallization mode selected by those compounds. For example, in (**I**) there are four intramolecular hydrogen bonds of different lengths, O3....H14= 2.17 Å, angle 120.12^0, Cl....H22= 2.26 Å, angle 173.73^0, O2....H21= 2.45 Å, angle 102.19^0, and O4....H21= 2.41 Å, angle 102.58^0. In (**II**), there are strong intramolecular hydrogen bonding between the terminal hydrogens and water oxygen group, the water hydrogens and -NO$_2$ oxygen groups, H5...Ow1= 2.11 Å, angle 156.16^0, H14....Ow1= 2.16 Å, angle 176.14^0, Hw1....O5= 2.17 Å, angle 136.40^0, and Hw1....O6= 2.91 Å, angle 139.19^0. However, in (**III**) and (**IV**), there are no such strong intramolecular hydrogen bonding contacts.

CONCLUSIONS

(a) In [*cis*–Co(abap)(NO$_2$)$_2$]Cl (**I**) and [*cis*–Co(abap)(NO$_2$)$_2$]ClO$_4$·H$_2$O (**II**) (conglomerates), behave oppositely to the tren series with same counter ions; we also noted a remarkable change in the conformation of the -NO$_2$ groups which probably comes from the stereo hindrance — the formation of the two six-membered rings other than five-membered rings is one possible reason to form the conglomerates, as suggested by hydrogen bonding and by the torsional angles of the -NO$_2$ ligands.

(b) Finally, in [*cis*–Co(abap)(NO$_2$)$_2$]Cl (**I**) and [*cis*–Co(abap)(NO$_2$)$_2$]ClO$_4$·H$_2$O (**II**) (conglomerates), there are very strong intermolecular hydrogen bonding between the anions and one hydrogen, between -NO$_2$ oxygens and other terminal hydrogens thereby increasing the rigidity of the cations, whose symmetry is now lowered to C_i. We believe these factors also influence compounds (**I**) and (**II**) to select the conglomerate pathway, as has been suggested in many previous studies.[1-24,44-45]

NOTES AND REFERENCES

* Author to whom correspondence should be addressed.
+ Fellow of the Robert A. Welch Foundation.
1. I. Bernal, *Inorg. Chim. Acta* 96:99 (1985).
2. I. Bernal, *Inorg. Chim. Acta* 101:175 (1985).
3. I. Bernal, E.O. Schlemper and C.K. Fair, *Inorg. Chim. Acta* 115:25 (1985).
4. I. Bernal and J. Cetrullo, *Inorg. Chim. Acta* 120:109 (1986).
5. I. Bernal, *Inorg. Chim. Acta* 121:1 (1986).
6. I. Bernal and J. Cetrullo, *Inorg. Chim. Acta* 122:213 (1986).
7. I. Bernal, *J. Coord. Chem.* 15, 337 (1987).
8. I. Bernal, *Inorg. Chim. Acta* 131:53 (1987).
9. I. Bernal and J. Cetrullo, *Inorg. Chim. Acta* 131:201 (1987).
10. I. Bernal and J. Cetrullo, *Inorg. Chim. Acta* 134:105 (1987).
11. I. Bernal and J. Cetrullo, *Inorg. Chim. Acta* 144:227 (1988).
12. I. Bernal, *Inorg. Chim. Acta* 142:21 (1988).
13. I. Bernal and J. Cetrullo, *Inorg. Chim. Acta* 150:75 (1988).
14. I. Bernal and J. Cetrullo, *J. Coord. Chem.* 20:247 (1989).

15. I. Bernal and J. Cetrullo, *J. Coord. Chem.* 20:259 (1989).
16. I. Bernal and J. Cetrullo, *J. Coord. Chem.* 20:237 (1989).
17. I. Bernal and J. Cetrullo, *Struct. Chem.* 1:227 (1990).
18. I. Bernal and J. Cetrullo, *Struct. Chem.* 1:235 (1990).
19. I. Bernal, J. Cetrullo and S. Berhane, *Struct. Chem.* 1:361 (1990).
20. I. Bernal, J. Myrczek and J. Cai, *Polyhedron* 12:1149 (1993).
21. I. Bernal and J. Cetrullo, *Inorg. Chim. Acta* 142:235 (1988).
22. I. Bernal, J. Cai, S.S. Massoud, S.F. Watkins and F.R. Fronczek, *J. Coord. Chem.* in press (1995).
23. I. Bernal, J. Cetrullo and J. Myrczek, *Mater. Chem. and Phys.* 35:290 (1993).
24. I. Bernal, J. Cai and J. Myrczek, *Inorg. Chem.*, accepted (1995).
25. J. Chin, M. Drouin and A.G. Michel, *Acta Cryst.* 46:1022 (1995).
26. M. Laing, S. Baines and P. Sommerville, *Inorg. Chem.* 10:1057 (1971).
27. B. Nubert, H. Siebert, K. Wiedenheimmer, Weiss and M.I. Ziegler, *Acta. Crystallogr.* B75:1020 (1979)
28. R.L. Fanshawe and A.G. Blackman, *Inorg. Chem.* 34, 421 (1995).
29. P.D. Streater, Taylor, R.C. Hider, and J. Porter, *J. Med. Chem.* 33:1749 (1990).
30. TEXRAY-230 is a modification of the SDP-Plus31 set of X-ray crystallographic programs distributed by Molecular Structure Corporation, 3200 Research Forest Dr., The Woodlands, TX 77386, for use with their automation of the CAD-4 diffractometer. Version of 1985.
31. SDP-Plus is the Enraf-Nonius Corporation X-ray diffraction data processing programs distributed by B.A. Frenz & Associates, 209 University Dr. East, College Station, TX 77840. Version of 1985.
32. R.B. Roof, "A Theoretical Extension of the Reduced Cell Concept in Crystallography," Report LA-4038, Los Alamos Scientific Laboratory (1969).
33. D.T. Cromer, J.P. Waber, "International Tables for X-ray Crystallography," the Kynoch Press, Birmingham, England (1975); vol. IV, Tables 2.2.8 and 2.3.1, respectively, for the scattering factor curves and the anomalous dispersion values.
34. A.C. Larson, F.L. Lee, Y. LePage, M. Webster, J.P. Charland and E. J. Gabe, "The NRCVAX Crystal Structure System" as adapted for PC use by Peter S. White, University of North Carolina, Chapel Hill, N.C., 27599-3290.
35. I. Bernal, J. Cai and S.S. Massoud, *J. Coord. Chem.* accepted (1995).
36. I. Bernal, J. Cai and J. Myrczek, *Acta Chemica Hungarica – Models in Chemistry* 132:451 (1995).
37. I. Bernal, J. Cai, J. Cetrullo, S. Massoud, S.F. Watkins and F.R. Fronczek, *J. Coord. Chem.*, in press (1995).
38. I. Bernal, X. Xia and F. Somoza, unpublished data (1995).
39. (a) V.G. Albano, P. Bellon and M. Sansoni, *J. Chem. Soc., D.,* 899 (1969). (b) V.G. Albano, G.M. Ricci and M. Sansoni, *Inorg. Chem.* 8:2109 (1969). (c)V. G. Albano, P. Bellon and M. Sansoni, *J. Chem. Soc. A* 2420 (1971).
40. M. Hanack, "Conformation Theory," p. 44, Academic Press, New York (1965).
41. (a) I. Bernal, J. Cetrullo, J. Myrczek and S.S. Massoud, *J. Coord. Chem.* 29:287 (1993). (b) I. Bernal, J. Cetrullo, J. Myrczek and S.S. Massoud, *J. Coord. Chem.* 29:319 (1993).
42. N.L. Allinger and L.A. Freiberg, *J. Am. Chem. Soc.* 83:2393 (1960).
43. I. Bernal, J. Cetrullo and W.G. Jackson, *Struct. Chem.* 4:235 (1993).

LIST OF CONTRIBUTORS

Frank Allen
Cambridge Crystallographic Data Centre
12 Union Road
Cambridge
CB2 1EZ
England
Tel. +44 1223 336425
Fax: +44 1223 336033
e-mail allen@chemcrys.cam.ac.uk

Anton Amann
Laboratorium für Physikalische Chemie
ETH–Zentrum
CH–8092 Zürich
Tel. +41 1 632 4405
Fax: +41 1 632 1021
e-mail amann@phys.chem.ethz.ch

Ivan Bernal
Department of Chemistry
University of Houston
Houston TX 77204-5641
USA
Tel. 713 743 2775
Fax: 713 743 2709
e-mail ibernal@uh.edu

Jan C. A. Boeyens
Department of Chemistry
University of the Witwatersrand
P.O. Wits
2050 Johannesburg
South Africa
Tel. +27 11 716 4097
Fax: +27 11 716 3826
e-mail jan@hobbes.gh.wits.ac.za

Peter Comba
Anorganisch-Chemisches Institut
Universität Heidelberg
Im Neuenheimer Feld 270
D–69120 Heidelberg
Tel. +49 6221 54 8453
Fax: +49 6221 54 6617
e-mail comba@akcomba.oci.uni-heidelberg.de

Gastone Gilli
Dipartimento di Chimica
Universitá di Ferrara
Via Borsari 46
I–44100 Ferrara
Italy
Tel. +39 532 210 370/291 141
Fax: +39 532 240 709
e-mail M38A@icineca.cineca.it

Leslie Glasser
Department of Chemistry
University of the Witwatersrand
P.O. Wits
2050 Johannesburg
South Africa
Tel. +27 11 716 2070
Fax: +27 11 339 7967
e-mail glasser@aurum.chem.wits.ac.za

Alajos Kálmán
Budapest–114
POB 17
H1525
Hungary
Tel. +36 1 212 4790
Fax: +36 1 212 5020
e-mail akalman@cric.chemres.hu

Tibor Koritsánszky
Institut für Kristallographie
Freie Universtiät Berlin
Takustr. 6
D–14195 Berlin
Tel. +49 30 838 6787
Fax: +49 30 838 3464
e-mail tibor@chemie.fu-berlin.de

John Ogilvie
Department of Chemistry
Oregon State University
Corvallis
OR 97331-4003
USA
Tel. 541 737 6716
Fax: 541 737 2062
e-mail ogilviej@ccmail.orst.edu

Eiji Ōsawa
Dept. Knowledge–Based Information Engineering
Toyohashi University of Technology
Toyohashi 441
Japan
Tel. 81 532 48 5588
Fax: 81 532 48 5588
e-mail osawa@cochem.tutkie.tut.ac.jp

Brian T. Sutcliffe
Department of Chemistry
University of York
York
Y01 5DD
England
Tel. +44 1904 432 515
Fax: +44 1904 432 516
e-mail bts1@mailer.york.ac.uk

INDEX

α-hydroxy-ω-carboxylic acid, 189
Ab-initio calculation, 6
Adiabatic effects, 46, 48, 50
Ammonia
   MASER transition, 58
   pyramidal structure, 56
Araki perturbation theory, 87
Asparagic acid, 60
Attractors, 90

Bloch sphere, 63
Bond, bonds, 13, 57
Bond ellipticity, 147
Bond path, 145
Born-Oppenheimer approximation, 46, 60, 167
Bragg scattering, 148

C-Cl···O=C interactions, 115
CAHB, 135
Cambridge Structural Database (CSD), 8, 106, 124, 128
Canonical decomposition, 70, 72, 83
Charge neutralization, 181
Chemical structure, 61
Chirality, 60
Chrysanthemic acid, 108
Clamped nuclei, 17, 18, 30
Classical limit, 89
Classical structures, 70
CLFSE/MM, 173
Close packing, 200
Closed-shell interaction, 146
Cluster analysis, 109, 205
Cluster distance, 193
Clusters, 192, 205
Cobalt(III/II) hexaamine couples, 179
Complementarity, 210, 212, 214, 215
Complexity, 122
Compressed atom, 100
Conformational analysis, 175, 185
Conformational distance, 192
Conformational interconversion pathways, 109
Conformational preferences, 108
Conformational space search, 191
Conglomerate crystallization, 223
Constructionism, 122

Convolution approximation, 149
Coordinate systems
   fixed in the body, 24
   translationally invariant, 19
Copper(II) hexaamine complexes, 173
Crisp sets, 90
Critical point, 144
Crystal, 199
Crystal engineering, 105
Crystal structure databases, 124
Crystallographic databases, 105
Crystallography, 2, 227
CSD information content, 106
CSD software system, 106
Curie-Weiss magnet, 72, 83
Cyclopropyl-carbonyl substructure, 108

d-orbitals
   partly filled, 171
Decoupling, 16, 17, 21
Degrees of freedom, 205
Density operator, 62
Designing, 107
Dicopper(I) complex, 181
Dicopper(II) systems
   weakly coupled, 176
   weakly dipolar coupled, 181
Dipole-dipole interactions, 115
Directionality, 112
Doob-Meyer bracket, 92
Dressing procedures, 63

Effective hamiltonian, 43
Eigenstates, 61
Electric dipolar moment, 47, 48, 49, 50
Electron density, 143
   topology of, 144
   Laplacian of, 145
Electron spin, 99, 100
Electronic coordinates, 22
Electronic effects, 168, 176
Electronic motion, 21, 28
Emergent laws, 123
Entropy, 85
Environment, 63, 99, 168
Environmental effects, 174

EPR spectra, 168, 176
EPR spectroscopy, 176
Equidistribution 66, 82
Ergodicity, 90
Erythromycin, 191
   seco acid for total synthesis of, 189
Eulerian angles, 24, 25, 28, 33
Ewald summation, 202
External perturbations, 60

Fermi's Golden Rule, 74
Flexible coordination geometries, 171
Force fields, 169, 201, 202, 206
   electronically doped, 174
Frequencies of spectral lines, 41, 51, 52
Furan, 117
Fuzzy classical observables, 87, 90
Fuzzy classical structures, 88
Fuzzy sets, 90

Gas phase calculations, 174
Generic force constants, 171
Genetic algorithm, 206

Handedness, 59, 60, 88, 90
Harmonic sine function, 172
Holism, 121
Hydrogen bonds, 56, 69
   C-H$\cdots$X, 115
   charge-assisted, 135
   covalent nature of, 135
   homonuclear and heteronuclear, 124, 139
   in β-diketone enols, 125
   in β-ketoester enols, 129
   in δ- and ζ-diketone enols, 130
   in DNA and proteins, 132
   in selecting the crystallization pathway, 224
   modeling of, 124
   O-H$\cdots$O-, 135, 139
   O$\cdots$H$^+\cdots$O, 135, 139
   resonance-assisted, 112, 124
   resonance-induced, 112
   resonant N-H$\cdots$O, 132
   resonant O-H$\cdots$O, 124, 129
Hydrogen bonding, 176
   at 'univalent' sulphur, 112
Hydrophobicity, 176

Individual quantum theory, 63
Intensities of spectral lines, 41, 51
Interactions
   1,3 nonbonded, 171
Interatomic surface, 145
Internal coordinates, 24–27, 29, 31, 32
Interpolation, 170
Intermolecular perturbation theory, 109
Ion pairing, 174, 176
Isolated molecules, 174
Isomerism, 59, 61, 90
Isomers, 70
Isostructurality, 210–220

Itô equation, 78

Jahn-Teller distortion, 173

Karplus relations, 174
Kinematic scattering theory, 148
KMS state, 71
Knot type, 59, 61
Knowledge acquisition, 105

Large-deviation theory, 85
Lattice energy, 202, 203
Level splitting, 60
Lifetimes, 76
Ligand field based perturbation, 172
Ligand field spectra, 168
Ligand field theory, 176
Ligand-ligand repulsion model, 170
Lyapunov coefficients, 91

Macrocyclization, 190
   transition-state of, 196
Magnetic dipolar moment, 47, 52
Martingale, 78, 79, 91
Maximum entropy decomposition, 83, 84, 89
Mean molecular dimensions, 107
Metal ion selectivities, 175
MM2(91), 194
MM-AOM, 176, 185
MM-EPR, 176, 181
MM-Redox, 175, 179
Models, 120
Molecular frame
Molecular mechanics, 8, 168
Molecular mechanics force fields, 170
Molecular modeling, 1, 121, 167
Molecular properties, 167
Molecular shape, 8 12, 13, 32, 102
Molecular structure, 2, 16, 32, 168
Monodeuteroaniline, 60
Multiple minimum, 199, 202
Multipole refinement, 151

Natural laws, 120
Natural product
   total synthesis of, 189
Newton-Raphson procedure, 174
Nitro groups, 115
NOE, 174
Nonadiabatic rotational effects, 42–44, 48–51
Nonadiabatic vibrational effects, 43, 44, 46, 48, 52
Non-covalent interactions, 109
Non-pure states, 61
Nuclear coordinates, 21
Nuclear motion, 17, 29
Nuclear structure, 88, 90
Nucleophilicity, 177

Optimization, 202, 206
Optional stopping theorem, 92

p-terphenyl crystal, 74
π-systems as H-bond acceptors, 115
Packing, 8, 199
Patterns, 167
Phase transition, 60, 102
Plane groups, 201
PM3 modification of MNDO, 194
Polar lattice, 223
Polarity of diatomic molecules, 48, 50
Potential energy function, 168
Potential energy (hypersurface), 8, 18, 31, 36
Potential energy operator, 26
Principal component analysis, 109
Protein Data Bank, 107, 124
Pure states, 61, 73, 91
    equidistribution of, 66, 82
    stable distribution of, 82

Quantum deviations, 56
Quantum fluctuations, 88

Racemization, 87
RAHB, 124
Rational molecular design, 105
Reaction pathway,
Reaction dynamics, 115
Redox properties, 168
Redox potential, 175
Reductionism, 122
Regular dynamics, 75
Rotating dipole, 47, 52
Rotational *g* factor, 45, 46, 48, 49

Scattering amplitude, 148
    thermal average of, 148
Scientific theories, 120
Semimartingale, 93
Shape, 11, 14
Shared interaction, 146
Simulated annealling, 205
Single molecule, 74, 86
Single molecule spectroscopy, 74
Solution structure, 174
Solvation, 174, 176

Space groups, 201–206
Spectroscopy, 5
    of single molecules, 74
Spin-equilibrium compounds, 174
Statistical quantum mechanics
Stereospecifities, 175
Stochastic differential equations, 77
Stochastic dynamics, 74, 76, 91
Stopping time, 77, 92
Strain energy, 169
Stratonovich stochastic equation, 78
Structurally stable, 90
Supramolecular chemistry, 109
Supramolecules, 209, 210, 214, 215, 220
Symmetry breaking, 91

Template reaction, 181
Term energies, 176
Terrylene defect, 74
Thioether, 116
Thiophene, 117
Trajectories, 76
Transient states, 90
Transition metal compounds, 168
Transition probabilities, 76, 79
Transitions, 76
Tyrosinase, 181

Unique labeling problem, 181
UV-vis-NIR, 176

VALBOND, 173
Valence/valency, 13, 15
Valence shell charge concentration, 146
Valence state, 101
Vibrational frequencies, 171, 174

Weakly coupled dicopper(II) systems, 176
    EPR spectra of, 176
Weakly dipolar coupled dicopper(II), 181
Wiener process, 77

X-ray crystallography, 227
X-ray diffraction, 143